Hasan Padamsee, Jens Knobloch, and Tom Hays

RF Superconductivity for Accelerators

Related Titles

Wangler, T. P.

RF Linear Accelerators

2nd Edition

420 pages with approx. 120 figures

2007

Hardcover

ISBN: 978-3-527-40680-7

Buckel, W., Kleiner, R.

Superconductivity

Fundamentals and Applications

475 pages with approx. 247 figures

2004

Hardcover

ISBN: 978-3-527-40349-3

Clarke, J., Braginski, A. I. (eds.)

The SQUID Handbook

Volume I: Fundamentals and Technology of SQUIDs and SQUID Systems

409 pages with approx. 75 figures

2004

Hardcover

ISBN: 978-3-527-40229-8

Clarke, J., Braginski, A. I. (eds.)

The SQUID Handbook

Volume II: Applications of SQUIDs and SQUID Systems

653 pages with approx. 50 figures

Hardcover

ISBN: 978-3-527-40408-7

Hasan Padamsee, Jens Knobloch, and Tom Hays

RF Superconductivity for Accelerators

Second Edition

WILEY-VCH Verlag GmbH & Co. KGaA

The Authors

Hasan Padamsee
Laboratory of Nuclear Studies
Cornell University
Ithaca, USA
Hsp3@cornell.edu

Jens Knobloch
BESSY GmbH
Berlin, Germany
knobloch@bessy.de

Tom Hays
Cosmic Consulting
Gore, USA
trhays@cosmicconsulting.org

Cover

Printed with kind permission of Professor Hasan Padamsee.

1st Edition 2008
1st Reprint 2008

All books published by Wiley-VCH are carefully produced. Nevertheless, authors, editors, and publisher do not warrant the information contained in these books, including this book, to be free of errors. Readers are advised to keep in mind that statements, data, illustrations, procedural details or other items may inadvertently be inaccurate.

Library of Congress Card No.:
applied for

British Library Cataloguing-in-Publication Data
A catalogue record for this book is available from the British Library.

Bibliographic information published by the Deutsche Nationalbibliothek
Die Deutsche Nationalbibliothek lists this publication in the Deutsche Nationalbibliografie; detailed bibliographic data are available in the Internet at <http://dnb.d-nb.de>.

© 2008 WILEY-VCH Verlag GmbH & Co. KGaA, Weinheim

All rights reserved (including those of translation into other languages). No part of this book may be reproduced in any form - by photoprinting, microfilm, or any other means - nor transmitted or translated into a machine language without written permission from the publishers. Registered names, trademarks, etc. used in this book, even when not specifically marked as such, are not to be considered unprotected by law.

Printing Strauss GmbH, Mörlenbach
Binding Litges & Dopf Buchbinderei GmbH, Heppenheim

Printed in the Federal Republic of Germany

Printed on acid-free paper

ISBN: 978-3-527-40842-9

Contents

PREFACE xiii

I BASICS

1 Introductory Overview 3

1.1 Radio Frequency Cavities for Accelerators / 3
1.2 Attractiveness of RF Superconductivity / 6
1.3 Basics of RF Superconductivity for Accelerators / 8
    1.3.1 Surface Resistance / 8
    1.3.2 Field Limits / 9
    1.3.3 Units for Magnetic Field Quantities / 10
    1.3.4 Accelerator Physics Issues for Structure and Couplers / 11
1.4 The State of the Art in Gradients / 13
1.5 Historical Foundations of RF Superconductivity / 14
    1.5.1 Electrons, Velocity of Light Particles / 14
    1.5.2 Protons and Heavy Ions (Low-Velocity Particles, $v/c < 0.3$) / 18
    1.5.3 Other Early Applications / 21
1.6 State of the Art for Accelerators Based on RF Superconductivity / 21
    1.6.1 Heavy Ion Linacs / 22
    1.6.2 Storage Rings / 27
    1.6.3 Recirculating Linacs / 31
    1.6.4 Free Electron Lasers / 33
1.7 A Summary of the State of the Art / 34

2 Cavity Fundamentals and Cavity Fields 37

2.1 Radio-Frequency Fields in Cavities / 37
    2.1.1 The Pill-Box Cavity / 40
    2.1.2 The Accelerating Voltage / 42
    2.1.3 Peak Surface Fields / 43

2.2 Figures of Merit / 43
  2.2.1 Power Dissipation and the Cavity Quality / 44
  2.2.2 Shunt Impedance / 47
  2.2.3 Refrigerator Requirements / 48
2.3 Application of RF Codes / 49
  2.3.1 Numerical Techniques / 49
  2.3.2 Using Symmetry / 51
  2.3.3 More Examples / 51
  2.3.4 Code Comparison / 55

3 Superconductivity Essentials 57

3.1 Introduction / 57
3.2 The Free Electron Model / 57
  3.2.1 Success of the Classical Free Electron Model / 57
  3.2.2 Quantum Mechanical Description / 62
3.3 Enter Superconductivity / 66
3.4 Electrical Properties, DC and RF Resistance / 72
3.5 Thermal Conductivity in the Superconducting State / 75

4 Electrodynamics of Normal and Superconductors 77

4.1 Introduction / 77
4.2 Skin Depth and Surface Resistance of Normal Conductors / 77
4.3 The Anomalous Skin Effect / 79
4.4 Perfect Conductors / 80
4.5 Meissner Effect / 82
4.6 Surface Impedance of Superconductors in the Two-Fluid Model / 85
4.7 BCS Treatment of Surface Resistance / 88

5 Maximum Surface Fields 91

5.1 Introduction / 91
5.2 The Thermodynamic Critical Field / 91
5.3 Positive Surface Energy Superconductors (Type I) / 93
5.4 Negative Surface Energy Superconductors (Type II) / 96
5.5 The RF Critical Magnetic Field / 99
5.6 Maximum Surface Electric Field / 102

II PERFORMANCE OF SUPERCONDUCTING CAVITIES

6 Cavity Fabrication and Preparation 105

6.1 Introduction / 105
6.2 Niobium / 105

6.3 Forming Sheet Niobium / 108
    6.3.1 Deep Drawing / 108
    6.3.2 Spinning / 111
6.4 Trimming / 114
6.5 Electron-Beam Welding / 115
6.6 Postpurification / 118
6.7 Tuning / 119
6.8 Surface Preparation / 120
    6.8.1 Chemical Treatment / 120
    6.8.2 Rinsing / 121
6.9 Clean Assembly / 123
6.10 Summary / 125

7 Multicell Field "Flatness" Tuning   129

7.1 Introduction / 129
7.2 Circuit Model / 130
    7.2.1 Compensating for Beam Tubes / 131
    7.2.2 Eigenvectors / 132
    7.2.3 Eigenvalues / 133
    7.2.4 Dispersion Diagram / 133
7.3 Modeling an Out-of-Tune Cavity / 133
7.4 Refresher on Perturbation Techniques / 134
7.5 Applying The Perturbation / 136
7.6 "Bead Pulling" to Measure the Field Profile / 137
7.7 Constructing the Model from Measurements / 140
7.8 Two-Cell Worked Example / 140
7.9 Five-Cell Cavity Example / 142

8 Cavity Testing   145

8.1 Introduction / 145
8.2 RF Measurements / 145
    8.2.1 Undriven Cavity / 146
    8.2.2 Driven Cavity with One Coupler / 148
8.3 Cavity Behavior Examples / 154
    8.3.1 Steady State / 154
    8.3.2 Switch RF Off / 155
    8.3.3 Switch RF On / 156
8.4 Rectangular Pulses / 156
8.5 Frequency Domain Measurements / 157
8.6 RF Equipment and Electronics / 160
8.7 Measuring Q_0 Versus E / 160
8.8 Strongly Coupled Input / 164

8.9 Temperature Mapping / 164
    8.9.1 A Cavity Test Using Thermometry / 167

9 Residual Resistance — 171

9.1 Introduction / 171
9.2 Typical Residual Losses / 171
9.3 Trapped Magnetic Flux / 173
9.4 Residual Losses From Hydrides / 175
9.5 Residual Loss From Oxides / 177

10 Multipacting — 179

10.1 Introduction / 179
10.2 Experimental Observation of Multipacting in Cavities / 179
10.3 Multipacting Basics / 181
10.4 Secondary Electron Emission / 182
    10.5 Common Multipacting Scenarios / 184
        10.5.1 One-Point Multipacting / 185
        10.5.2 Two-Point Multipacting / 189
    10.6 Numerical Multipacting Simulations / 192
        10.6.1 Multipacting Thresholds Determined with Electron Tracking / 192
10.7 Avoiding Multipacting / 196

11 Thermal Breakdown — 199

11.1 Introduction / 199
11.2 Thermal Breakdown of Superconductivity / 199
11.3 Examples of Defects / 201
11.4 A Simple Model for Thermal Breakdown / 205
11.5 Solutions to Thermal Breakdown / 207
    11.5.1 Guided Repair / 207
    11.5.2 Raising the Thermal Conductivity of Niobium / 208
    11.5.3 Thin Films of Niobium on Copper / 209
11.6 Heat Transport at the Helium Interface / 210
11.7 Thermal Model Simulations / 213
11.8 Methods to Improve Niobium Purity / 217
11.9 Quench Suppression with High-Purity Niobium / 220
11.10 Defect-Free Cavities / 223

12 Field Emission — 227

12.1 Introduction / 227
12.2 Diagnosing Field Emission / 228

CONTENTS

- 12.3 Theory of Field Emission / 230
- 12.4 Field Emitters in Superconducting Cavities / 235
- 12.5 DC Studies of Field Emission / 242
- 12.6 A Brief Look at the Impact of Field Emission Studies on Cavity Performance / 247
- 12.7 Nature of Field Emitters / 250
  - 12.7.1 The Tip-on-Tip Model / 251
  - 12.7.2 The Role of the Interface / 251
  - 12.7.3 The Metal–Insulator–Metal Model / 252
  - 12.7.4 Condensed Gas and Adsorbates / 256
- 12.8 Investigations on Processed Emitters in RF Cavities / 257
  - 12.8.1 Dissecting Single-Cell Test Cavities / 258
  - 12.8.2 Demountable Mushroom Cavity Studies / 261
  - 12.8.3 Copper Cavity Studies / 263
  - 12.8.4 Emitter Processability and Fowler–Nordheim Properties / 264
- 12.9 DC Voltage Breakdown Studies / 265
- 12.10 The Role of Gas in Processing / 268
- 12.11 Summary—A Picture for Field Emission and Processing / 270
- 12.12 Simulating Field Emission Heating / 272

13 The Quest for High Gradients   281

- 13.1 Introduction / 281
- 13.2 A Review of the State of the Art / 281
- 13.3 A Statistical Model for the Performance of Field Emission Dominated Cavities / 283
- 13.4 Overcoming Thermal Breakdown / 285
- 13.5 Early Methods for Overcoming Field Emission / 287
  - 13.5.1 Helium Processing / 287
  - 13.5.2 Heat Treatment of Niobium Cavities / 289
- 13.6 High-Pressure Rinsing to Avoid Field Emission / 293
- 13.7 High-Power Pulsed RF Processing / 296
  - 13.7.1 RF Power Supply and High-Power Test Stand / 297
  - 13.7.2 HPP Results / 300
  - 13.7.3 The Controlling Parameter for RF Processing / 302
  - 13.7.4 Limitations to HPP / 307
  - 13.7.5 Stability of Processing Benefits and Recovery from Vacuum Accidents / 309
- 13.8 Closing Remarks on the Gradient Quest / 312

14 Alternate Materials to Solid Niobium   315

- 14.1 Introduction / 315
- 14.2 Sputtered Niobium on Copper / 316

14.3 Nb_3Sn / 319
14.4 High-Temperature Superconductors / 325

III COUPLERS AND TUNERS

15 Mode Excitation and Its Consequences — 331

15.1 Introduction / 331
15.2 Monopole Mode Excitation by a Point Charge / 331
15.3 Monopole Mode Excitation by a Bunch / 334
15.4 Monopole Mode Excitation by a Train of Bunches / 335
    15.4.1 Cryogenic Losses / 338
15.5 Dipole Mode Excitation / 340
15.6 Instabilities from Beam Cavity Interactions / 342
    15.6.1 Single-Bunch Effects / 343
    15.6.2 Coupled-Bunch Instabilities / 345
15.7 Code Examples for HOM Studies / 349

16 Higher Order Mode Couplers — 355

16.1 Introduction / 355
16.2 Preliminary Design Considerations / 355
16.3 Waveguide Couplers / 357
    16.3.1 Performance of Waveguide HOM Couplers / 360
16.4 Coaxial Couplers / 361
    16.4.1 Performance of Coaxial Couplers / 372
16.5 Beam Tube Couplers for High-Current Applications / 374
    16.5.1 Performance of Beam Pipe HOM Couplers / 379

17 Coupling Power to the Beam — 381

17.1 Introduction / 381
17.2 The Equivalent Circuit / 382
17.3 Beam Loading / 383
17.4 Resonant Operation / 386
    17.4.1 Optimal Coupling in the Presence of Beam Loading / 388
    17.4.2 Current and Frequency Fluctuations / 389
17.5 Nonsynchronous Operation / 392
    17.5.1 Phase Stability in the Presence of Little Beam Loading / 392
    17.5.2 Cavity Detuning / 394
    17.5.3 Phase Stability in the Presence of Heavy Beam Loading / 396
17.6 Reexamination of the Circuit Model for Beam Loading / 398
17.7 Typical Parameters / 398
17.8 Special Considerations / 400

18 Input Power Couplers and Windows — 403

- 18.1 Introduction / 403
- 18.2 Couplers / 403
  - 18.2.1 Design Issues / 403
  - 18.2.2 Coaxial Couplers / 404
  - 18.2.3 Waveguide Couplers / 408
- 18.3 Windows / 410
  - 18.3.1 Design Issues / 410
  - 18.3.2 Windows for Coaxial Input Couplers / 412
  - 18.3.3 Windows for Waveguide Couplers / 413
  - 18.3.4 Materials Aspects for Windows / 415
- 18.4 Electron Activity in Couplers and Windows / 416
  - 18.4.1 Antimultipactor Measures / 417
  - 18.4.2 Conditioning and Diagnostics / 418
- 18.5 Performance of Input Couplers and Windows / 421
- 18.6 Couplers for High-Pulsed-Power Processing / 421

19 Tuners and Frequency Related Issues — 425

- 19.1 Introduction / 425
- 19.2 Requirements for Tuners / 425
- 19.3 Microphonics / 427
- 19.4 Lorentz Force Detuning and Pondermotive Oscillations / 428
- 19.5 Tuner Designs / 431
- 19.6 Tuner Examples / 432
  - 19.6.1 Mechanical Tuners / 432
  - 19.6.2 Thermal Tuner / 434

IV FRONTIER ACCELERATORS

20 High-Current Accelerators — 439

- 20.1 The Need for Frontier Accelerators in High-Energy Physics / 439
- 20.2 High-Current Storage Rings / 440
- 20.3 The Benefits of Superconducting RF for High-Current Storage Rings / 442
- 20.4 Systems Under Development / 446
- 20.5 Crab Cavities for Bunch Rotation / 450
- 20.6 Intense Proton Accelerators / 453
- 20.7 Pulsed Neutron Sources for Materials Research / 454
- 20.8 Transmutation Applications / 455
  - 20.8.1 Reduction of Nuclear Waste / 455
  - 20.8.2 Tritium Production / 455
- 20.9 Accelerator Based Fission Reactors / 455

20.10 Advantages of the Superconducting Approach to High-Intensity Proton Linacs / 456
20.11 Progress in Superconducting Cavities for High-Current Proton Accelerators / 457

21 High-Energy Accelerators   459

21.1 Introduction / 459
21.2 Issues in Optimizing the Design Parameters of Linear Colliders / 460
21.3 The Superconducting Linear Collider (TESLA) / 463
21.4 Attractive Features of TESLA / 466
21.5 Design Flexibility and Energy Upgrades / 470
21.6 The Two-Beam Accelerator with Superconducting Linac / 473
21.7 Muon Colliders / 474
21.8 Concluding Remarks on Future Prospects / 475

PROBLEMS   477

REFERENCES   491

INDEX   515

Preface

Radio frequency superconductivity has become an important technology for particle accelerators. Superconducting cavities have been operating routinely for many years in a variety of accelerators for high-energy physics, low-energy to medium-energy nuclear physics research, and free-electron lasers. Many hundreds of meters are installed and operated in state-of-the-art machines. As a result, there are a large number of workers in the field at national accelerator laboratories such as ANL, TJNAF (formerly CEBAF), CEN-Saclay, CERN, DESY, FNAL, KEK, and LANL, as well as at universities such as Stanford, SUNY at Stony Brook, Washington State, Florida State, Cornell, Wuppertal, and others. Over the course of the development of this field, much progress has been made in understanding the physics of the phenomena that influence the achievable gradient and Q. Techniques have been invented and successfully implemented to continually advance the performance of superconducting cavities. As a result, many new applications are on the horizon. The field is growing steadily. Laboratories are seeking to expand research and development into advanced accelerators such as TeV linear colliders, intense proton accelerators, and muon colliders.

Many review articles are now available covering the state of the art in rf superconductivity and its application to particle accelerators. There have been seven International Workshops on RF Superconductivity. The proceedings from these workshops carry detailed information on the physics, technology, and applications of the field.

This book originated out of two series of lectures presented at the Joint US-CERN-Japan Accelerator School held in Hawaii during November 1994 and at the U.S. Particle Accelerator School held in San Diego during January 1995.

Our aim here is to introduce some of the key ideas of this exciting field, using a pedagogic approach, as well as to present an overview of the field. We therefore hope that this book will serve both as a review text for established workers in the field and as an introductory text for newcomers. Our overall treatment is biased toward accelerators for particles moving at nearly the velocity of light, although, in covering the state of the art, we present selected examples from the low-velocity applications to heavy-ion accelerators.

We divide the text into four parts. After an overview that introduces many of the key concepts of the field, the first part introduces the basic concepts of microwave cavities for particle acceleration, the elements of superconductivity, and rf superconductivity. The presentation leads to the theoretically expected performance of superconducting cavities. The second part is devoted to the observed behavior of superconducting cavities, as well as the reasons for the departures from the expected "ideal" performance; these topics are followed by an in-depth presentation of past and present efforts to overcome the performance limitations. At suitable junctures throughout, we present appropriate technological aspects such as cavity fabrication, tuning, surface treatment, clean environments, and niobium purification. In the third part we cover general issues connected with beam cavity interaction, and also cover the related issues for the critical components, such as input couplers, output couplers, and tuners, together with representative examples from various applications. In the final part we discuss applications of superconducting cavities to frontier accelerators of the future, drawing heavily on the examples that are in their most advanced stage, such as for high-current machines and for a TeV linear collider. Issues related to operation, commissioning, and availability of accelerators are beyond the scope of this treatment. Accordingly we do not delve deeply into topics such as rf controls, interlocks, and quench detectors. Similarly, we do not cover the subjects of cryostat, refrigeration, and cryogenic engineering as related to the development and application of superconducting cavities.

The book concludes with a problems section to illustrate and amplify text material as well as to draw upon example applications of superconducting cavities to existing and future accelerators.

For the most part, we use units dictated by common usage; where appropriate we present conversion factors to MKS units. The bulk of our discussion presumes that the reader is familiar with electricity and magnetism and electromagnetic waves at the level of *Fields and Waves in Communications Electronics* by Ramo, Whinnery, and Van Duzer [1], as well as the physics of electrons in metals at the undergraduate level—for example, *Fundamentals of Statistical and Thermal Physics* by Reif [2] or *Solid State Physics* by Ashcroft and Mermin. No background in superconductivity or accelerator physics is assumed. We recommend a list of recent review articles in the field [4–8] as well as the set of proceedings of the seven International Workshops on RF Superconductivity held between 1980 and 1996 [9–15]. The reader should consult these proceedings for detailed information on the activities at the various laboratories involved in the development and applications of rf superconductivity.

We are deeply indebted to Maury Tigner and Walter Hartung for their thorough review of the entire book, and we are grateful to David Whittum for his review of major sections of the book. Their many valuable suggestions are greatly appreciated. We also wish to thank Joseph Kirchgessner, Sergey Belomestnykh, Eric Chojnacki, Ernst Haebel, and David Pritzkau for reviewing particular chapters.

PREFACE

We are especially grateful to Mel Month for making possible our participation in the U.S. Particle Accelerator School as well as in the joint US-CERN-Japan Accelerator School.

HASAN PADAMSEE
JENS KNOBLOCH
TOM HAYS

Cornell University

Preface to Re-Printed Edition

It has now been nearly 10 years since the successful publication and enthusiastic reception of RF Superconductivity for Accelerators. It is a great pleasure to undertake a second printing of the successful book. A few minor text errors have been corrected. Some figures have been cleaned up. We hope that the re-printing will fulfill the need of the large numbers of new comers to the field.

In the meantime a new book *RF Superconductivity, Science, Technology and Applications* is under preparation, due to be released in 2008. Much has happened in the field since the appearance of the first work. There has been a substantial expansion in the number of accelerator applications and in the number of laboratories engaged in the field. RF superconductivity has become a major sub-field of accelerator science. The aim of the new book is to discuss advances in the science, technology and on-going applications of RF superconductivity over the years between 1998 and 2007. The content will be completely new.

Part I

Basics

CHAPTER 1

Introductory Overview

1.1 RADIO FREQUENCY CAVITIES FOR ACCELERATORS

A key component of the modern particle accelerator is the device that imparts energy to the charged particles. This is an electromagnetic cavity resonating at a microwave frequency. Consider first the case of a charged particle moving at nearly the velocity (v) of light ($v/c \approx 1$). As it traverses the half-wavelength ($\lambda/2$) accelerating gap in half a radio-frequency (rf) period, it sees the electric field pointing in the same direction for continuous acceleration. Figure 1.1 parts (a) to (c) show the evolution of the typical superconducting accelerating structure for a velocity-of-light particle. Starting from the single-cell, pill-box cavity (a) resonating in the TM_{010}, fundamental mode, the beam tubes are added and the cell shape changed to round the wall as is required for superconducting cavities (b). Part (c) shows the multicell structure with coupler ports and the relationship between the phase of the electric field in adjacent cells. A 4-cell, 350 MHz niobium structure developed at CERN for LEP [16] is shown in Figure 1.1(d). The properties of the accelerating mode will be discussed in Chapter 2.

Design considerations for such velocity-of-light structures will be discussed in various chapters throughout this book. The resonant frequency is typically between 350 MHz and 3000 MHz depending on the trade-offs involved in each specific application. The considerations involved in choosing the optimum rf frequency and the optimum aperture will be discussed in a later section. Here we address the size of the accelerating gap, i.e., the length of the cell, which, as already pointed out, is half a wavelength ($\lambda/2$) long. The phase of the electric field on the axis of each cell is shown for the accelerating mode in Figure 1.1. Ports outside the cell region are for input power couplers and higher-order mode power output couplers. The requirements for these couplers and the types of couplers that are inserted into the ports will be discussed in Part II.

Superconducting accelerating structures for low-velocity ($\beta = v/c < 0.3$) charged particles, such as heavy ions, have a very different geometry. Figure 1.2 parts (a) to (c) depict the evolution of a typical low-velocity resonator. This structure evolves from a coaxial transmission line a quarter wavelength long, resonating in the TEM mode. A drift tube is suspended from the end of the hollow center conductor. The structure has two accelerating cells between the

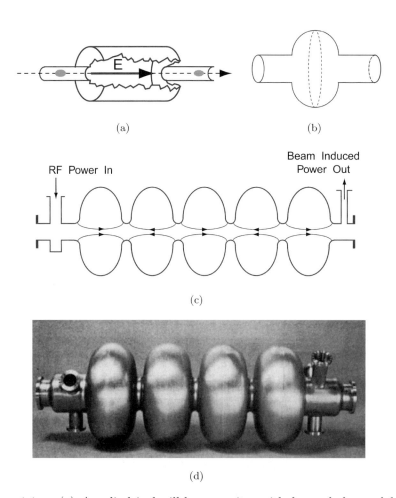

Figure 1.1: (a) A cylindrical pill-box cavity, with beam holes and beam pipes, resonating in the TM_{010} mode for which the electric field is maximum on the axis. (b) As was discovered later, the cylindrical shape is unsuitable for superconducting cavities because of multipacting. Rounding the curved wall eliminates the problem. (c) The typical accelerating structure consists of a chain of cavities. There are ports on the beam tubes to bring rf power in to establish the fields and to deliver power to the beam. There are additional ports for removing power induced by the beam in the higher-order resonant modes of the cavity. (d) A 4-cell cavity suitable for LEP at CERN. The resonant frequency is 350 MHz. The active cell length is 0.43 m. The cavity is made from the superconductor, niobium, and cooled in a bath of liquid helium at 4.2 K. (Courtesy of ACCEL.)

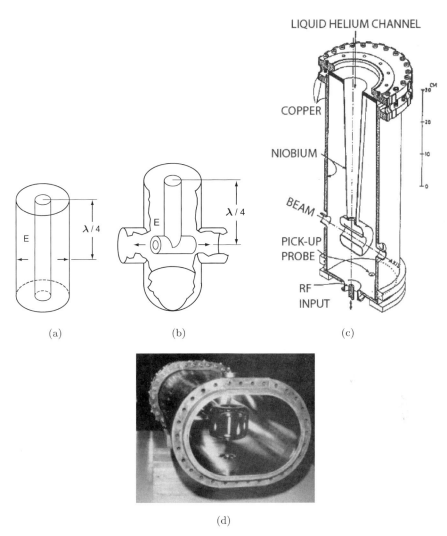

Figure 1.2: (a) A coaxial line, $\lambda/2$ in length, is shorted at both ends to form a resonator with maximum electric field at $\lambda/4$. (b) In a quarter-wave resonator, a field-free drift tube is suspended from the center conductor in the maximum electric field region. (c) Schematic of a complete resonator for the JAERI heavy-ion linac. The resonant frequency is 130 MHz. The height is about 60 cm. (d) A view inside the completed resonator. The inner conductor, which is made from niobium, is hollow and filled with liquid helium. The outer conductor is niobium clad with copper for conduction cooling. (Courtesy of JAERI.)

ends of the drift tube and the beam hole openings located in the outer conductor of the coax. A niobium structure developed for the JAERI (Tokai, Japan) heavy-ion linac [17] is shown in Figure 1.2(d).

The accelerating gap is $\beta\lambda/2$, where $\beta = v/c$. Since β is small, λ must be chosen to be large, to achieve a useful acceleration. Therefore a low resonant frequency is chosen, typically 100 MHz. For example, if $\beta = 0.1$ at $f = 100$ MHz, the accelerating gap of 15 cm yields a respectable accelerating voltage for a typical superconducting cavity gradient of several MV/m. The wavelength also sets the height of the quarter-wave resonator to a manageable 3/4 meter for the example of Figure 1.2. The other advantage of a low rf frequency is that beam bunches can occupy a smaller rf phase angle yielding less energy spread.

Ions with velocities between $0.01c$ and $0.2c$ are accelerated in structures with rf frequencies between 50 and 200 MHz [18]. A survey of the structures developed by various laboratories shows a strong correlation between the β value needed and the appropriate choice of rf frequency [19]. Other design considerations for low-velocity resonators will be discussed later.

1.2 ATTRACTIVENESS OF RF SUPERCONDUCTIVITY

The two most salient characteristics of a superconducting accelerating cavity are its average accelerating field, E_{acc}, and the quality factor Q_0, which is the intrinsic Q of the resonant cavity. The quality factor is a universal figure of merit for resonators and is defined in the usual manner as the ratio of the energy stored in the cavity (U) to the energy lost (P_c) in one rf period. It measures the number of oscillations a resonator will go through before dissipating its stored energy. The Q_0 depends on the microwave surface resistance of the metal. In general, one would like to have as high an accelerating field and as high a Q_0 as possible. The intrinsic bandwidth of a resonator is inversely proportional to Q_0. Q_0 values as high as 10^{11} have been reached. These are some of the highest Q values found in nature.

The strongest incentive to use superconducting cavities is in accelerators that operate in a continuous wave (cw) mode or at a high duty factor ($> 1\%$). For cw operation, the power dissipation in the walls of a copper structure is substantial. Here superconductivity comes to the rescue. The microwave surface resistance of a superconductor is typically five orders of magnitude lower than that of copper, and therefore the Q_0 is five orders of magnitude higher. As we shall see in Chapter 2, the dissipated power per unit length is given by

$$\frac{P}{L} = \frac{E_{\text{acc}}^2}{\frac{r_a}{Q_0} Q_0}. \tag{1.1}$$

Here r_a/Q_0 is the geometric shunt impedance in Ω/m, and it depends primarily on the geometry of the structure. These properties will be discussed thoroughly in Chapter 2. For now, we use the result to compute the required powers to illustrate one of the prime benefits of rf superconductivity. The typical

Table 1.1: AC power required to operate 500 MHz superconducting and normal conducting cavities at 1 MV/m and 5 MV/m

Option	Super-conducting	Normal Conducting
Q_0	2×10^9	2×10^4
r_a/Q_0 (Ohm/m), rf frequency = 500 MHz	330	900
P/L (Watt/m) for $E_{acc} = 1$ MV/m	1.5	56,000
AC Power (kW/m) for $E_{acc} = 1$ MV/m	0.54	112
AC Power (kW/m) for $E_{acc} = 5$ MV/m	13.5	2,800

quality factor of a superconducting cavity is in the range of 10^9 to 10^{10}.

Table 1.1 compares 500 MHz normal conducting (e.g., copper) and superconducting (e.g., niobium) cavities at $E_{acc} = 1$ MV/m to show that the dissipated power per meter is reduced by nearly a factor 4×10^4. The real gain of a superconducting cavity is, however, not as spectacular, since $P_c \approx 1.5$ W/m is dissipated at liquid helium temperature. The ac power required to operate the refrigerator must take into account the efficiency, which is the product of two terms. The first is the Carnot efficiency

$$\eta_c = \frac{T}{300 - T} = 0.014 \qquad (1.2)$$

at $T = 4.2$ K. The second part is the technical efficiency which is typically around 0.2, but can approach 0.3 for modern, large systems. Therefore the net refrigerator efficiency at 4.2 K is only 0.0028. For the copper cavity, the required wall plug power (we refer to this as the ac power) must also include the efficiency of the rf power source (e.g., the klystron), which is typically 0.5. Hence a superconducting cavity reduces the ac power by a factor of 200. Note in Table 1.1 that the typical r_a/Q_0 of the superconducting structure is chosen to be nearly a factor of 3 smaller than for the normal conducting case, for reasons which we discuss in Chapter 2.

For applications demanding high cw voltage, such as increasing the energy of storage rings, the advantage of superconducting cavities becomes clear. Since the dissipated power increases with the square of the operating field, only superconducting cavities can economically provide the needed voltage. For example, LEP requires 2.5 GV (gigavolts) to double its energy from 50 GeV to 100 GeV per beam. If copper cavities were to be used at the higher accelerating fields, both the capital cost of the klystrons and the ac power operating cost would become prohibitive. Nearly 3 MW/m of ac power would be required to operate a copper cavity at 5 MV/m. At 500 MHz, this is nearly 1 MW per cell. There are also practical limits to dissipating more than 100 kW in a copper cell to prevent the surface temperatures from exceeding 100 °C and causing vacuum degradation, stresses, and metal fatigue due to thermal expansion. Therefore the typical cw operating field for a copper cavity is usually kept below 1 MV/m. High fields (> 50 MV/m) can be produced in copper cavities, but only for microseconds, before the rf power needed becomes prohibitive.

Apart from the general advantages of reduced rf capital and reduced rf associated operating costs, superconductivity offers each arena of cw or high duty factor accelerators certain special advantages that stem from the low cavity wall losses. For example, as we shall see in Chapter 2, because of superconductivity one can afford to make the beam hole of a superconducting cavity much larger than for a normal conducting cavity. The resulting drop in geometric shunt impedance (i.e., r_a/Q_0) for the accelerating mode is not a significant concern because of the immensely larger intrinsic Q_0 of superconducting cavities. But the advantages of the larger beam hole are substantial. For example, in high-current storage rings, where multibunch instabilities stemming from the fundamental and higher-order modes are major worries, the large beam holes substantially reduce the impedance of the dangerous modes. For a linear collider, where emittance growth along the linac is a serious threat, the large beam hole significantly reduces short-range wakes. For the future intense proton linacs, where scraping of the proton beam tails is a major worry because of activation of the accelerator, the wide beam hole would greatly reduce the risk of beam-loss-induced radioactivity.

1.3 BASICS OF RF SUPERCONDUCTIVITY FOR ACCELERATORS

1.3.1 Surface Resistance

The remarkable properties of superconductivity are attributed to the condensation of charge carriers into Cooper pairs. At $T = 0$ K, all charge carriers are condensed. At higher temperatures, some carriers are unpaired; the fraction of unpaired carriers increases exponentially with temperature until none of the carriers are paired above T_c. The pairs move frictionlessly. In this simplified picture, known as the London two-fluid model, when a dc field is turned on, the pairs carry all the current, shielding the applied field from the normal electrons. Electrical resistance vanishes.

In the case of rf currents, however, dissipation does occur for all $T > 0$ K, albeit very small compared to the normal conducting state. A simple expression for the rf surface resistance is

$$R_s = A(1/T)f^2 \exp(-\Delta(T)/kT) + R_0. \tag{1.3}$$

Here A is a constant, dependent on the material parameters of the superconductor, such as the penetration depth, $\lambda_L(0)$, the coherence length, ξ_0, the Fermi velocity, v_F, and the mean free path, l. 2Δ is the energy gap of the superconductor, i.e., the energy needed to break the pairing. These fundamental material aspects of superconductivity are covered in Chapter 3. Equation 1.3 fits the experimental results for frequency and temperature dependence remarkably well for $T < 0.5T_c$. The quadratic frequency dependence and the exponential temperature dependence of the first term of Equation 1.3 are easy to see with the

BASICS OF RF SUPERCONDUCTIVITY FOR ACCELERATORS 9

two-fluid model. While the Cooper pairs move frictionlessly, they do have inertial mass. This means that for high-frequency currents to flow, forces must be applied to bring about alternating directions of flow. Hence ac electric fields will be present in the skin layer and they will continually accelerate and decelerate the normal carriers in the skin layer, leading to dissipation. The power dissipated is proportional to the internal electric field and to the normal component of the current. The interior electric field, induced by the changing magnetic field, is proportional to the rf frequency, f. The "normal" component of the current, being proportional to the interior electric field, gives a second factor of f. The normal component of the current also depends on the number of carriers thermally excited across the energy gap, 2Δ, and is given by the Boltzmann factor $\exp(-\Delta/kT)$.

The operating temperature of a superconducting cavity is usually chosen so that the first term in Equation 1.3 is reduced to an economically tolerable value. R_0, referred to as the residual resistance, is influenced by several factors. Some of the sources are extraneous to the superconducting surface, for example lossy joints between components of the structure. Other factors originate at the superconducting surface. Mechanisms for residual losses are discussed in Chapter 9, along with the appropriate references. Typical R_s values for Nb cavities fall in the range from 100 to 10 nΩ. The record for the lowest surface resistance is 1 nΩ.

1.3.2 Field Limits

The accelerating field, E_{acc}, is proportional to the peak electric (E_{pk}) as well as the magnetic field (H_{pk}) on the surface of the cavity. Therefore, besides the phenomenally low rf surface resistance, other important fundamental aspects of superconducting cavities are the maximum surface fields that can be tolerated without increasing the microwave surface resistance substantially or without causing a catastrophic breakdown of superconductivity. The ultimate limit to accelerating field is the theoretical rf critical magnetic field, which, as we shall see in Chapter 5, is the superheating field, H_{sh}. For the most popular superconductor, niobium, this is about 0.23 Tesla, while for lead, the other superconducting material used for accelerator cavities, it is 0.12 Tesla. (See the following section for a discussion of the various units used for magnetic field.) These surface fields translate to a maximum accelerating field of 55 MV/m for a typical $\beta = 1$ niobium structure and roughly 30 MV/m for a $\beta < 1$ niobium structure. The exact values depend on the detail structure geometries. A thorough discussion for the rf critical fields, along with the appropriate references, is found in Chapter 5.

Typical cavity performance is significantly below the theoretically expected surface field. One important phenomenon that limits the achievable rf magnetic field is "thermal breakdown" of superconductivity, originating at sub millimeter-size regions of high rf loss, called defects. When the temperature outside the defect exceeds the superconducting transition temperature, T_c, the losses increase, as large regions become normal conducting. Thermal breakdown is the subject of extensive discussion, along with the appropriate references, in Chapter 11. Here we also discuss measures developed to overcome thermal breakdown, such as (a) improving the thermal conductivity of niobium by purification or (b) using thin films of niobium (or lead) on a copper substrate cavity.

In the early stages of the development of superconducting cavities, a major performance limitation was the phenomenon of "multipacting." This is a resonant process in which a large number of electrons builds up within a small region of the cavity surface. When the resonant conditions are met, the avalanche absorbs the rf power making it impossible to raise the fields by increasing the incident rf power. The electrons impact the cavity walls leading to a large temperature rise and eventually to thermal breakdown. With the invention of the proper cavity shape, multipacting is no longer a significant problem for velocity-of-light structures. Multipacting is still a nuisance, and in some cases an impediment, for structures for low-velocity particles, as well as for coupling devices. We will discuss the general problem of multipacting, along with the appropriate references, in Chapter 10. Here we also cover the techniques developed to minimize and avoid multipacting.

In contrast to the magnetic field limit, we know of no theoretical limit to the tolerable surface electric field. Fields up to 220 MV/m have been imposed on a superconducting niobium cavity without any catastrophic effects [20]. However, at high electric fields, an important limitation to the performance of superconducting cavities arises from the emission of electrons from high electric field regions of the cavity. Power is absorbed by the electrons and deposited as heat upon impact with the cavity walls. If the emission grows intense at high electric fields it can even initiate thermal breakdown. We have learned much about the nature of field emission sites and made progress in techniques to avoid them as well as to destroy them. These subjects, along with the appropriate references, are treated fully in Chapters 12 and 13.

As the limiting mechanisms of multipacting, thermal breakdown and field emission were each understood, and in turn overcome, cavity performance has improved steadily over time. Consider the evolution in gradients for the velocity-of-light structures. The lowest panel of Figure 1.3 shows the state of the art in off-line accelerating gradients around 1974, when multipacting was the chief impediment. The middle panel shows the improvement after the multipacting problem was solved, and thermal breakdown became the dominant limitation. The upper panel shows the state of the art in gradients around 1991, when field emission was the dominant mechanism.

1.3.3 Units for Magnetic Field Quantities

In the MKS system the units for magnetic field (H) have the dimensions of A/m. The magnetic flux density (B) has the dimensions of Vs/m^2. In the cgs system the units for B are gauss and for H are oersted. Since free space permeability, $\mu = 1$, in the cgs system, the numerical values for B and H are the same. To convert from MKS to cgs units, use 1 A/m = $4\pi \times 10^{-3}$ Oe. Finally 1 T = 10^4 Oe.

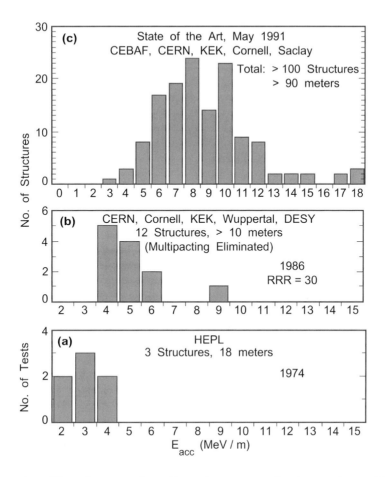

Figure 1.3: The advances of gradients in superconducting cavities with improved understanding of the limitations and invention of cures. (a) Gradients limited by multipacting. (b) The multipacting problem was solved by developing the proper shape, but the gradients were limited by thermal breakdown. (c) The thermal breakdown problem was alleviated by using high-purity, high-thermal-conductivity niobium. Now the gradients are predominantly limited by field emission, although thermal breakdown is also encountered.

1.3.4 Accelerator Physics Issues for Structure and Couplers

A number of accelerator physics considerations influence the size of the aperture, the choice of the rf frequency, the number of cells for a superconducting structure and the coupling strength of input and output power couplers. A trade-off must usually be made between accelerator physics requirements and those of cavity performance.

Besides accommodating the beam diameter, there are several considerations that dictate a large aperture. By increasing the energy coupling between cells, a large aperture will reduce the degree of field nonuniformity from cell to cell caused by errors in the shape or frequency of individual cells. A large aperture will improve the capability of power to propagate from the input coupler to the cells, where it is transferred to the beam. As we have already mentioned, a larger aperture, as well as a lower rf frequency, will reduce the short- and the long-range wake fields, allowing better beam quality and higher beam current. On other hand, a large aperture will increase the peak surface electric and magnetic fields, lowering the best achievable accelerating field. It will also decrease the shunt impedance of the fundamental mode, raising the refrigerator load.

Since the surface resistance is proportional to the square of the frequency, this consideration favors low rf frequencies. Minimization of beam instabilities caused by long- and short-range wake fields also favors low frequencies. On the other hand, for a given accelerator structure length, a low-frequency unit will have a larger surface area with an increased probability for encountering defects that may cause thermal breakdown and for encountering emitters that cause field emission.

A small number of cells is desirable to maintain a flat electric field profile from cell to cell in the accelerating mode. When a large amount of power is to be coupled into the beam per meter of structure, the length of the structure (or the number of cells) may have to be limited in order to keep the power passing through the rf window below the state-of-the-art limit for windows on accelerating cavities. On the other hand, a large number of cells is desirable to minimize system costs, minimize the effects of fringing fields, and minimize the wasted space between cavities.

The input coupling factor is normally chosen so that no incident power is reflected when the beam current is at its design value. Depending on the beam current desired for the application, this usually translates to a bandwidth for the input coupler-cavity system of the order of 0.1–10 kHz, much larger than the intrinsic bandwidth of a high-Q_0 superconducting cavity (typically 0.1–1 Hz). However, in the absence of beam, there is a severe mismatch between the generator and the cavity, so that most of the power is reflected. Therefore the window and input coupler must be designed to tolerate large values of reflected power.

In high-current accelerators, the fields excited by the beam in the higher-order modes (HOM) of the cavity have to be kept small to avoid excessive power dissipation at liquid helium temperature. Since superconducting cavities store energy well, the energy deposited by the beam into the HOMs linger for long periods of time. When a particle passes through a cavity more than once, HOMs can give rise to multibunch instabilities. The HOM fields induced in the cavity on one pass act on a particle on successive passes and can cause the beam to become unstable. The best way to avoid cryogenic losses and beam instabilities from HOMs is to damp the HOMs by adding HOM couplers to the structure.

An important consideration associated with damping HOMs is that of "trapped modes." Modes can be trapped if the shape of the individual cells is such that a HOM has negligible fields in the end cells, where the HOM couplers are usually located to damp these fields. In this respect, a small number of cells is desirable to avoid trapped modes.

For low-velocity accelerators superconducting cavities are extended, loaded structures (for example, drift tubes supported by pipes) and generally have reduced mechanical stability. Since the ion beam currents are low (nA), the input coupler bandwidths are of the order of 10–100 Hz. Ambient acoustic noise (microphonics) excites mechanical vibrational modes of the cavity, causing the eigenfrequency to vary. The range of variation depends on the mechanical properties of the cavity and ambient acoustic conditions. For use in an accelerator, the cavity rf phase must be synchronized with an rf clock. This requires either (a) driving the cavity at a frequency different than the resonant frequency or (b) rapidly tuning the cavity to cancel the effects of acoustically induced mechanical distortions. Either method requires the control system to provide (reactive) rf power in proportion to the cavity stored energy. The microphonics problem is less severe for $\beta = 1$ applications because of the larger bandwidth of the input coupler-cavity system.

1.4 THE STATE OF THE ART IN GRADIENTS

Even at the modest fraction of the ultimate potential, many attractive applications are now in place and new ones are forthcoming. We will discuss these applications in the subsequent sections of this overview chapter, as well as in the chapters of Part IV which are devoted to future applications. As our understanding of field emission and thermal breakdown continues to improve, new techniques emerge to further advance gradients, such as postpurified high-purity niobium to further raise the thermal conductivity, high pressure rinsing to provide cleaner, field emission free surfaces, and high-pulsed-power processing (HPP), to destroy residual emitters.

If these advanced techniques are not used, the state of the art for gradients achieved in laboratory tests of structures is shown in Figure 1.4. The average off-line gradient in bulk niobium cavities is 9 MV/m and the highest value is 18 MV/m. (Gradients achieved in thin-film niobium-on-copper cavities are not included in this histogram but are presented in Chapter 14.) The new techniques for bulk niobium cavities have demonstrated that gradients can be improved to between 20 and 30 MV/m in multicell structures. In Figure 1.5 we compare the recent data for about 20 structures prepared by the most advanced techniques with the results of the previous histogram (Figure 1.4). The reason for using a log scale in Figure 1.5 is that the total number of structures prepared by the new techniques is still small relative to the existing technology base. But the progress to date is indeed encouraging. If gradients 20–30 MV/m can be reliably achieved, exciting new applications are on the horizon, such as for a TeV linear collider or a multi-TeV muon collider.

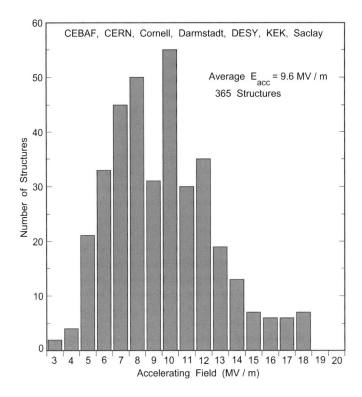

Figure 1.4: Gradients achieved in laboratory tests for multicell structures by a variety of laboratories.

1.5 HISTORICAL FOUNDATIONS OF RF SUPERCONDUCTIVITY

The pioneer laboratory in the exploration of rf superconductivity for its application to accelerators was the High-Energy Physics Lab (HEPL) at Stanford University. The applications to nuclear physics and free electron lasers that emerged from their early efforts are now realized, some of them at other locations. The efforts at Stanford soon expanded to heavy-ion linacs and to high-energy beams for particle physics. We will see the evolution of various structure geometries, and we will note their performance and some of the early limitations.

1.5.1 Electrons, Velocity of Light Particles

Exploration of rf superconductivity for particle accelerators began at Stanford University in 1965 with the acceleration of electrons in a lead-plated resonator [21]. Figure 1.6 shows single-cell and seven-cell niobium cavities of the HEPL shape. In 1977 HEPL completed the first Superconducting Accelerator (SCA) providing 50 MV with a 27 meter linac [22]. Multicell HEPL sections are shown

HISTORICAL FOUNDATIONS OF RF SUPERCONDUCTIVITY

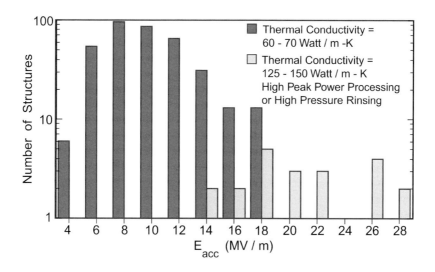

Figure 1.5: Improvement in gradients achieved with the combined application of the most recently developed techniques such as postpurification to reach higher thermal conductivity, high-pressure water rinsing to prepare cleaner surfaces, and high-pulsed-power rf processing to destroy field emission sites.

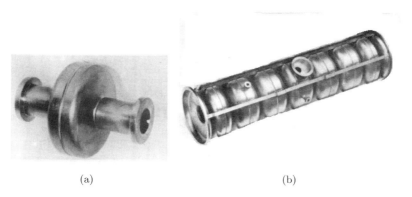

Figure 1.6: (a) Single-cell HEPL niobium cavity, resonant frequency 1.3 GHz, active cell length 11.5 cm. (b) HEPL 7-cell subsection. The niobium cavities were fabricated by sheet metal forming and electron beam welding. (Courtesy of Stanford University.)

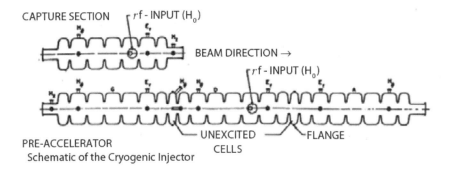

Figure 1.7: (a) Seven-cell subsection with five cells and two half end-cells. This arrangement is used to obtain a flat electric field profile along the beam axis in the accelerating mode. (b) Multicell subsection for the HEPL SCA. In the long structure, several subsections are joined together at the unexcited cells. (Courtesy of Stanford University.)

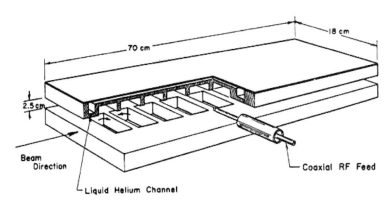

Figure 1.8: Schematic of the 11-cell muffin-tin cavity showing the input power coupler. The two halves are supported by a ring (not shown) with beam and coupler holes.

in Figure 1.7. The University of Illinois completed construction in 1977 of a microtron using sections of the SCA that provided 13 MV [23]. These accelerators aimed to provide a cw high-current ($\approx$ several $\times 10$ μA) beam for precision nuclear physics research. Both at Stanford and at Illinois the 1.3-GHz structures operated at an accelerating gradient of about 2 MV/m. The phenomenon of multipacting limited the achievable gradient. The HEPL accelerator continues to operate as a free electron laser (FEL).

The desire to increase the energy of the Cornell 12-GeV electron synchrotron with high gradient superconducting cavities gave rise [24] to the noncylindrically symmetric, muffin-tin structure (Figure 1.8). In aiming for a high

HISTORICAL FOUNDATIONS OF RF SUPERCONDUCTIVITY

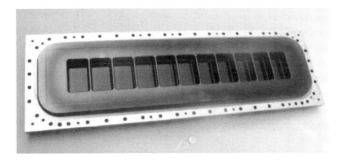

Figure 1.9: One half of the 11-cell niobium, muffin-tin cavity, machined from a solid piece. The active cell length is 60 cm.

Figure 1.10: A 1.5-GHz niobium muffin-tin cavity, active length 0.5 m. On the left, entering the cavity from the top half, is the waveguide input coupler. Two waveguide HOM couplers are attached to the end cells on the bottom half. Two such cavities were used for the first test of a superconducting cavity in an electron-positron colliding beam accelerator.

gradient, an important idea was to use a higher rf frequency (2.86 GHz). This would push the fields at which multipacting would occur to correspondingly higher values. Because of the smaller beam aperture at the higher frequency it was necessary to invent a structure open in the midplane so that the horizontal fan of synchrotron radiation could escape the cavity without hitting the cold walls. The rectangular geometry was a natural way to open the midplane.

In 1975 a 60-cm-long, 2.86-GHz muffin-tin structure (Figure 1.8 and Figure 1.9) successfully accelerated an electron beam of 110 μA to 4 GeV in the Cornell Electron Synchrotron [25]. This was the first application of rf superconductivity to a high-energy-physics accelerator. At a Q_0 of 10^9, the gradient reached was 4 MV/m, limited by thermal breakdown.

Continuing the exploration of rf superconductivity for high-energy physics, the first test of a superconducting cavity in a high-energy-physics storage ring

Table 1.2: Performance of superconducting cavities in storage ring beam tests carried out at several laboratories

Lab	Yr	Accel.	Cells	Freq. MHz	E_{acc} MV/m	Limit
Cornell [26]	82	CESR	2×5	1,500	1.9	MP[a]
Karlsruhe [27]	82	PETRA	1	500	2.3	MP[a]
CERN [28]	83	PETRA	5	500	2.1	Low Q_0[b]
KEK [29]	84	TAR[c]	3	508	3.5	Vacuum
KEK [30]	84	TAR[c]	3	508	4.1	IC[d]
Cornell [31]	84	CESR	2×5	1,500	6.5	FE[e]
					2.4	Q_0
DESY [32]	85	PETRA	9	1,000	2.5	Q_0
KEK [33]	86	TAR	5	508	3.5	FE[e]

[a] MP = multipacting
[b] refrigerator capacity imposed limit
[c] TAR = Tristan Accumulation Ring
[d] IC = input coupler
[e] FE = field emission

was carried out in 1982 at the Cornell Electron Storage Ring (CESR) using two 1.5-GHz muffin-tin structures (Figure 1.10), operating at an average gradient of 1.9 MV/m and a Q_0 of 10^9. The maximum current stored was 12 mA [26].

This test was soon followed by a series of successful storage ring beam tests as described in Table 1.2. Continued development of superconducting cavities for electron accelerators is covered later in this chapter.

1.5.2 Protons and Heavy Ions (Low-Velocity Particles, v/c < 0.3)

The Institute of Nuclear Physics at Karlsruhe (KFK) began in the late 1960s to explore a superconducting linear accelerator for protons using helically loaded niobium cavities [34]. A similar exploration started at Argonne National Labs [35]. Niobium pipes carrying liquid helium were wound into a helix and suspended from the cylindrical wall of the vacuum tank (Figure 1.11). This type of structure exhibited poor mechanical stability so that vibrations made it very difficult to control the rf phase.

In 1974, the California Institute of Technology [36] successfully tackled the mechanical stability problem by developing a split-ring cavity geometry, shown in Figure 1.12. Fabricated from a thin film of lead electroplated on an underlying copper structure, this resonator became the basis of the booster accelerator at the State University of New York at Stony Brook [37], which began operation in 1985. The typical operating gradient is 2.5 MV/m.

In 1975 Argonne National Lab [38] started development of a niobium split-ring structure (Figure 1.13) which eventually became the basis of ATLAS (Argonne Tandem Linear Accelerator System). Commissioned in 1978, ATLAS

Figure 1.11: Helical niobium resonator developed at Argonne National Lab for a heavy-ion linac. The helix was later supplanted by the split-ring structure. (Courtesy of Argonne National Labs.)

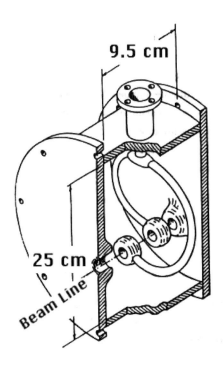

Figure 1.12: The three-gap split loop, lead-on-copper resonator originated at California Institute of Technology, now used in the Stony Brook heavy-ion linac. It is suitable for accelerating charged particles with $\beta = 0.1$; the rf frequency is 150 MHz.

Figure 1.13: The niobium split-ring resonators used in the heavy-ion accelerator facility, ATLAS. From left to right, the structure properties are: 97 MHz, $\beta = 0.1$; 145.5 MHz, $\beta = 0.16$; 97 MHz, $\beta = 0.06$. The niobium components are hollow, and filled with liquid helium at 4.2 K. The outer housing is made from a niobium-copper composite and cooled by conduction. (Courtesy of Argonne National Labs.)

holds the world record for the longest running accelerator based on rf superconductivity [39]. It has accumulated more than 100,000 hours of beam on target. The on-line operating gradient for the split-ring structures is 2.5–3.5 MV/m. Throughout its illustrious career, ATLAS has seen several upgrades and continues to operate as a world-class nuclear physics research facility.

To improve the mechanical properties of the helix, KFK [40] modified the geometry into the tapered helix structure (Figure 1.14). This resonator later became the basis of the heavy-ion booster accelerator at Saclay [41] that began operation in 1988. The typical operating gradient was 2.5 MV/m. This accelerator was retired in 1995 after a successful career of nuclear physics research.

A structure with mechanical stability superior to both the split-ring and the modified helix is the coaxial quarter-wave resonator, conceived at Stony Brook [42]. Such a resonator (Figure 1.2), which uses straight inductors, also has the advantage of lower peak surface magnetic field at the expense of a larger transverse dimension. The quarter-wave resonator became the basis for several new projects, such as at the University of Washington [43], JAERI (Tokai, Japan) [44], and INFN Legnaro, Italy [45]. Further development of rf superconductivity for low-velocity particles is covered in the state-of-the-art section.

STATE OF THE ART FOR ACCELERATORS

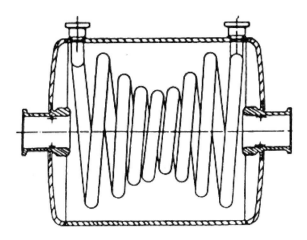

Figure 1.14: Tapered-helix structure developed at KFK and used in the Saclay heavy-ion linac. This is a $\lambda/2$-long, two-gap equivalent resonator with rf frequency 81 MHz and $\beta = 0.09$. Courtesy of CEN, Saclay.

1.5.3 Other Early Applications

In the early 1970s, efforts were started by a CERN-Karlsruhe collaboration [46] and by Brookhaven National Lab (BNL) [47] to use superconducting cavities to separate K mesons from the beam line of proton synchrotrons. The CERN group successfully completed a 2.86-GHz, 5.5-meter separator structure (Figure 1.15 and Figure 1.16) which operated in the CERN beam line for more than 3,000 hours [48]. K mesons between 8 and 26 GeV/c and antiprotons between 3.3 and 37 GeV/c were successfully separated. An enrichment factor of 10 with respect to unwanted particles was achieved. The average deflecting field achieved was 1.3 MV/m (limited by thermal breakdown) at a Q_0 above 2×10^9.

1.6 STATE OF THE ART FOR ACCELERATORS BASED ON RF SUPERCONDUCTIVITY

Based on the success of the pioneering efforts described in the previous sections, as well as on steady advances in achievable gradients, large-scale application of superconducting cavities to electron and ion accelerators is now established at many laboratories around the world. These accelerators provide high-energy electron and positron beams for elementary particle research, medium-energy electron beams for nuclear physics research, low-energy, heavy-ion beams for nuclear research, and high-quality electron beams for free electron lasers.

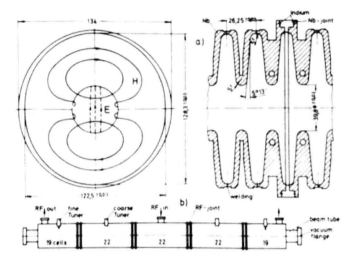

Figure 1.15: (a) Field lines for the deflecting mode. (b) Geometry of a deflecting cell and a joint cell. (c) Schematic layout of one separator assembled from five sections. The various ports are for couplers and tuners. (Courtesy of CERN.)

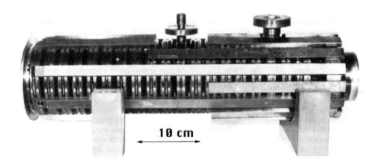

Figure 1.16: A 19-cell niobium separator section, overall length 60 cm. The section is reinforced by niobium bars. (Courtesy of CERN.)

1.6.1 Heavy Ion Linacs

Heavy ions, from helium to uranium, are accelerated to energies from a few to 20 MeV/nucleon and then used in experiments as projectiles to bombard other nuclei. Above $\approx$ 5 MeV/nucleon, the ions have sufficient energy to overcome the Coulomb barrier and penetrate the nucleus. The collisions cause energy, mass, and angular momentum to be transferred between the projectile and target nuclei, enabling structure research on the evolution of nuclear shape as a function of excitation energy and other aspects, such as spin.

Superconducting linacs providing precision beams of heavy ions have consistently been one of the most successful applications. These machines are energy boosters for tandem Van de Graaff accelerators, which are typically limited to 10 MV by the large capital cost of a Van de Graaff. RF superconductivity provides cw operation at gradients > 2 MV/m, making it economically viable to increase the energy of the tandem. Many university-based laboratories have found that the capital cost of a superconducting booster is much lower than the equivalent voltage electrostatic tandem. While other accelerating systems, such as cyclotrons, produce high ion beam energies, a superconducting linac preserves the excellent beam quality provided by the tandem accelerator. CW operation possible with superconducting resonators eliminates transient effects and thereby provides the beam quality essential for good physics.

Heavy-ion accelerators must efficiently accelerate particles whose velocity changes along the accelerator. They must also be able to accelerate a variety of ions with different velocity profiles. Several structures are therefore needed, each of which must be optimized for a particular velocity range. A major advantage of superconducting resonators is that a high voltage can be obtained in a short structure. The booster linac can therefore be formed as an array of independently phased resonators, making it possible to vary the velocity profile of the machine. The superconducting booster is therefore capable of accelerating a variety of ion species and charge states. The independently phased array forms a system which provides a high degree of operational flexibilty and tolerates variations in the performance of individual cavities. Superconducting boosters therefore have excellent transverse and longitudinal phase space properties, and they excel in beam transmission and timing characteristics. Because of the intrinsic modularity, there is also the flexibility to increase the output energy by adding higher β sections at the output, or to extend the mass range by adding lower β resonators at the input.

The peak surface fields to accelerating field ratios are significantly higher in the low-velocity structures. For a gradient of 1 MV/m, the peak surface electric field typically ranges from 4 to 6 MV/m (compared to 2 to 2.6 MV/m for $\beta = 1$ structures) and the peak surface magnetic field ranges from 60 to 200 Oe (compared to 40 to 47 Oe for $\beta = 1$ structures).

Besides the pioneer accelerators at Argonne and Stony Brook discussed in the historical introduction, eight heavy-ion accelerator facilities are now operating, utilizing over 225 resonators made of niobium or lead-on-copper. These facilities are located at the University of Washington, Florida State University, Kansas State University, CEN Saclay (France), JAERI (Tokai, Japan), and INFN (Legnaro, Italy). These accelerators provide beams with mass up to 100 atomic units and energies up to 25 MeV per nucleon. New heavy-ion accelerator facilities utilizing superconducting structures are being planned or under construction at Australian National University, Sao Paolo (Brazil), New Delhi and Bombay. Figure 1.17 shows an assembly of quarter wave resonators ready for installation into the cryostat for the JAERI heavy-ion linac. Table 1.3 taken from a recent review [19], summarizes the achieved performance of the structures at the various institutions.

While ATLAS at Argonne has been operating routinely since 1978, a major

Figure 1.17: An assembly of four quarter-wave resonators for JAERI; $\beta = 0.1$, frequency = 130 MHz. The overall height of each resonator is about 60 cm. The resonators correspond to the sketch of Figure 1.2. (Courtesy of JAERI.)

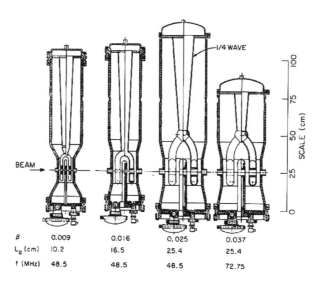

Figure 1.18: Interdigital structures for the ATLAS injector linac. The niobium drift tubes and supporting center conductors are hollow and filled with liquid helium. The housing is made from niobium-copper composite. (Courtesy of Argonne.)

Table 1.3: Performance of superconducting resonators for heavy ion linacs

Laboratory	Structure	β	RF Freq. (MHz)	$\Delta E/q$ (MV)	Dissipated Power (Watt)
Stony Brook	Pb/Sn quarter wave	0.068	150	0.51	6
Stony Brook	Pb/Sn split loop	0.1	150	0.67	6
JAERI	Nb quarter wave	0.08	129	0.51	4
U. Washington	Pb quarter wave	0.1	149	0.54	8
U. Washington	Pb quarter wave	0.2	149	0.86	16
Argonne	Nb interdigital	0.009	49	0.45	6
Argonne	Nb interdigital	0.016	49	0.48	6
Argonne	Nb interdigital	0.025	49	0.76	6
Argonne	Nb interdigital	0.037	73	0.76	6
Argonne	Nb split ring	0.1	145	1.08	4
Argonne	Nb half wave	0.12	355	1.26	40
Argonne	Nb half wave	0.12	355	1.05	10
Argonne	Nb quarter wave	0.15	400	0.82	21
Argonne	Nb spoke	0.3	855	0.43	24
INFN-Legnaro	Nb quarter wave	0.056	80	0.76	5
INFN-Legnaro	Nb quarter wave	0.056	80	0.72	1
INFN-Legnaro	Nb quarter wave	0.11	160	0.9	10
INFN-Legnaro	Nb quarter wave	0.17	240	0.85	6

upgrade [49] has been to replace the combination of negative ion source and tandem accelerator by an electron cyclotron resonance (ECR) ion source and a superconducting injector linac with a series of superconducting resonators spanning β values between 0.009 and 0.037. The ECR is a plasma device designed to provide highly charged ions at low velocities. A solenoidal magnetic field provides radial confinement of electrons and ions, while an array of permanent magnets gives azimuthal confinement. The ions within the trap region are bombarded by electrons excited by microwaves into electron cyclotron resonance. Located on a high-voltage platform, the ECR source can produce ions with high charge states and sufficient velocity to be injected directly into the upgraded superconducting linac. Argonne has successfully incorporated low-β quarter-wave resonators into an interdigital structure (see Figure 1.18) to accelerate the very-low-velocity ions.

The gradients obtained during operation of heavy-ion accelerators are in the range of 2.5 to 3.5 MV/m for most structures. The niobium interdigital structures perform appreciably above these levels. For example, the lowest-velocity section ($\beta = 0.009$) operates at 6 MV/m.

The highest-β resonator developed is the niobium two-gap spoke resonator with $\beta = 0.28$ (Figure 1.19(a) and Figure 1.19(b)). In laboratory tests a cw accelerating field of 11 MV/m was obtained. A record accelerating field of 18 MV/m was achieved in a 355 MHz coaxial half-wave resonator (Fig-

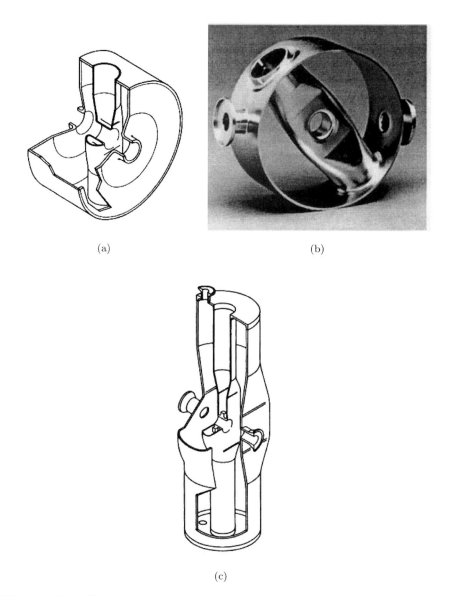

Figure 1.19: The spoke resonator with $\beta = 0.28$, frequency 855 MHz. (a) Schematic, (b) photograph. (c) A record accelerating field was obtained in this niobium half-wave resonator developed for future high current proton linacs ($\beta = 0.12$, frequency = 355 MHz). (Courtesy of Argonne.)

ure 1.19(c)) with $\beta = 0.12$. The associated E_{pk} and H_{pk} were respectively, 58 MV/m and 936 Oe [50].

1.6.2 Storage Rings

To study the fundamental properties of matter, high-energy physicists have built colliding beam storage rings of steadily increasing energies. Beams of charged particles circulating in opposite directions collide at one or more interaction points, where a detector is placed to track and measure secondary particles produced in the collisions. The colliding beams are electron-positron, (e.g., CESR at Cornell, TRISTAN at KEK in Japan, and LEP at CERN in Switzerland) proton-antiproton (e.g., SPS at CERN and the TEVATRON at Fermilab), or electron-proton (e.g., HERA at DESY in Germany). Several of these machines have already benefited from the use of superconducting rf technology.

For several years, TRISTAN with superconducting cavities held the record for the highest energy e^+e^- collider at 32 GeV per beam. Located at CERN, the LEP has a circumference of 27 km and operates at an energy above 50 GeV per beam. We will discuss the program presently underway to raise the energy toward 100 GeV/beam. With the higher energy, LEP-II has confirmed the existence of the W meson (one of the carriers of the weak force) and plans to measure the mass with a high accuracy. Such studies rigorously test the consistency of the Standard Model of high-energy physics which embodies our present understanding of the fundamental nature of matter. Deviations from the Standard Model at LEP and elsewhere would signal the presence of new physics. LEP also hopes to shed light on the mass and the nature of the Higgs boson to provide an understanding of the wide disparity between the masses of the elementary quarks and leptons, the basic building blocks of all matter. (See also Chapter 20)

Another frontier high-energy physics accelerator that uses superconducting rf is HERA, which provides electron-proton collisions at 300 GeV in the center of mass (CM). Located in a common tunnel, an electron ring operates at a beam energy of 30 GeV and a superconducting-magnet-based proton ring operates at a beam energy of 800 GeV. One of the objectives of HERA is to probe the structure of the proton by deep inelastic scattering to very high momentum transfer.

Electrons in storage rings lose energy in the form of synchrotron radiation. Because the energy loss increases as the fourth power of the beam energy, the electron storage rings TRISTAN, HERA and LEP have needed high gradient superconducting cavities to increase their beam energies. There is another important reason to use superconducting cavities in such machines. As we shall discuss in Chapter 2, with superconducting cavities it is possible to choose a cavity geometry with large beam holes, as compared to normal conducting structures. Therefore superconducting cavities present a much lower impedance to the beam. This helps to avoid multibunch beam instabilities and permits the operation of storage rings at higher currents to reach higher luminosity.

The structures for TRISTAN, HERA and LEP are chains of 4- or 5-cell

Table 1.4: Achieved beam energy increase for TRISTAN, HERA, and LEP

Accelerator	Meters Installed	Volts Installed (MV)	Original Beam Energy (GeV)	Upgraded Beam Energy (GeV)
TRISTAN	48	200	27	32
HERA	20	75	28	30
LEP	246	1320	50	80

cavities with resonant frequencies between 350 MHz and 500 MHz. Table 1.4 gives for each storage ring the length of superconducting structures installed, the voltage delivered, and the upgraded beam energy. Each structure is equipped with input couplers that must deliver 50–100 kW to the beam. Output HOM couplers must remove 10–100 W of beam-induced power, and must also damp dangerous modes to Q_0 values of 10^3 to 10^4, so that the beam-induced energy will not disturb subsequent beam bunches. The principles and techniques of input and HOM couplers will be covered in Part III. Altogether, more than 300 meters of superconducting cavities have been installed in TRISTAN, HERA and LEP. These structures have been operated at gradients between 3 and 6 MV/m.

Figure 1.20(a) shows the layout of two 5-cell cavities with couplers, housed inside the cryostat as used in TRISTAN [51]. Figure 1.20(b) shows a 5-cell niobium TRISTAN cavity [52]. TRISTAN was one of the first high-energy physics machines to utilize superconducting cavities on a large scale. Forty-eight meters were operated for more than 6 years [53]. In the absence of beam, these cavities have been tested at 6–7 MV/m in the accelerator tunnel. With beam current up to 14 mA, they were run at 4–5 MV/m. For comparison, the copper rf systems in the same machine run at 0.5–1 MV/m. The maximum power delivered to the beam by the superconducting cavities was 70 kW/coupler while 10–100 W were removed via the HOM couplers. More than 40,000 hours of running time with beam have been accumulated in TRISTAN without any degradation of cavity performance over many years on line. The cavities at the end of the each 24-meter string were located close to the machine dipoles and quadrupoles so that, due to the higher incident synchrotron radiation, the end cavities showed a tendency to trip more frequently than the rest of the cavities. There is evidence that trips usually came from excess arcing in the HOM couplers. The TRISTAN superconducting rf system is now removed for construction of the B-factory, KEK-B. A new superconducting rf system under development for possible use in the high-energy ring of KEK-B will be discussed in Chapter 20 [54].

HERA has been operating with 20 meters of superconducting cavities for more than 4 years [55]. Figure 1.21 shows a 4-cell niobium cavity for the HERA electron ring. After installation in the tunnel, these cavities were run at 6–7 MV/m without beam and 3–5 MV/m with beam. In many cases, the operating gradient with beam has been limited by the available rf power for the beam. The maximum beam current accelerated is 50 mA in HERA.

STATE OF THE ART FOR ACCELERATORS

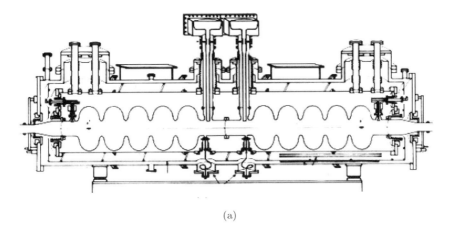

(a)

(b)

Figure 1.20: (a) TRISTAN 5-cell, 500-MHz cavities inside their cryostat. The active length of each structure is 1.5 m. Input couplers enter at the center of the cryostat. HOM couplers are visible at the outer ends of the structures. (b) TRISTAN niobium, 5-cell cavity, active length 1.5 m. Niobium stiffening rings are welded to each cell. Coupler ports are welded on the beam tube. Thirty-two such cavities were operated in TRISTAN for more than 6 years. (Courtesy of KEK.)

Figure 1.21: (Front left) A 500 MHz, 4-cell niobium cavity for HERA, active length 1.2 m. (Front right) A niobium cavity inside its liquid helium vessel. (Back) One of eight complete cryomodules that have been operating in HERA for more than 4 years. (Courtesy of DESY.)

Figure 1.22: A 10-meter-long cryomodule for LEP, housing four 4-cell, 350-MHz cavities made from a niobium film sputtered onto a copper cavity. More than 50 such cryomodules will be used to upgrade the energy from 50 GeV to 100 GeV per beam. Thirty-six cryomodules have already been installed and operated successfully. (Courtesy of CERN.)

A major superconducting rf installation is in progress at CERN to raise the beam energy of LEP from 50 to 100 GeV per beam [56]. The energy loss per turn is 1860 MeV per electron at 90 GeV, and the power lost due to synchrotron radiation is about 6 MW per beam for a current of 3 mA. As of July 1996, 246 meters of cavities were installed in LEP to increase the energy of LEP to 80 GeV per beam. Figure 1.22 shows a cryomodule housing four 4-cell LEP cavities.

Several of the installed modules have been operated at 6 MV/m, at a $Q > 3 \times 10^9$, with a beam current of 7 mA. The superconducting rf system boosts the voltage supplied by 120 copper cavities (280 MV) to more than 1600 MV. When completed, LEP-II will have installed a total of 465 meters of superconducting rf cavities. Instead of using cavities fabricated out of bulk sheet niobium, a unique approach adopted for LEP-II is to sputter a thin film of niobium on to a cavity fabricated out of copper sheet. The techniques and results for this new technology are discussed in more detail in Chapter 14.

LEP-type cavities have also been installed in the CERN SPS for the lepton acceleration cycle and have operated routinely for several years at 5 MV/m [57]. These are single-cell 400 MHz cavities. More than 25,000 hours of operation have been accumulated. A special feedback system was developed to zero out the shunt impedance of the cavity during the proton acceleration cycle, so that the cavities do not disturb the proton beam. The average proton current is 50 mA and the peak current is 300 mA.

1.6.3 Recirculating Linacs

Medium energy electron accelerators are desired to improve the basic understanding of nuclear matter, in particular to elucidate the quark and gluon structure of protons and neutrons. Superconducting cavities offer special advantages to electron accelerators for nuclear physics in the 1 to 10 GeV energy range: high average current, low peak current, continuous beam, and excellent beam quality. For precise measurement of small electromagnetic cross sections and for coincidence detection of reaction particles, a high beam duty factor, preferably cw, and a high average beam current (100–200 μA) are desirable. In addition, the beam must have a high quality for adequate resolution of closely spaced nuclear states, low energy spread ($< 10^{-4}$) and low transverse emittance ($< 10^{-5}$ m-rad, normalized) to reduce background arising from the beam halo. Because of the highly stable operation possible with a cw superconducting linac, the rf phase and amplitude can be controlled very precisely, yielding a very low energy spread. In cw operation, the desired average beam current can be achieved with a low peak current made possible by the superconducting linac. Therefore the interaction of the beam with the cavity and the vacuum chamber is weak and the small emittance of the beam can be preserved through the linac.

Superconducting cavities for recirculating linacs have rf frequencies between 1300 and 3000 MHz. Stanford University (HEPL) began operating the first such machine in 1977 [22]. They installed 27 meters of niobium structures at 1300 MHz, and they reached an output energy of 84 MeV in two passes.

Figure 1.23: A 1500-MHz, 5-cell niobium cavity pair for TJNAF (formerly CEBAF). The active length is 1 m. The cavities have waveguide input couplers (center) and waveguide HOM couplers (at ends). (Courtesy of TJNAF.)

Figure 1.24: A pair of 1500-MHz 5-cell, niobium cavities tested in CESR at Cornell before being adopted by TJNAF (formerly CEBAF).

Darmstadt has been operating the DALINAC [58] for more than 5 years with 10 meters of 3-GHz cavities at gradients between 4 and 10 MV/m (average 6 MV/m). They have achieved a beam energy of 120 MeV in three recirculating passes, and they have obtained beam currents up to 60 μA, with an energy spread of 2.5×10^{-4}, and emittance of $2\pi \times 10^{-6}$. The first structures for the S-DALINAC were designed, constructed and prepared by the University of Wuppertal [59]. Since that time, the structures, couplers and tuners have been upgraded to reach the desired level of performance.

The 4-GeV recirculating linac, TJNAF (formerly CEBAF), is one of the two biggest superconducting rf installations to date (July 1996) with 170 meters of 1.5-GHz cavities. The cavity design (Figure 1.23) chosen for CEBAF was based on the Cornell design (Figure 1.24) because it met gradient and Q_0 requirements, damped higher-order modes well, and was proven in a beam test at CESR with a beam current of 26 mA at an average $E_{\text{acc}} = 4.5$ MV/m [31].

TJNAF has achieved the design beam energy of 4 GeV in five recirculating passes with a cw beam current of 25 μA, as chosen for the commissioning stage. The normalized emittance achieved was $< 1.6 \times 10^{-5}$ m-rad and the energy spread was 2.8×10^{-5}. In the accelerator, the cavities average gradients above 6 MV/m at a $Q_0 > 5 \times 10^9$ [60]. Once all the klystrons are operating at their design value of 5 kW, it is expected that the cavities will run at an average gradient of 7–8 MV/m to yield a significant energy upgrade. The maximum gradient reached on-line was 14 MV/m. In early commissioning tests, the design current of 200 μA was accelerated through the first 9 meters to 45 MeV. At present the maximum current is 125 μA.

1.6.4 Free Electron Lasers

Free electron lasers (FELs) offer many desirable characteristics over conventional lasers: wavelength tunability, high average power, and high efficiency of conversion of ac to laser power. Because the lasing principle does not involve transitions between fixed atomic or molecular levels, as for chemical lasers, the FEL is wavelength tunable (by adjusting the beam energy). High peak power and high average power infrared and ultraviolet FELs serve as valuable research tools in solid state physics, chemistry, biology and medicine. They offer a variety of applications in high-power microwaves, materials processing, surface processing, micromachining, surgery, and defense [61].

There are many candidate drivers for an FEL: direct current (dc) electrostatic accelerators, storage rings, induction linacs and rf linacs. The rf linac is suitable for a variety of FELs designed to generate infrared light or ultraviolet light. A linac driver offers high extraction efficiency and therefore higher power. To achieve lasing, it is necessary to focus the electron beam inside the laser beam so that there is adequate spatial overlap between the two beams. Therefore good beam quality (i.e., low energy spread and low emittance) is essential for FEL operation. A superconducting linac is ideally suited to provide the excellent beam quality for the same reasons discussed in connection with a cw linac for precision nuclear physics.

It is possible to envision both high power and high efficiency in the superconducting linac-driven FEL. In high-power lasers, most of the energy input into the laser is rejected as waste heat, so that the power limitation is eventually set by the ability to quickly remove the waste heat from the gas, as in the conventional high-power CO_2 lasers. In an FEL, the electron beam can carry a large power, and the waste heat in the form of the electron beam moves out very fast. Therefore, the electron beam energy not converted to photon energy is largely recoverable.

The first FEL beam was demonstrated nearly two decades ago with a 50-MeV beam from the SCA at Stanford [62]. In the SCA experiment at 1.6 microns, more than 1% of the beam energy was converted to laser energy. Energy recovery was also demonstrated at the SCA [63]. A 50-MeV beam was decelerated to 5 MeV by recirculating through the linac at the appropriate phase. In this experiment, the power required to maintain the operating field level was less than 10% of the power required in the nonrecovery mode. Thus the overall system efficiency, defined as photon power out/rf power supplied to the beam, was increased by a factor of 10. The Stanford Picosecond FEL Center is now the home of SCA. The FEL produces intense picoseond pulses of light in the mid- and far- infrared regions of the spectrum (3–50 microns wavelength). The wavelength is continuously and rapidly tunable.

New FEL projects based on superconducting linacs are underway at Darmstadt[58], JAERI[64] and TJNAF[65]. At Darmstadt, a 50-MeV beam from the S-band DALINAC will be diverted through a wiggler and mirror section to lase at 2.5 and 7.0 microns. Spontaneous emission in the IR has already observed. JAERI has completed a 13-MeV superconducting linear accelerator using 500-MHz cavities. The laser is designed for 10.3 μm. The future FEL facility at TJNAF aims to provide 1–3 kW average power in the IR and UV ranges. In the first phase, a beam energy of 100 MeV will be reached at the end of a 24-cavity linac. A recirculation leg for energy recovery is planned. In the second phase, the beam will be recirculated to double the energy for the UV light.

1.7 A SUMMARY OF THE STATE OF THE ART

All together more than 500 meters of superconducting cavities have been installed worldwide and sucessfully operated at accelerating fields up to 6 MV/m to provide a total of more than 2.5 GV for a variety of accelerators. Figure 1.25 shows the rapid growth in the voltage provided by superconducting cavities over the last decade. For ion accelerators, there has been steady growth since the early 1970s. One can see that the application to electron accelerators was nearly at a standstill in the decade between 1974 and 1984. During this decade, research and development in the technology was progressing to overcome the field limiting problems of multipacting and thermal breakdown.

Upon installation of superconducting structures into an accelerator, the accelerating gradients reached are usually found to be 20–30% lower, as shown in Figure 1.26, which summarizes the experience of KEK [66] and TJNAF [67]. The degradation is attributed to two factors. The first reason is that during the accelerator commissioning stage the full rf power is not available. For example, at present, TJNAF operates its klystrons at 2 kW instead of the design 4 kW. As mentioned before, the average on-line accelerating gradient for TJNAF is expected to increase to 8 MV/m when the full power is applied. Secondly, field emitting dust particles are introduced into the cavity during assembly of the power couplers and during installation of the cavity into the accelerator beam line. We will see in Chapter 13 that high-pulsed-power rf processing (HPP) promises to combat the degradation.

In closing, we have seen that practical structures with attractive performance levels have been developed for a variety of accelerators, namely, for linacs for electrons and ions as well as for electron and proton storage rings. These structures have been installed in the intended accelerators and operated at the desired beam current for many years. To reach the design gradients, substantial progress has been made in understanding field and Q_0 limitations and in inventing new techniques for cavity preparation. As we will see in later chapters, the progress of gradients continues, and with it new applications are emerging.

A SUMMARY OF THE STATE OF THE ART

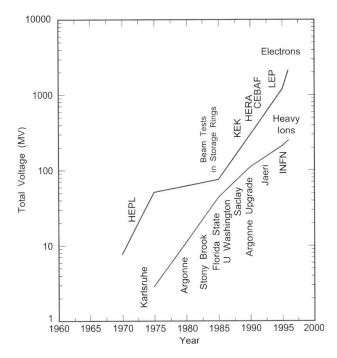

Figure 1.25: Total installed voltage capability with superconducting cavities for electron and heavy-ion accelerators. The growth in application to electron accelerators leveled out during the decade between 1974 and 1984 because of the difficulties presented by multipacting and thermal breakdown. Once these limitations were overcome, the accumulated installed voltage began to grow exponentially.

In considering new applications of superconducting cavities, the gradient and aperture advantages must be balanced against the added cryogenic complexity of the refrigerator and cryogen distribution system, as well as the demands for clean surface preparation. With the present niobium-based technology, superconducting cavities are limited to accelerating fields less than 60 MV/m. At higher fields, the corresponding surface magnetic field will exceed the superheating critical field so that a phase transition to the normal state will occur. Another factor to bear in mind is that the useful length to active length ratio is typically 50% due to the filling factor of active cavities in the cryostat as well as to the need for other accelerator components in the beam line. In the future, better cryomodule designs hope to improve the filling factor to seventy per cent.

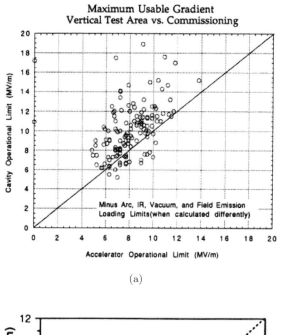

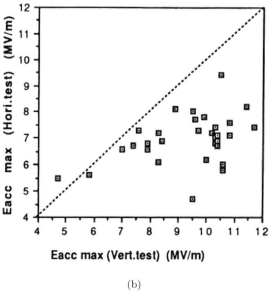

Figure 1.26: A comparison of the on-line performance of structures with their performance in the acceptance test carried out in the laboratory in a vertical test cryostat. (a) TJNAF: on-line results on horizontal axis. (b) TRISTAN: on-line results on vertical axis. (Courtesy of KEK and TJNAF.)

CHAPTER 2

Cavity Fundamentals and Cavity Fields

This chapter outlines how the fields in a cavity can be calculated. In the process we determine various important quantities such as the accelerating field, the peak electric and magnetic fields, the cavity quality factor, and the shunt impedance. These figures of merit shed light on a cavity's effectiveness in accelerating charges and permit a comparison between different shapes.

To illustrate these concepts we will often revert to an idealized cavity — the pill-box shape. Although it is not used for practical applications due to multipacting limitations, it is a useful example because it lends itself to analytic calculations. Nevertheless, the quantities developed here apply equally well to more practical cavity shapes. At the end of this chapter we describe computer codes developed to handle cavity shapes which are not amenable to an analytic treatment. This review is only intended to apprise the reader of the codes that are available. It does not, however, provide an exhaustive treatment.

2.1 RADIO-FREQUENCY FIELDS IN CAVITIES

Consider the electromagnetic fields near a perfect conductor. According to Maxwell's equations, outside the conductor the electric field ($\mathbf{E}$) is perpendicular to the surface whereas the magnetic field ($\mathbf{H}$) must be parallel to the surface. This is expressed by the boundary conditions

$$\hat{n} \times \mathbf{E} = 0, \quad \hat{n} \cdot \mathbf{H} = 0, \tag{2.1}$$

where $\hat{n}$ is the unit vector normal to the surface of the conductor. Consider next the fields, excited by a radio-frequency (rf) source, in an infinite waveguide of uniform cross-section bounded by a perfect conductor (Figure 2.1). We assume that the spatial and temporal variations of the fields are given by

$$\mathbf{E}(\mathbf{x}, t) = \mathbf{E}(\rho, \phi) e^{ikz - i\omega t} \tag{2.2}$$
$$\mathbf{H}(\mathbf{x}, t) = \mathbf{H}(\rho, \phi) e^{ikz - i\omega t}. \tag{2.3}$$

Here ω is the angular frequency and k is the wave number. Maxwell's equations combine to yield the wave equation

$$\left(\nabla^2 - \frac{1}{c^2} \frac{\partial^2}{\partial t^2} \right) \left\{ \begin{array}{c} \mathbf{E} \\ \mathbf{H} \end{array} \right\} = 0, \tag{2.4}$$

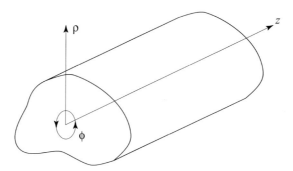

Figure 2.1: A constant cross-section waveguide.

where c is the speed of light and we are assuming that the relative permeability and permittivity are both unity. Substituting the fields in (2.2) and (2.3) into (2.4) reduces the wave equation to

$$\nabla_\perp^2 + \left(\frac{\omega^2}{c^2} - k^2\right) \left\{ \begin{array}{c} \mathbf{E} \\ \mathbf{H} \end{array} \right\} = 0, \qquad (2.5)$$

where

$$\nabla_\perp^2 = \nabla^2 - \frac{\partial^2}{\partial z^2}. \qquad (2.6)$$

The solutions to the eigenvalue equation (2.5) form an orthogonal set with eigenvalues $\gamma^2 = \omega^2/c^2 - k^2$.

Maxwell's equations can be combined to express the transverse fields in the waveguide ($E_\perp$ and $H_\perp$) as a function of the longitudinal components (E_z and H_z) [68]. However, E_z and H_z are independent of each other. Furthermore, the boundary conditions in (2.1) can be recast as

$$E_z|_S = 0, \qquad \left.\frac{\partial H_z}{\partial n}\right|_S = 0. \qquad (2.7)$$

The different boundary conditions imposed on E_z and H_z and the fact that E_z and H_z are independent imply that the solutions to (2.5) form two sets of modes, generally with different eigenvalues. The two families are denoted *transverse magnetic* (TM) modes and *transverse electric* (TE) modes. For TM modes a longitudinal component of $\mathbf{E}$ exists whereas $\mathbf{H}$ is transverse everywhere (with respect to the z axis). Conversely, for TE modes a longitudinal component of $\mathbf{H}$ exists and $\mathbf{E}$ is transverse everywhere. In both cases the transverse magnetic and electric fields are related by

$$\mathbf{H}_\perp = \pm \frac{\hat{z} \times \mathbf{E}_\perp}{Z}, \qquad (2.8)$$

where

$$Z = \frac{k}{\epsilon_0 \omega} \qquad (2.9)$$

RADIO-FREQUENCY FIELDS IN CAVITIES

is the wave impedance for TM modes and

$$Z = \frac{\mu_0 \omega}{k} \tag{2.10}$$

for TE modes. The positive case applies to waves traveling in the $+\hat{z}$ direction, while the negative case refers to backward traveling waves. The transverse components can be derived from the longitudinal components by

$$\mathbf{E}_\perp = \pm \frac{ik}{\gamma^2} \boldsymbol{\nabla}_\perp E_z \quad \text{(TM modes)} \tag{2.11}$$

$$\mathbf{H}_\perp = \pm \frac{ik}{\gamma^2} \boldsymbol{\nabla}_\perp H_z \quad \text{(TE modes),} \tag{2.12}$$

where E_z and H_z satisfy (2.5).

However, we are interested in cavities, not waveguides. A simple way of converting a waveguide into a cavity is to add conducting faces at $z = 0$ and $z = d$. Standing waves are created due to the reflections at the ends, where the boundary conditions (2.1) also need to be satisfied. We can combine forward and backward traveling waves to do this explicitly. One finds

$$E_z(\mathbf{x}, t) = \psi(\rho, \phi) \cos\left(\frac{p\pi z}{d}\right) e^{i\omega t}, \quad p = 0, 1, 2 \ldots \quad \text{(TM modes)} \tag{2.13}$$

$$H_z(\mathbf{x}, t) = \psi(\rho, \phi) \sin\left(\frac{p\pi z}{d}\right) e^{i\omega t}, \quad p = 1, 2, 3 \ldots \quad \text{(TE modes)} \tag{2.14}$$

with $k = p\pi/d$. When substituting the above expressions into the eigenvalue equation (2.5), we see that the fields $\psi(\rho, \phi)$ are the solutions of the eigenvalue equation

$$\left(\nabla_\perp^2 + \gamma_j^2\right) \psi(\rho, \phi) = 0, \tag{2.15}$$

where

$$\gamma_j^2 = \left(\frac{\omega_j}{c}\right)^2 - \left(\frac{p\pi}{d}\right)^2 \tag{2.16}$$

is the jth eigenvalue.

The transverse field components can still be obtained by applying (2.11), (2.12) and (2.8) to the forward and backward traveling components of (2.13) or (2.14) separately [68]. One finds for TM modes

$$\mathbf{E}_\perp = -\frac{p\pi}{d\gamma_j^2} \sin\left(\frac{p\pi z}{d}\right) \boldsymbol{\nabla}_\perp \psi(\rho, \phi) \tag{2.17}$$

$$\mathbf{H}_\perp = \frac{i\omega_j}{\eta c \gamma_j^2} \cos\left(\frac{p\pi z}{d}\right) \hat{z} \times \boldsymbol{\nabla}_\perp \psi(\rho, \phi) \tag{2.18}$$

and for TE modes

$$\mathbf{E}_\perp = -\frac{i\eta \omega_j}{c \gamma_j^2} \sin\left(\frac{p\pi z}{d}\right) \hat{z} \times \boldsymbol{\nabla}_\perp \psi(\rho, \phi) \tag{2.19}$$

$$\mathbf{H}_\perp = \frac{p\pi}{d\gamma_j^2} \cos\left(\frac{p\pi z}{d}\right) \boldsymbol{\nabla}_\perp \psi(\rho, \phi), \tag{2.20}$$

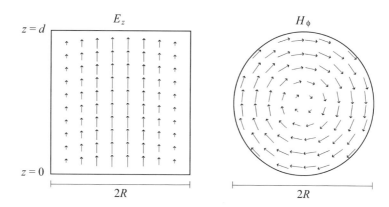

Figure 2.2: Vector plots of the electric and magnetic fields in the TM$_{010}$ mode of the pill-box cavity. Left: E_z in the ρ–z plane. Right: H_ϕ in the ρ–ϕ plane.

where

$$\eta = \sqrt{\frac{\mu_0}{\epsilon_0}} \qquad (2.21)$$

is the impedance of free space.

2.1.1 The Pill-Box Cavity

As an example of a TM mode, consider a cylindrical pill-box cavity of length d and radius R. The solutions to the eigenvalue equation (2.15) are Bessel functions. The lowest frequency TM mode is

$$E_z = E_0 J_0\left(\frac{2.405\rho}{R}\right) e^{-i\omega t} \qquad (2.22)$$

$$H_\phi = -i\frac{E_0}{\eta} J_1\left(\frac{2.405\rho}{R}\right) e^{-i\omega t}, \qquad (2.23)$$

and all other field components vanish. J_0 and J_1 are zeroth- and first-order Bessel functions, respectively. This mode is denoted TM$_{010}$ (the nomenclature is explained below) and its field pattern is shown in Figure 2.2.

The resonant frequency is given by

$$\omega_{010} = \frac{2.405c}{R}, \qquad (2.24)$$

which is independent of the cavity length. This is not surprising, considering that there is no variation of the fields in the $\hat{z}$ direction.

Normally the modes are classified by the nomenclature TM$_{mnp}$. The integer indices m, n, and p are measures of the number of sign changes E_z undergoes in the ϕ, ρ, and z directions, respectively.

For the pill-box cavity the complete set of TM_{mnp} modes is

$$E_z = E_0 \cos\left(\frac{p\pi z}{d}\right) J_m\left(\frac{u_{mn}\rho}{R}\right) \cos(m\phi) \qquad (2.25)$$

$$E_\rho = -E_0 \frac{p\pi R}{du_{mn}} \sin\left(\frac{p\pi z}{d}\right) J'_m\left(\frac{u_{mn}\rho}{R}\right) \cos(m\phi) \qquad (2.26)$$

$$E_\phi = E_0 \frac{mp\pi R^2}{\rho d u_{mn}^2} \sin\left(\frac{p\pi z}{d}\right) J_m\left(\frac{u_{mn}\rho}{R}\right) \sin(m\phi) \qquad (2.27)$$

$$H_z = 0 \qquad (2.28)$$

$$H_\rho = iE_0 \frac{m\omega_{mnp} R^2}{\eta c \rho u_{mn}^2} \cos\left(\frac{p\pi z}{d}\right) J_m\left(\frac{u_{mn}\rho}{R}\right) \sin(m\phi) \qquad (2.29)$$

$$H_\phi = iE_0 \frac{\omega_{mnp} R}{\eta c u_{mn}} \cos\left(\frac{p\pi z}{d}\right) J'_m\left(\frac{u_{mn}\rho}{R}\right) \cos(m\phi) \qquad (2.30)$$

$$\gamma_{mn} = \frac{u_{mn}}{R}, \qquad \omega_{mnp} = c\sqrt{\gamma_{mn}^2 + \left(\frac{p\pi}{d}\right)^2}, \qquad (2.31)$$

where u_{mn} is the nth root of $J_m(x)$ and $J'_m(x)$ is the derivative of $J_m(x)$ with respect to x.

A set of TE modes similar to those in (2.25)–(2.30) exists as well. However, in an exact pill-box cavity, TE modes have no longitudinal electric field and thus cannot accelerate the beam. Conversely, the beam cannot excite these modes and thus TE modes are of little interest to us until we consider real cavities with beam holes. We will therefore restrict our development to the TM modes.

To accelerate a beam, the longitudinal component of **E** must not vanish at the beam axis. Of all the Bessel functions, only J_0 satisfies this condition and all possible accelerating modes are of the type TM_{0np}. They are also called *monopole* modes because of their field distribution. The TM_{010} mode is usually chosen for acceleration because it has the lowest eigenfrequency.

Modes of the type TM_{1np} have a net deflecting field on axis. These are referred to as *dipole* modes, and they are undesirable in accelerating cavities because they disrupt the beam. Modes with $m = 2$ are *quadrupole* modes, and so on (see Problem 1).

To allow a beam through the cavity we need to add beam pipes at either end. In doing so, one has to ensure that the cutoff frequency of the beam pipes is above the frequency of the accelerating mode, so that the cavity fields cannot propagate out of the cavity.

The addition of these holes makes it very difficult to calculate the fields analytically. Techniques such as mode matching [69] can sometimes be used to yield analytical results. However, more commonly, computer codes such as SUPERFISH [70], URMEL [71], and SuperLANS [72] are used for numerical analyses. They are discussed at the end of this chapter.

In cases where the cavity is not too severely perturbed from the cylindrical case, one can identify modes similar to the TM and TE modes discussed above so that one maintains the same nomenclature. However, due to the perturbative effect of the beam tubes, TM modes have a finite H_z and TE modes acquire a longitudinal electric field. They are called *hybrid modes*. When considering the excitation of modes by the beam (discussed in Chapter 15) thus both TE and TM modes need to be considered.

2.1.2 The Accelerating Voltage

Assume an electron travels at the speed of light (c). This is a reasonable approximation for e^+e^- accelerators with energies greater than 10 MeV. The charge enters an accelerating cavity on axis at time $t = 0$ and leaves at a time $t = d/c = T_{\text{cav}}$. During the transit, it sees a time-varying electric field. The time it takes the electron to traverse the cavity needs to equal one-half an rf period for the charge to receive the maximum kick from the cavity; that is,

$$T_{\text{cav}} = \frac{d}{c} = \frac{\pi}{\omega_0}, \quad (2.32)$$

where ω_0 is the angular frequency of the accelerating mode. The electron always sees a field pointing in the same direction provided the charge enters when the electric field changes sign. We now want to calculate the energy received by the electron. We can define the accelerating voltage (V_c)[1] for a cavity by the following:

$$V_c = \left| \frac{1}{e} \times \text{maximum energy gain possible during transit} \right|. \quad (2.33)$$

This voltage is given by the line integral of E_z as seen by the electron (E_{el}); that is,

$$V_c = \left| \int_0^d E_{\text{el}} \, dz \right|. \quad (2.34)$$

Since

$$E_{\text{el}} = E_z(\rho = 0, z) e^{i\omega_0 z/c + i\varphi}, \quad (2.35)$$

where φ is some arbitrary phase, we find

$$V_c = \left| \int_0^d E_z(\rho = 0, z) e^{i\omega_0 z/c} dz \right|. \quad (2.36)$$

Consider, for example, the pill-box cavity operating in the TM_{010} mode. Using the field in (2.22) we find

$$V_c = E_0 \left| \int_0^d e^{i\omega_0 z/c} dz \right| = dE_0 \frac{\sin\left(\frac{\omega_0 d}{2c}\right)}{\frac{\omega_0 d}{2c}} = dE_0 T. \quad (2.37)$$

The quantity T is known as the *transit factor*. Provided that (2.32) is satisfied, $T = 2/\pi$ for the pill-box TM_{010} mode.

---
[1] Note that in many texts V_c is referred to as V_{acc}.

FIGURES OF MERIT

Often one quotes the average accelerating electric field (E_acc) that the electron sees during transit. This is given by

$$E_\text{acc} = \frac{V_c}{d} \tag{2.38}$$

which, for the TM$_{010}$ pill-box mode, evaluates to $E_\text{acc} = 2E_0/\pi$.

2.1.3 Peak Surface Fields

When applying superconducting cavities to accelerators, the question arises as to how large the accelerating field can be. This depends on the maximum surface fields the cavity can maintain. We therefore introduce two further fields, the peak surface electric field (E_pk) and the peak surface magnetic field (H_pk). H_pk is important because a superconductor will quench above the rf critical magnetic field (H_c^rf) — see Chapter 5. Thus H_pk may not exceed H_c^rf. E_pk, on the other hand, is important because of the danger of field emission in high electric field regions (see Chapter 12). To maximize the accelerating field it is therefore important to minimize the ratios of the peak fields to the accelerating field. For the TM$_{010}$ mode we have been studying, we see from the fields in (2.22) and (2.23) that

$$E_\text{pk} = E_0 \tag{2.39}$$

$$H_\text{pk} = \frac{E_0}{\eta} J_1(1.84) = \frac{E_0}{647\,\Omega}. \tag{2.40}$$

Thus we obtain the following ratios:

$$\frac{E_\text{pk}}{E_\text{acc}} = \frac{\pi}{2} = 1.6$$

$$\frac{H_\text{pk}}{E_\text{acc}} = 2430\,\frac{\text{A/m}}{\text{MV/m}} = 30.5\,\frac{\text{Oe}}{\text{MV/m}}.$$

By referring to Figure 2.2, we see that the peak electric and peak magnetic field regions both lie on the end faces of the cavity, where E_pk is on axis and H_pk is at $\rho = 0.77R$.

2.2 FIGURES OF MERIT

When designing a cavity for a certain application, the shape and size will often go through many iterations before a satisfying design is found. It is therefore important to develop figures of merit to be able to compare different designs and to identify the superior shape.

We already encountered two such figures in the ratios E_pk/E_acc and H_pk/E_acc which are useful in determining the maximum accelerating field that can be achieved in a cavity.

In this section we will discuss several other figures of merit. In particular, it is important to find quantities that are independent of the eigenfrequency so that we can compare cavities of different shapes without needing to take their overall size into consideration.

To this end, consider scaling a cavity's linear dimensions by a factor a. It seems intuitive that the fields in this new cavity are unaffected by the operation, except that we need to replace the argument $\mathbf{x}$ by $a\mathbf{x}$ — that is,

$$\left\{ \begin{array}{c} \mathbf{E}(\mathbf{x}) \\ \mathbf{H}(\mathbf{x}) \end{array} \right\} \to \left\{ \begin{array}{c} \mathbf{E}(a\mathbf{x}) \\ \mathbf{H}(a\mathbf{x}) \end{array} \right\}. \tag{2.41}$$

Certainly, these new fields do satisfy the boundary conditions at the scaled cavity's walls. However, we also need to show that the fields satisfy the eigenvalue equation (2.15). Substituting the scaled field in (2.15) and using cartesian coordinates we find

$$\left(\frac{\partial^2}{\partial x^2} + \frac{\partial^2}{\partial y^2} + \gamma_j^2 \right) \psi(a\mathbf{x}) = 0, \tag{2.42}$$

or

$$\left(\frac{\partial^2}{\partial (ax)^2} + \frac{\partial^2}{\partial (ay)^2} + \frac{\gamma_j^2}{a^2} \right) \psi(a\mathbf{x}) = 0. \tag{2.43}$$

By defining $\mathbf{X} = a\mathbf{x} = (X, Y, Z)$ and $\Gamma_j = \gamma_j/a$ we can simply write

$$\left(\frac{\partial^2}{\partial X^2} + \frac{\partial^2}{\partial Y^2} + \Gamma_j^2 \right) \psi(\mathbf{X}) = 0. \tag{2.44}$$

Equation 2.44 is identical with (2.15), thereby confirming that the fields in the scaled cavity are indeed unchanged, except for the scaling factor a in the argument. However, we see that the eigenfrequencies changed. By (2.16) and the definition of Γ_j, one finds

$$\omega_j \propto \frac{1}{a}; \tag{2.45}$$

that is, the cavity's mode spectrum is inversely proportional to the cavity size. The importance of this result will become clear shortly.

2.2.1 Power Dissipation and the Cavity Quality

So far we have talked about cavities with perfectly conducting walls. In order to support the cavity fields, currents flow within a thin surface layer of the walls. Even a superconductor has a small resistance at rf frequencies, so that the wall currents that sustain the fields dissipate energy (see Chapter 3). The losses can be characterized by the surface resistance (R_s), which is defined via the power dissipated per unit area (dP_c/ds) due to Joule heating:

$$\frac{dP_\text{c}}{ds} = \frac{1}{2} R_\text{s} |\mathbf{H}|^2, \tag{2.46}$$

FIGURES OF MERIT

where $\mathbf{H}$ is the local magnetic field. Typically, R_s is several tens of nanohms for a well-prepared niobium superconducting surface, whereas for copper the value is in the milliohm range.

Theoretically, the fields in the cavity are modified by the dissipative material. However, because R_s is so small, we can neglect this perturbation and use the fields calculated with perfectly conducting walls.

An important figure of merit for accelerating cavities is the quality factor (Q_0), which is related to the power dissipation and is defined as

$$Q_0 = \frac{\omega_0 U}{P_c}, \qquad (2.47)$$

where U is the stored energy and P_c is the power dissipated in the cavity walls. The Q_0 is roughly 2π times the number of rf cycles it takes to dissipate the energy stored in the cavity.

Since the time averaged energy in the electric field equals that in the magnetic field, the total energy in the cavity is given by

$$U = \frac{1}{2}\mu_0 \int_V |\mathbf{H}|^2 \, dv = \frac{1}{2}\epsilon_0 \int_V |\mathbf{E}|^2 \, dv, \qquad (2.48)$$

where the integral is taken over the volume of the cavity. Furthermore, Equation 2.46 yields the dissipated power

$$P_c = \frac{1}{2} R_s \int_S |\mathbf{H}|^2 \, ds, \qquad (2.49)$$

where the integration is taken over the interior cavity surface.[2] Thus we find for Q_0:

$$Q_0 = \frac{\omega_0 \mu_0 \int_V |\mathbf{H}|^2 \, dv}{R_s \int_S |\mathbf{H}|^2 \, ds}. \qquad (2.50)$$

The Q_0 is frequently written as

$$Q_0 = \frac{G}{R_s}, \qquad (2.51)$$

where

$$G = \frac{\omega_0 \mu_0 \int_V |\mathbf{H}|^2 \, dv}{\int_S |\mathbf{H}|^2 \, ds} \qquad (2.52)$$

is known as the *geometry constant*.[3] Some information about the geometry factor can be gained by scaling cavity's linear dimensions by a factor a. Using the approach outlined earlier, one finds that

$$\frac{\int_V |\mathbf{H}|^2 \, dv}{\int_S |\mathbf{H}|^2 \, ds} \propto a \propto \frac{1}{\omega_0}. \qquad (2.53)$$

---

[2] We are assuming that the surface resistance does not vary over the cavity surface.
[3] If R_s is not position independent then $Q_0 = G/\bar{R}_s$, where $\bar{R}_s = \int_S R_s |\mathbf{H}|^2 ds / \int_S |\mathbf{H}|^2 ds$.

From (2.52) and (2.53) we see that the geometry factor depends on the cavity shape but not its size. This quantity therefore is very useful for comparing different cavity shapes, irrespective of their size and wall materials. The quality factor, on the other hand, varies with the cavity size due to the frequency dependence of R_s (see Chapters 3 and 4).

For the pill-box TM$_{010}$ mode we find

$$U = E_0^2 \pi d \epsilon_0 \int_0^R \rho J_1^2\left(\frac{2.405\rho}{R}\right) d\rho \tag{2.54}$$

$$P_c = \frac{R_s E_0^2}{\eta^2} \left\{ 2\pi \int_0^R \rho J_1^2\left(\frac{2.405\rho}{R}\right) d\rho + \pi R d J_1^2(2.405) \right\} \tag{2.55}$$

which we can simplify using [1]

$$\int \rho J_\nu^2(\alpha \rho) \, d\rho = \frac{\rho^2}{2} \left[J_\nu^2(\alpha \rho) - J_{\nu-1}(\alpha \rho) J_{\nu+1}(\alpha \rho) \right] \tag{2.56}$$

to give

$$U = \frac{\pi \epsilon_0 E_0^2}{2} J_1^2(2.405) dR^2 \tag{2.57}$$

$$P_c = \frac{\pi R_s E_0^2}{\eta^2} J_1^2(2.405) R(R+d) \tag{2.58}$$

$$G = \frac{\omega_0 \mu_0 dR^2}{2(R^2 + Rd)} = \eta \frac{2.405d}{2(R+d)} = \frac{453 \frac{d}{R}}{1 + \frac{d}{R}} \Omega. \tag{2.59}$$

We see that for this specific example, G is indeed independent of the cavity's size. By combining (2.24) and (2.32), one finds that the acceleration is maximized if

$$\frac{d}{R} = \frac{\pi}{2.405} \tag{2.60}$$

in which case $G = 257$ Ω. If $R_s = 20$ nΩ for a superconducting cavity, then

$$Q_0 = \frac{G}{R_s} = 1.3 \times 10^{10}. \tag{2.61}$$

A typical length of $d = 10$ cm requires a cavity radius R of 7.65 cm or, equivalently, a resonant frequency of 1.5 GHz. For operation at $V_c = 1$ MV the following results are found to apply:

$$E_{\text{acc}} = \frac{V_c}{d} = 10 \text{ MV/m}$$

$$E_{\text{pk}} = E_0 = \frac{\pi}{2} E_{\text{acc}} = 15.7 \text{ MV/m}$$

$$H_{\text{pk}} = 30.5 \frac{\text{Oe}}{\text{MV/m}} E_{\text{acc}} = 305 \text{ Oe}$$

$$U = E_0^2 \frac{\pi \epsilon_0}{2} J_1^2(2.405) dR^2 = 0.54 \text{ J}$$

$$P_c = \frac{\omega U}{Q_0} = 0.4 \text{ W}.$$

2.2.2 Shunt Impedance

Another important quantity used to characterize the losses in a cavity is the shunt impedance (R_a), which is defined as

$$R_a = \frac{V_c^2}{P_c} \tag{2.62}$$

in units of ohms per cell. Equation 2.62 is the accelerator definition. Other, similar expressions exist for the shunt impedance and it is important not to confuse them. In circuit theory, for example, one uses

$$R_a^c = \frac{V_c^2}{2P_c}, \tag{2.63}$$

and a common definition for linacs is

$$r_a = \frac{V_c^2}{P_c'}, \tag{2.64}$$

where P_c' is the power dissipated per unit length. The linac shunt impedance is in ohms per meter.

Ideally we want the shunt impedance to be large for the accelerating mode so that the dissipated power is minimized. This is particularly important for copper cavities, where power dissipation in the walls is a major issue.

For the pill-box TM$_{010}$ mode we have (see (2.37) and (2.58))

$$R_a = \frac{4\eta^2 d^2}{\pi^3 R_s J_1^2(2.405) R(R+d)} = 2.5 \times 10^{12} \, \Omega. \tag{2.65}$$

Note that the ratio of R_a/Q_0 is given by

$$\frac{R_a}{Q_0} = \frac{V_c^2}{\omega_0 U}, \tag{2.66}$$

which is independent of the surface resistance. The scaling relation used earlier can also be applied to the expressions for V_c in (2.34) and U in (2.48). The ratio V_c^2/U can thus be seen to scale inversely with the cavity's linear dimensions, just like the frequency. Therefore R_a/Q_0 is independent of the cavity size. This quantity is frequently quoted as a figure of merit and is also used for determining the level of mode excitation by charges passing through the cavity (see Chapter 15). For the TM$_{010}$ mode we have

$$\frac{R_a}{Q_0} = 150 \, \Omega \, \frac{d}{R} = 196 \, \Omega, \tag{2.67}$$

provided that d/R satisfies (2.60).

Table 2.1 summarizes the various parameters we just discussed for the Cornell/CEBAF 5-cell cavity in Figures 1.23 and 1.24 as computed by numerical

Table 2.1: Figures of merit for the Cornell/CEBAF 5-cell cavity compared with the pill-box cavity

Quantity	CEBAF 5-cell	Pillbox
G	290 Ω	257 Ω
R_a/Q_0	480 Ω/5 cells	196 Ω/cell
$E_\mathrm{pk}/E_\mathrm{acc}$	2.6	1.6
$H_\mathrm{pk}/E_\mathrm{acc}$	47 Oe/(MV/m)	30.5 Oe/(MV/m)

programs such as SUPERFISH or URMEL. Note that due to the presence of the large beam holes, the shunt impedance is reduced and the peak surface fields are enhanced relative to the pill-box case. For a realistic cavity shape, R_a/Q_0 typically is lowered by a factor of 2 due to the presence of the beam holes.

In a normal conducting cavity, where power dissipation is a major concern, the R_a/Q_0 must be maximized by using a small beam hole. In many cases it is further enhanced by making the cell shape re-entrant (cf. Figure 20.1(a)) to lower the magnetic field in the equator region, thereby reducing P_c. These measures tend also to increase the R/Q_0 values of higher-order modes, which we will discuss in Chapter 15. As a result, the beam interacts more strongly with the higher-order modes, thereby degrading the beam quality and limiting the maximum possible bunch charge. With the phenomenally low R_s of superconducting cavities, the dissipated power is no longer an overriding issue, even after taking into account the refrigerator efficiency. Superconducting cavities thus can have large beam holes. The penalty in lower R_a/Q_0 for the accelerating mode is not serious. However, the reduced R/Q_0 for higher-order modes and the accompanying reduced beam–cavity interaction pay a big dividend.

2.2.3 Refrigerator Requirements

The power dissipated in the superconducting pill-box cavity as calculated in Section 2.2.1 is very small. However, these losses heat the liquid helium bath that surrounds the cavity. Together with the static heat leak to the cryostat, these losses comprise the cryogenic losses. Typically, the required refrigeration power exceeds the power dissipated in a 2 K helium bath by a factor of 750. Part of this factor is due to the technical efficiency (η_t) of the refrigerator; generally, $\eta_\mathrm{t} = 0.2$ for a large system. The other part is due to the Carnot efficiency η_c which at 2 K is

$$\eta_\mathrm{c} = \frac{T_\mathrm{He}}{T_\mathrm{rm} - T_\mathrm{He}} = \frac{2}{300 - 2} = 6.7 \times 10^{-3}, \qquad (2.68)$$

where T_He is the bath temperature and T_rm is room temperature (300 K).

We found that the 1.5-GHz pill-box cavity dissipates 0.4 W at $E_\mathrm{acc} = 10$ MV/m. The required refrigerator ac power due to these rf losses would be 300 W. On the other hand, the power dissipation in a copper cavity of the same geometry would be 60 kW for a typical $R_\mathrm{s} = 3$ mΩ! Due to klystron effi-

APPLICATION OF RF CODES 49

ciencies on the order of 0.5, the actual ac wall power required can be as high as 120 kW. Thus the ac power cost of running a copper cavity in cw mode would be several hundred times the cost for a niobium cavity. This illustrates one of the principal attractive features of superconducting cavities, which we outlined in Chapter 1.

2.3 APPLICATION OF RF CODES

Since only very simple accelerating structures can be solved analytically, we must, in general, use computers to apply numerical methods to determine the cavity mode frequencies and their electromagnetic fields. One can then compute the R_a/Q_0, the ratios $E_{\rm pk}/\sqrt{U}$ and $H_{\rm pk}/\sqrt{U}$, the ratios $E_{\rm pk}/E_{\rm acc}$ and $\mu_0 c H_{\rm pk}/E_{\rm acc}$, and the geometry factor $G = Q_0/R_s$ from the numeric solutions for the fields. Sometimes one computes Q values with mode damping materials and/or couplers present.

A generic computer code requires that the user specify the cavity geometry, properties of the materials used (metal, dielectric, lossy), and the boundary conditions. The boundary conditions are usually Neumann ($E_\parallel = 0$, "electric") or Dirichlet ($H_\parallel = 0$, "magnetic"). Often one can exploit symmetry in the problem and find the full solutions by computing only part of the structure if the appropriate boundary conditions are chosen. The cavity geometry is usually specified in terms of a sequence of arcs and straight lines. For 3-D codes such as MAFIA [73], the input is expanded to rectangular solids (brick shapes), cylinders, spheres, and solids of rotation.

In addition, the code may also require the user to specify the number of mesh points or even the exact arrangement of the mesh. Some codes also require an initial guess for the frequency sought.

To give a feel for the way these calculations are done in practice, we will present a few examples of code use. The examples presented here are ones that have been useful at various times for SRF applications. This list is by no means complete.[4] The online Code Compendium maintained by the Los Alamos Accelerator Code Group [74] is a useful place to learn of more computer codes.

We will discuss the 2-D codes URMEL [71], SUPERFISH [70], and SuperLANS [72, 75]. All of these codes numerically solve the eigenvalue problem constructed from Maxwell's equations. In a source free region

$$\left(\nabla^2 + \mu\epsilon\frac{\omega^2}{c^2}\right)\left\{\begin{array}{c}\mathbf{E}\\ \mathbf{H}\end{array}\right\} = 0. \qquad (2.69)$$

2.3.1 Numerical Techniques

Two types of numerical techniques available are finite differencing and finite element analysis. URMEL and SUPERFISH are finite difference codes while

---
[4]The presence of a code in this discussion is not meant as an endorsement of that code. Similarly a code's absence should not be interpreted as a judgment.

SuperLANS is a finite element code. To give a brief comparison, in finite differencing, there are fewer calculations per mesh point so the code runs faster for the same number of points. For the programmer, the formulation is simpler and there is less bookkeeping involved. In general, the finite element code does more calculations per mesh point but requires fewer mesh points than finite differencing to achieve the same accuracy. Often the finite element formulation is better at reducing numerical roundoff error.

Finite differencing is based on a Taylor expansion

$$f(x_{j+1}) = f(x_j) + f'(x_j) h + f''(x_j) \frac{h^2}{2} + \cdots \qquad (2.70)$$

where

$$h \equiv x_{j+1} - x_j. \qquad (2.71)$$

The problem is discretized with mesh spacing h. Derivatives are written in terms of discrete points — for example:

$$f'(x_j) = \frac{f(x_{j+1}) - f(x_{j-1})}{2h} \qquad (2.72)$$

and

$$f''(x_j) = \frac{f(x_{j+1}) - 2f(x_j) + f(x_{j-1})}{(2h)^2}. \qquad (2.73)$$

These derivatives are used in the differential equations of (2.69) to yield the solutions at the discrete mesh points.

In the finite element method, we start with the differential equation of the problem to be solved,

$$\mathcal{L}f(x) = g(x) \qquad (2.74)$$

where $\mathcal{L}$ is a differential operator. Expand the solution $f(x)$ in some convenient finite set of basis functions $N_0(x) \ldots N_J(x)$:

$$\tilde{f}(x) = N_0(x)f(x_0) + N_1(x)f(x_1) + \cdots + N_J(x)f(x_J). \qquad (2.75)$$

The goal is to find a function $\tilde{f}(x)$ that minimizes the residual, $R(x)$, defined by

$$R(x) \equiv \mathcal{L}\tilde{f}(x) - g(x). \qquad (2.76)$$

This is done by multiplying by a weighting function W and integrating over the domain. Then require that

$$(\mathcal{L}\tilde{f}, W) - (g, W) = (R, W) = 0 \qquad (2.77)$$

where $(A, B) = \int AB \, dx$ is the inner product. By a suitably clever choice of weighting function, the problem turns into a manageable sequence of algebraic operations. Once the expansion coefficients are known, a piecewise continuous solution is available over the whole domain. This is a significant difference from the finite difference method, where the solution is only determined at discrete points.

APPLICATION OF RF CODES 51

2.3.2 Using Symmetry

The 2-D code URMEL models the cavity shape with a rectangular mesh. Only axisymmetric structures can be modeled. The use of a uniform mesh may result in a jagged or chunky shape. One can achieve a smoother fit to the cavity shape if a variable mesh density is used. Figure 2.3 shows a comparison of these meshing strategies. [5]

If the shape has reflection symmetry as well as axial symmetry, as in the last example, one need only model half the cell. When doing so, one has to change the boundary condition at the symmetry plane to find all the modes. The properties of a cavity with an infinite number of cells (periodic structure) can be computed by modeling a half-cell with different boundary conditions.

Figure 2.4 shows the four choices for the boundary conditions at the edges of the half-cell and the lowest frequency mode that satisfies each of these. Note that when computing the most popular accelerating mode, the π mode, only the case in Figure 2.4(b) is used.

When computing the properties of a cavity with a finite number of cells, the full structure (not just a single cell) needs to be modeled since the beam tubes may play a significant role. In Chapter 7, we will discuss multicell cavities in more detail. Figure 2.5 shows URMEL results for a 2-cell cavity. In this case the reflection symmetry was not used.[6] A multicell cavity is just a set of coupled single-cell cavities. Since there are N modes of oscillation for a system of N coupled oscillators, the two modes shown are the only two modes of the system in the fundamental (TM_{010}) passband. Figure 2.6 shows three of the modes of a 6-cell cavity when its reflection symmetry is exploited. The symmetry plane is taken to have a magnetic-type boundary. The other three modes for this structure would be obtained by using an electric-type boundary at the symmetry plane. (This structure has been tuned so that in the π mode, there is an equal amount of stored energy in each cell.)

2.3.3 More Examples

The code SUPERFISH is another 2-D analysis tool. It carries out essentially the same function as URMEL but has some different features. Instead of a rectangular mesh, SUPERFISH uses a triangular mesh to perform its finite differencing. The triangular mesh allows an excellent fit to most shapes without the need for large variations in mesh density. Figure 2.7 shows an example of the triangular mesh for half of a cavity cell. Unlike URMEL, which computes many modes of the structure in one run and can compute multipole modes, SUPERFISH computes only one mode at a time and is restricted to monopole modes. The user must supply an initial guess for the frequency and a drive point for the magnetic field. The code will excite the structure at the drive

---

[5] The authors of URMEL do not recommend varying the mesh density by more than a factor of five.

[6] In practice, the reflection symmetry would be used to minimize the number of meshpoints needed. The full structure is shown here to make it easier to see the mode characteristics.

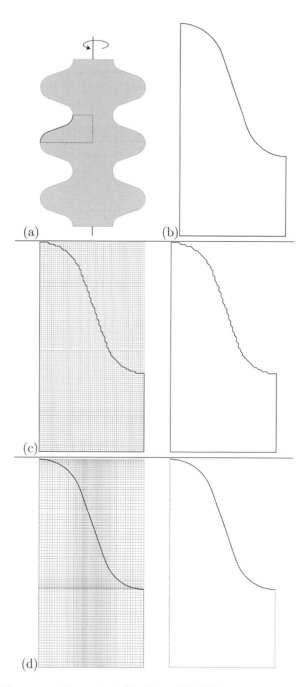

Figure 2.3: When modeling the half-cell in URMEL, the variable mesh density allows a closer fit to the shape. (a) Three cells of an infinite periodic structure. The part modeled is indicated. (b) Input shape to model. (c) Uniform mesh. (d) Variable mesh density.

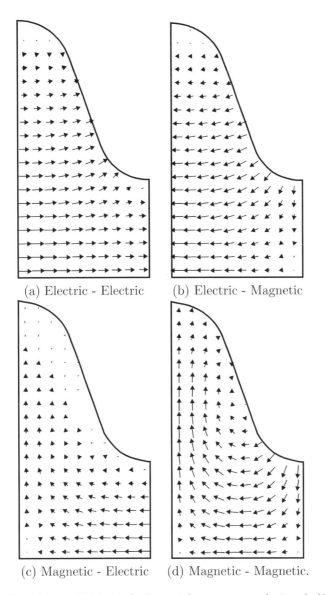

Figure 2.4: The electric fields in the lowest frequency mode in a half-cell cavity for the four possible sets of boundary conditions.

point and converge upon the resonant frequency. Unlike URMEL, SUPERFISH allows the user to model dielectric materials in addition to perfect conductors and vacuum.

SuperLANS solves the same class of problems as SUPERFISH but uses a finite element method instead of a finite difference method. The contour fitting mesh is composed of arbitrary quadrilaterals instead of triangles. Unlike

CAVITY FUNDAMENTALS AND CAVITY FIELDS

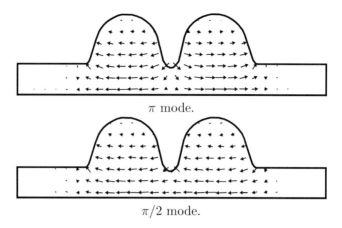

Figure 2.5: The electric field patterns of both modes in the fundamental passband of a 2-cell cavity as computed by URMEL.

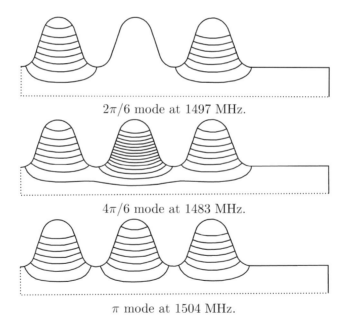

Figure 2.6: Three modes in the fundamental passband computed by URMEL on a structure designed to have a "flat" π mode. (See Chapter 7.) Here all symmetries of the structure are exploited and the symmetry plane is taken to have a "magnetic" boundary. To obtain the other three modes for this structure, an "electric" boundary would be used instead.

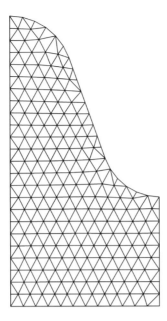

Figure 2.7: The SUPERFISH contour fitting triangular mesh for a cavity half-cell.

SUPERFISH, no drive point or frequency guess is required, and up to 10 modes can be computed in one run.

2.3.4 Code Comparison

Because of the very different ways in which the codes mesh and compute the problems, it is difficult to make meaningful cross-comparisons of performance. One code might take longer to solve a problem than another, but it may give more modes in its answer or be more accurate. Figure 2.8 shows an example of the meshing of a spherical cavity (using appropriate symmetries) as done by three codes. The spherical cavity is used as the test problem because it offers a reasonable challenge for meshing and its modes have an analytical solution.

For this case SUPERFISH and URMEL offer very similar performance. SuperLANS takes longer to solve the problem than the other codes, but yields better accuracy in computing the resonant frequency of the fundamental mode when the same number of meshpoints are used [76].

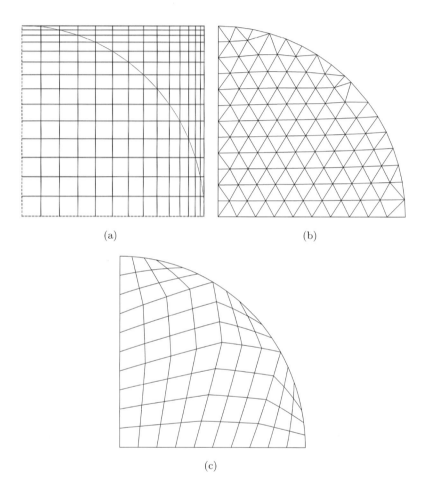

Figure 2.8: Meshing for a spherical cavity using three different 2-D codes. (a) Variable density rectangular mesh for URMEL. (b) Contour fitting triangular mesh of SUPERFISH. (c) Contour fitting quadralateral mesh of SuperLANS.

CHAPTER 3

Superconductivity Essentials

3.1 INTRODUCTION

Many aspects of radio frequency (rf) superconductivity, both fundamental and technological, are deeply connected with the physics of electrons in metals. Apart from the fascinating phenomenon of superconductivity, the physics of electrons in metals is important in understanding how heat is transported across the wall of the cavity, how electrons tunnel out of metal surfaces to contaminate the vacuum in the rf cavities (field emission), or how secondary electrons are generated to produce multipacting. It is therefore worthwhile to discuss briefly the fundamental theory of conduction electrons in metals. We refer the reader to Ashcroft and Mermin [3] or Reif [2] for a rigorous development of the concepts presented here. The intuitive approach adopted here allows us to stage the introduction of some of the important physical quantities that underlie the field of rf superconductivity, such as the Fermi energy, the Fermi velocity, the London penetration depth, the energy gap, the coherence length and the mean free path.

3.2 THE FREE ELECTRON MODEL

In the simple free electron picture for the behavior of a metal (Figure 3.1), electrons in the outer shells of atoms in a metal are easily detached from the core and wander away from the parent ions that define the solid lattice. These valence electrons move freely through the lattice and are responsible for the excellent electrical and thermal conductivity of metals.

3.2.1 Success of the Classical Free Electron Model

In the earliest approach, called the Drude model, the conduction electrons were treated by analogy to an ideal gas. Once released from the atoms, the electrons move freely through the lattice. The Coulomb attraction of the positive ions is represented by a flat potential well, and the electrons are treated as non-interacting particles. Their mutual repulsion is treated globally by suitably choosing the depth of the potential well. When traveling through a perfect crys-

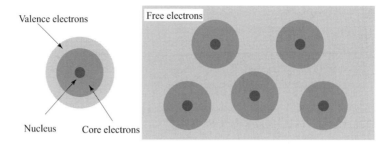

Figure 3.1: A simple free electron picture.

tal lattice, electrons suffer no scattering. However, they do get scattered when they suffer frequent collisions with the lattice imperfections, such as missing atoms, interstitial atoms, and chemical impurities. Electrons also get scattered by the vibrating lattice ions when the perfect periodicity of the lattice is disturbed by thermally excited sound waves. The word phonon is used here to remind us that the lattice vibrations are quantized. The thermal motion of the lattice increases with temperature. If the temperature is low enough the number of thermally excited phonons is small compared to the number of impurities, so that impurity scattering is dominant. Between collisions, electrons move freely for a mean free time τ, called the relaxation time.

By applying the kinetic theory to such a "gas" of free electrons, it is possible to construct a simple model for the electrical conductivity (σ) and thermal conductivity (κ) of metals, as we shall now see. If there are n electrons per unit volume, and they move with a drift velocity $\Delta \mathbf{v}$ under the influence of an electric field $\mathbf{E}$, the electrical current is given by

$$\mathbf{j} = -ne\Delta \mathbf{v}. \tag{3.1}$$

Since electrons emerge from collisions in a random direction there is no contribution to the average electron velocity from the velocity before the collision. The additional velocity, $\Delta \mathbf{v}$, is simply the acceleration, $e\mathbf{E}/m$, times the average time τ between collisions:

$$\Delta \mathbf{v} = e\mathbf{E}\tau/m. \tag{3.2}$$

We also assume that the increment of the electron speed due to the electric field is negligible compared to the thermal speed. After collisions, the directed velocity gain is lost due to random thermal motion. The conductivity equation in a metal then becomes

$$\mathbf{j} = \frac{ne^2\tau}{m}\mathbf{E} \quad \text{or} \quad \mathbf{j} = \sigma \mathbf{E}, \tag{3.3}$$

where σ is the dc electrical conductivity given by

$$\sigma = \frac{ne^2\tau}{m}. \tag{3.4}$$

THE FREE ELECTRON MODEL

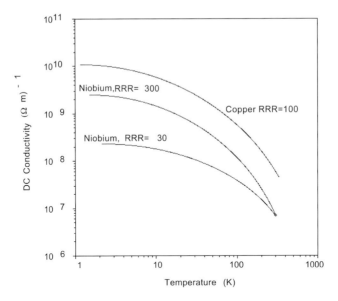

Figure 3.2: Temperature dependence of the dc conductivities for copper, low-purity niobium, and high-purity niobium.

Typically, $n = 10^{22}$ cm^{-3}, and $\tau = 10^{-14}$ s, from the observed resistivity. In the Drude model, a simple estimate for the average electron speed is made from the classical equipartition of energy theorem

$$\frac{1}{2}mv^2 = \frac{3}{2}k_\mathrm{B}T, \qquad (3.5)$$

to give $v = 10^7$ cm/s for room temperature. From this we see that the electron mean free path, $\ell = v\tau$, is typically 10^{-7} cm.

The electrical conductivity of metals increases with decreasing temperature because the scattering term from lattice vibrations decreases. But the improving conductivity eventually saturates to a value determined by impurity scattering, as shown in Figure 3.2. Because electrons are also the dominant heat carriers, this limitation will turn out to be an equally important factor in understanding the heat-carrying capability of the wall of a niobium cavity. For example, for relatively impure Nb, the dc resistivity ($\rho \propto 1/\sigma$) drops by a factor of 30 from its room temperature value of 1.4×10^{-8} Ω-m to its residual value. The factor 30 by which resistivity drops to the residual value is called the "residual resistance ratio", or RRR. High-purity niobium has higher RRR values, the theoretical limit being 35,000 [77], determined by scattering of electrons from lattice vibrations.

We turn now to thermal conductivity. In a metal, heat is transported by electrons and phonons. At low temperature, the number of phonons per unit volume $\propto T^3$, by analogy with photons. (This behavior basically reflects the 3-D properties of space via the volume of phase space for phonons.) Because

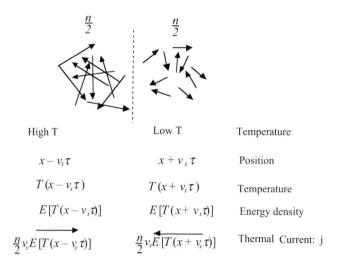

Figure 3.3: To calculate the thermal current we evaluate the transport of energy from the high-temperature region to the low-temperature region.

of the T^3 dependence, the predominant mechanism for heat transport at low temperatures is by the conduction electrons, i.e., the free-electron gas. In a simple one-dimensional model, illustrated in Figure 3.3, there is a net flow of heat energy from high temperature to low temperature. The thermal current density, j_q (watt/area), flowing across an area is given by

$$j_q = -\kappa \frac{dT}{dx}. \tag{3.6}$$

The heat current density can be estimated by considering the number density $(n/2)$ of electrons that flow across the area, times the velocity (v_x) of the electrons, times the energy which each electron carries. Thus

$$j_q = n v_x E. \tag{3.7}$$

From Figure 3.3 we see that an electron coming from the high-temperature region (on the left) has an energy $E\left[T(x - v_x\tau)\right]$ which is larger than the energy of an electron coming from the lower-temperature region (on the right). Note that the energy of electrons is also a function of the temperature. Thus there is a net transport of energy from the high-temperature region to the low-temperature region. The thermal current is

$$j_q = \frac{n}{2} v_x E\left[T(x - v_x\tau)\right] - \frac{n}{2} v_x E\left[T(x + v_x\tau)\right]. \tag{3.8}$$

The first term is the contribution to the thermal current from electrons arriving at x from the high-temperature side. The second term is the contribution

THE FREE ELECTRON MODEL

from the low-temperature side. Using

$$E[T(x - v_x\tau)] - [T(x + v_x\tau)] = -\frac{dE}{dT}\frac{dT}{dx}2v_x\tau, \tag{3.9}$$

the thermal current becomes

$$j_q = -nv_x^2\tau\frac{dE}{dT}\frac{dT}{dx}. \tag{3.10}$$

To go to the 3-D case, we can write from the equipartition theorem of the kinetic theory of gases

$$v_x^2 = \frac{1}{3}v^2. \tag{3.11}$$

The quantity dE/dT is related to the heat capacity from the definition of specific heat at constant volume:

$$C_\text{v} = n\frac{dE}{dT}. \tag{3.12}$$

Using the specific heat, along with the definition of thermal conductivity (Equation 3.6), we derive

$$\kappa = \frac{1}{3}v^2\tau C_\text{v}. \tag{3.13}$$

In this simple kinetic view, the salient result is that the ratio of the thermal to the electrical conductivity becomes

$$\frac{\kappa}{\sigma} = \frac{mv^2 C_\text{v}}{3ne^2}. \tag{3.14}$$

Now we can further apply some of the key results of the classical kinetic theory of gases, such as the relation between temperature and mean square velocity, as well as the specific heat contribution from each degree of freedom:

$$\frac{1}{2}mv^2 = \frac{3}{2}k_\text{B}T \tag{3.15}$$

and

$$C_\text{v} = \frac{3}{2}nk_\text{B}, \tag{3.16}$$

where k_B is Boltzmann's constant. In determining the specific heat we assume, as in an ideal gas, that the electrons obey Maxwell-Boltzmann statistics, i.e., the number of electrons with energy between ε and $\varepsilon + d\varepsilon$ is given by the energy distribution function

$$f(\varepsilon) = e^{-\varepsilon/k_\text{B}T}. \tag{3.17}$$

Using Equations 3.15 and 3.16, the relation between thermal and electrical conductivities then becomes

$$\frac{\kappa}{\sigma} = \frac{3}{2}\left(\frac{k_\text{B}}{e}\right)^2 T. \tag{3.18}$$

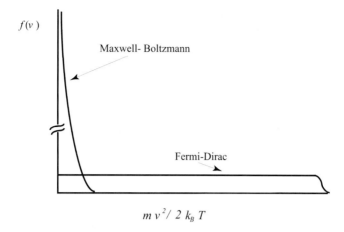

Figure 3.4: A comparison between classical and quantum velocity distributions for free electrons.

The greatest success of the classical free electron gas model was this "explanation" of the empirically established Wiedemann-Franz (WF) law, which states that

$$\frac{\kappa}{\sigma} = LT, \qquad (3.19)$$

where L, the Lorenz number, is independent of temperature and defect concentration. L is also observed to be approximately the same for most metals. If electrons are the predominant carriers of both heat and electrical current in metals, then the thermal conductivity of a metal is proportional to the electrical conductivity. As with electrical conductivity, the thermal conductivity of a metal is limited by impurity scattering of electrons. We will see in Chapter 6 that by removing the appropriate impurities from niobium, the thermal conductivity of the wall of a niobium cavity increases, thereby providing better cooling. The accompanying increase in electrical conductivity, or the RRR value, serves as a convenient measure of the purity of the metal. The formal definition of RRR is

$$\text{RRR} = \frac{\text{resistivity 300 K}}{\text{residual resistivity at low temperature (normal state)}}. \qquad (3.20)$$

3.2.2 Quantum Mechanical Description

There comes now a serious difficulty with the classical free electron model. The electron "gas" does not appear to contribute at all to the heat capacity of a metal. From the equipartition of energy theorem, we expect an additional contribution of $(3/2)k_B$ per electron, but the observed electron contribution is nearly zero. Another contradiction to observed behavior is that the classical treatment predicts that the specific heat of a metal is temperature-independent.

THE FREE ELECTRON MODEL

The source of these problems is that in the classical description the electron velocities and the energies are given by the Maxwell–Boltzmann distribution as shown in Figure 3.4. In the quantum mechanical description the velocity distribution for the electrons is given the by Fermi–Dirac function (Figure 3.4), which is vastly different.

In quantum mechanics, an electron traveling through a metal is represented by a plane wave. The wavelength is inversely proportional to the velocity according to the de Broglie relation, $\lambda = h/p$. The amplitude of the wave is proportional to the density of electrons. The wave passes through a perfectly periodic crystal lattice without being scattered or losing its momentum. The wave function of the electron must satisfy the boundary conditions that keep it confined within the metal. These conditions lead to a finite number of states in a given energy range, referred to as the density of states, $D(\varepsilon)$. The occupation of these states is governed by the Pauli exclusion principle which introduces an important statistical correlation between different electrons. Electrons with parallel spins avoid each other. In building up the ground state for N electrons in a volume V, we may place only two electrons (with opposite spins) in each allowed quantized energy level. The next two electrons have to be placed at a higher energy level. As we continue to place electrons we use up higher and higher energies. The energy level at which the last two electrons are placed is called the Fermi level, and the highest energy is known as the Fermi energy, ε_F. We have

$$\varepsilon_F = \frac{\hbar^2 (3\pi^2 n)^{2/3}}{2m_e}. \tag{3.21}$$

Here $\hbar$ is Planck's constant and m_e is the mass of the electron. The velocity of electrons at the Fermi level is referred to as the Fermi velocity v_F:

$$v_F^2 = \frac{2\varepsilon_F}{m}. \tag{3.22}$$

It is the Fermi velocity which plays the major role in the quantum theory of metals, rather than the thermal velocity of the classical electron gas:

$$v = \left(\frac{3k_B T}{m}\right)^{1/2}. \tag{3.23}$$

Because of the Pauli exclusion principle, the Fermi velocity is very high, typically 1% of the velocity of light. For niobium, the Fermi energy, ε_F, is 5.32 eV, the Fermi temperature, $T_F = \varepsilon_F/k_B$, is 6.18×10^4 K, and the Fermi velocity, v_F, is 1.37×10^6 m/s. Without the exclusion principle, the classical electron velocity would be much lower, i.e., $k_B T \approx 0.02$ eV. Note also that the mean free path of the electrons, $\ell = v_F \tau$, is very much larger than the distance between neighboring atoms.

The Pauli exclusion principle and the resulting Fermi–Dirac (FD) statistics resolve the equipartition energy problem and explain why the observed specific heat of the electron gas is very low.

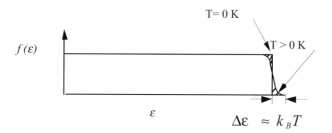

Figure 3.5: The step function Fermi–Dirac distribution at $T = 0$ K, and the Fermi-Dirac distribution function for $T > 0$ K.

Let us make a simple derivation of the electronic specific heat. The probability that a given state of energy ε is occupied by an electron is given by the FD distribution:

$$f(\varepsilon) = \frac{1}{\exp\left(\frac{\varepsilon - \varepsilon_F}{k_B T}\right) + 1}. \tag{3.24}$$

As shown in Figure 3.5, at $T = 0$ K, the FD distribution is a step function. It is equal to unity for $\varepsilon < \varepsilon_F$ and is equal to zero for $\varepsilon > \varepsilon_F$. At a finite temperature, some electrons, those within $k_B T$ below the Fermi energy, are excited to energy states within $k_B T$ above the Fermi energy. Only these electrons contribute to the specific heat and the thermal conductivity. The electrons with $\varepsilon \ll \varepsilon_F$ have very little effect on the macroscopic properties of a metal because they are in completely filled states. A small change in temperature (compared to T_F) does not affect these electrons since all the nearby states are completely full. The mean energy of these electrons is also unaffected, and they do not contribute to the heat capacity. Only the small number of electrons in the region $k_B T \approx \varepsilon_F$ contribute to the heat capacity. In the tail end of this region, the electron energies obey the Maxwell-Boltzmann (MB) distribution. Therefore each electron in this region contributes $(3/2)k_B$ to the heat capacity. Only a fraction $k_B T/\varepsilon_F$ of the total number of electrons are in the tail region of the MB distribution. All these excited electrons move with nearly the Fermi velocity v_F. The number of excited electrons per unit volume is the width $k_B T$ times the density of states per unit volume, $D(\varepsilon)$. The excitation energy is also of the order $k_B T$. Therefore, the total thermal energy density is

$$D(\varepsilon)(k_B T)^2. \tag{3.25}$$

The heat capacity, which is the temperature derivative of the energy, is therefore proportional to T. The exact expression for free electrons is

$$C_v = \frac{\pi^2}{2}\left(\frac{k_B T}{\varepsilon_F}\right) n k_B = \gamma T, \tag{3.26}$$

and for a metal with density of states $D(\varepsilon)$ it is

$$C_v = \frac{\pi^2}{3} k_B^2 T D(\varepsilon) = \gamma T. \tag{3.27}$$

THE FREE ELECTRON MODEL

Here γ, the coefficient of the electronic specific heat, is a measure of the available electronic density of states. Since at ordinary temperatures $k_B T \ll \varepsilon_F$, the electronic contribution to the specific heat is greatly depressed; at room temperature it is only 1% of the total. This is one of the most important predictions of Fermi–Dirac statistics, and it is a simple consequence of the Pauli exclusion principle. Note also that the specific heat is now temperature-dependent, in agreement with the observed behavior.

Another important consequence of the FD distribution is that when an electric field is applied, most of the electrons are unable to accept the momentum because they do not find empty states in their vicinity. Only those electrons in the tail of the distribution have free states accessible.

Now let us revisit the basis of the Wiedemann–Franz law:

$$\frac{\kappa}{\sigma} = \frac{mv^2 C_v}{3ne^2}. \tag{3.28}$$

In the quantum mechanical treatment, we have seen that the specific heat is reduced by a factor of 100. But the mean square electron velocity is also increased by the same factor, because of the high Fermi velocity. Therefore, using Equation 3.22 for v_F and Equation 3.26 for C_v, we get

$$\frac{\kappa}{\sigma T} = \frac{\pi^2}{3}\left(\frac{k_B}{e}\right)^2 = 2.44 \times 10^{-8} \frac{W\Omega}{K^2}. \tag{3.29}$$

The quantum mechanical treatment correctly predicts the WF law. The agreement between the classical electron gas theory and the WF law was therefore entirely fortuitous. However, it was useful to discuss the classical electron gas to appreciate the impact of quantum mechanics on electron velocity, mean free path, and electron temperature. It is remarkable that, apart from MB statistics, the free electron gas picture can still be used to describe the main features of electrical and thermal conduction as well as heat capacity, provided that FD statistics replaces the classical MB statistics.

More generally, the phonons, which we have largely ignored up to now, also play a role in both the thermal conductivity and the specific heat of a metal. As we pointed out earlier, the number of phonons is proportional to T^3. At low temperature, electron-impurity scattering dominates, and the electronic component of thermal conductivity, $\kappa_{electron} \propto T$. As the temperature rises, phonon scattering becomes more important. The electron mean free path, which is inversely proportional to the number of phonons, will therefore become $\propto T^{-3}$. For phonon scattering, $\kappa_{phonon} \propto T(1/T^3) \propto T^{-2}$. To a first approximation the scattering probabilities of the two mechanisms (impurities and phonons) can be considered independently. To obtain the combined contributions we add the two thermal resistivities components — that is, the reciprocals of the respective conductivities — as follows:

$$\frac{1}{\kappa} = \frac{B}{T} + AT^2. \tag{3.30}$$

When the phonon contribution is added, the total specific heat of metal has the form

$$C_\text{v} = \gamma T + AT^3. \tag{3.31}$$

Here $A = 12\pi^4 N_\text{A} k_\text{B}/5\theta_\text{D}^3$, where θ_D is the Debye temperature, e.g., 275 K for niobium, and corresponds to the highest possible phonon energy $k_\text{B}\theta_\text{D} = \hbar\omega_\text{D} \approx 0.01\text{--}0.02$ eV. The lattice contribution depends on the mean number of phonons $\propto T^3$ and the mean energy $\propto T^4$. Correspondingly, the specific heat $\propto dE/dT \propto T^3$. At low temperature, the phonon contribution to the specific heat is much small than linear electron contribution (γT).

3.3 ENTER SUPERCONDUCTIVITY

The most fascinating aspect of superconductivity is that the electrical resistance of a conductor goes to zero when the material enters the superconducting state. No experiment has yet been able to detect a measurable resistance in this state. The discoverer of this marvelous phenomenon, Kammerlingh-Onnes [78, 79], dramatically demonstrated the unmeasurable resistance by showing that the persistent current in a loop of superconducting wire he was carrying with him was completely stable even for the long period of time it took him to travel from his home in Leiden, Holland to his destination in Cambridge, England. By carefully monitoring the stability of a persistent current in a loop of superconducting wire, the upper limit for the resistivity of a superconductor (ρ_s) has now been established to be

$$\frac{\rho_\text{s}}{\rho_\text{n}} < 10^{-12}. \tag{3.32}$$

As a practical application, superconducting magnets can be operated in the persistent current mode with no cryogenic losses for an indefinitely long period.

Aside from infinite conductivity, another hallmark effect of superconductivity is perfect magnetic flux exclusion. When a superconducting material is cooled through its transition temperature in the presence of an external magnetic field, the magnetic flux is abruptly expelled. This spectacular magnetic behavior, which was first discovered by Meissner and Ochsenfeld [80], is called the Meissner effect. When we discuss this effect and its consequences in Chapter 4 we will see that it cannot be explained in terms of Maxwell's equations.

Many materials become superconducting when cooled below a transition temperature, T_c. Besides several pure elements, many compounds and alloys also exhibit superconductivity; the new high-temperature superconductors comprise of four or more elements and are also brittle oxide ceramics. Among the elements, niobium has the highest transition temperature, $T_\text{c} = 9.2$ K. Relative to all other elements, this high transition temperature makes niobium the most suitable for accelerator cavities. More importantly, niobium is readily available in pure form and is a highly workable metal for forming cavities. At sufficiently high purity, niobium also has a high thermal conductivity that helps provide

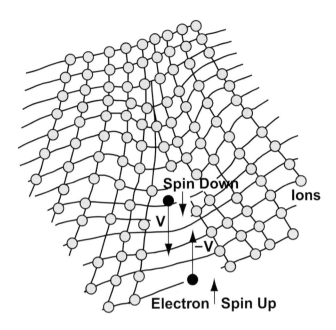

Figure 3.6: A simplified view of the electron–phonon interaction. An electron moving through the lattice distorts the lattice. The deformed lattice acts on a second electron moving in the opposite direction through the positive charge accumulation.

thermal stability of the rf surface. The other element used for superconducting cavities is lead, which has $T_c = 7.2$ K.

Forty years after the discovery by Kammerlingh-Onnes, the first theory of superconductivity was presented by Bardeen, Cooper, and Schrieffer [81], based on a pairing of electrons due to an attractive potential (the BCS theory). This theory has been enormously successful in providing a microscopic explanation for many aspects of superconductivity, as well as for predicting new behavior. Because of the interaction between the conduction electrons and the vibration of the atoms of the lattice, there is a small net *attraction* between the electrons. One way to think of the interaction is that an electron moving through the lattice distorts the lattice, and the deformed lattice in turn acts on a second electron because of the positive charge accumulation. As shown in Figure 3.6, the first electron passes through the lattice and attracts the positive ions. Because of their inertia, the ions cannot relax immediately. The shortest relaxation time, $2\pi/\omega_D$, corresponds to the highest possible lattice vibration frequency, ω_D. Therefore the first electron can influence a neighboring electron at a distance of $v_F 2\pi/\omega_D$, typically many tens of nanometers.

Another view of the attractive interaction is that each electron leaves behind a wake of phonons. Other electrons interact with the first electron if they travel

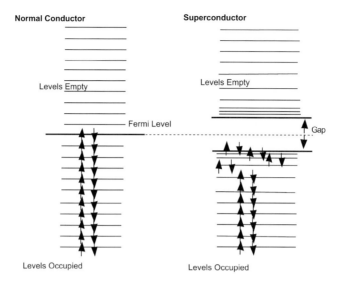

Figure 3.7: The change in the density of states that accompanies the superconducting state, showing the gap in the energy levels.

through this wake. The electron–electron attraction, mediated by phonons, is therefore analogous to the electromagnetic interaction between moving charges, which can be considered to be mediated by photons. Another analogy is provided by the nucleon interaction as mediated by mesons.

The importance to superconductivity of the interaction between electrons and lattice vibrations is confirmed by the "isotope effect" [82, 83]. Different isotopes of a given element have a different critical temperature according to

$$\frac{1}{T_\mathrm{c}} \propto \sqrt{M_\mathrm{atomic}}. \qquad (3.33)$$

The transition temperature depends on the mass of the ions comprising the lattice. This effect was predicted by Frolich [84], who first proposed the attractive mechanism. The electron–lattice interaction responsible for superconductivity is the very same interaction that determines the resistivity of metals at high temperatures. The higher transition temperature superconducting materials, such as niobium and lead, indeed have a high ideal resistivity in the normal state, because of their strong electron–phonon interaction. On the other hand, copper and gold have a weak electron–phonon interaction, and thus are excellent normal conductors, but do not show any superconducting transition.

Only a small fraction of the electrons cooperate in the interaction responsible for superconductivity, those within $k_\mathrm{B} T_\mathrm{c} \approx \hbar\omega_\mathrm{D}$ of the Fermi surface. Here T_c is the transition temperature, and $\hbar\omega_\mathrm{D}$ is a measure of the maximum phonon energy. In the presence of the attractive interaction, the normal Fermi distribution of electrons becomes unstable. The states near the Fermi surface get redistributed slightly. As shown in Figure 3.7, some electrons have their energy

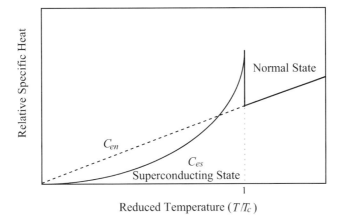

Figure 3.8: A comparison between the normal and superconducting state electronic specific heat.

pushed down from the Fermi level. The distribution of available states above the Fermi level is also altered. No states are available immediately above the Fermi level. An energy gap is created. The new ground state, the superconducting ground state, has a lower energy than the normal ground state.

One can view the energy gap as an absence of energy levels. As a result of this gap, the electronic component of the specific heat of a superconductor has the temperature dependence which is the characteristic thermal behavior of a system whose excited levels are separated from the ground state by an energy 2Δ,

$$C_{\text{es}} \propto \exp\left(-\frac{\Delta}{k_{\text{B}}T}\right). \tag{3.34}$$

As shown in Figure 3.8, the superconducting state electronic specific heat is quite different from the linear electronic specific heat of a normal metal.

The observed exponential behavior of the specific heat above $T = 0$ K provides the crucial evidence for the existence of the energy gap. This is clearly seen in Figure 3.9 for the data for a Pb–2%In alloy [85]. The slope of the measured curve is higher than the prediction of the BCS theory because lead has a higher value for $\Delta(0)/k_{\text{B}}T_{\text{c}}$ than the BCS predicted value (see below).

As we shall see later, the temperature dependence of the microwave surface resistance, $R_{\text{s}}(T)$, is another important manifestation of the presence of a gap in the electronic energy levels, and therefore it exhibits a behavior very similar to that of Figure 3.9.

A profound consequence of the attractive interaction is the condensation of electrons into Cooper pairs. The BCS theory considers what happens to two electrons near the Fermi surface in the presence of the attraction when each electron has an energy ε infinitesimally less than ε_{F}. At $T = 0$ K, the ground state is modified to give a lower energy by removing electrons from states with $\varepsilon < \varepsilon_{\text{F}}$ and allowing them to form pairs with electrons with energy $\varepsilon > \varepsilon_{\text{F}}$. The

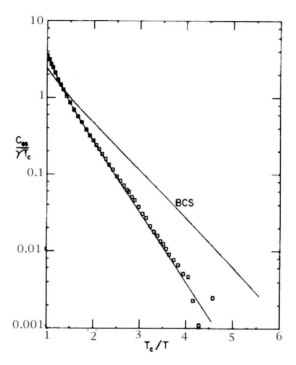

Figure 3.9: The ratio of the superconducting state electronic specific heat to the normal state electronic specific heat at T_c. The data are for a Pb–2%In alloy. The line is for the BCS theory using $\Delta(0)/k_B T_c = 1.76$. The line through the data is for the BCS theory with a modification that allows $\Delta(0)/k_B T_c = 2.43$.

two electrons form a bound state so that their total energy 2ε is less than $2\varepsilon_F$, because of the lowering of the potential energy from the mutual attraction.

These electron pairs are called Cooper pairs which can be regarded as new particles with twice the charge and twice the mass of an electron. The pairing energy is $2\Delta(T)$. The BCS theory predicts a relation between the pairing energy and the critical temperature. At $T = 0$ K,

$$2\Delta(0) = 2 \times (1.76 k T_c) = 3.12 \text{ meV}. \tag{3.35}$$

The actual size of the gap is different from one material to another due to differences in the electron–phonon interaction strength. For niobium and lead, $\Delta(0)/k_B T_c = 1.9$ and 2.4, respectively.

The electrons within a Cooper pair are not tightly bound to each other in the sense of electrons to protons within an atom. In a pair, the momenta and spin of the paired electrons are correlated as $(\mathbf{p}, \uparrow)$ $(-\mathbf{p}, \downarrow)$. It is this type of pairing which yields the minimum energy. Although individual electrons obey Fermi statistics, a Cooper pair is a boson. The Pauli exclusion principle does not apply; instead the pair obeys Bose–Einstein statistics. Therefore all the pairs can be in the same quantum state with the same center-of-mass energy.

When the electrons "condense" into Cooper pairs, the electrons involved are those within an energy range kT_c of the Fermi energy. We can use this range to find the momentum spread and the spatial extent of the electrons involved in the pairing interaction. We have

$$k_B T_c = \delta\left(\frac{p^2}{2m}\right) = \frac{p}{m}\delta p = v_F \delta p. \qquad (3.36)$$

Therefore

$$\delta p = \frac{k_B T_c}{v_F}. \qquad (3.37)$$

According to Heisenberg's uncertainty principle, the spatial extent of the pair, ξ_0, also known as the coherence length, is related to the momentum via

$$(\xi_0)(\delta p) \approx \hbar. \qquad (3.38)$$

From Equation 3.37 and Equation 3.38, the coherence length is

$$\xi_0 = \frac{\hbar v_F}{k_B T_c}. \qquad (3.39)$$

For niobium $\xi_0 = 39$ nm and for lead $\xi_0 = 83$ nm. The large distance indicates that the superconducting electrons possess long-range order and that the Cooper pairs overlap each other. The coherence length also gives the spatial extent of the superconducting state wavefunction. At the boundary between a normal and a superconducting region, the superconducting state wavefunction increases from zero to the maximum value over the coherence length. Since the BCS theory predicts a relation between the pairing energy and the critical temperature, another expression for ξ_0 can be obtained in terms of Δ:

$$\xi_0 = \frac{\hbar v_F}{\Delta}. \qquad (3.40)$$

At $T > 0$ K, the excited state of a superconductor can be constructed in a one-to-one correspondence with the normal state. In the first excited state, one electron would have momentum **p** and spin up, but its partner state with spin down would be *unoccupied*. The unpaired electron can be considered to behave as a "normal" or "free" electron. All other momentum states would still be occupied in pairs. In this excited state an energy gap still exists but the value of Δ is lower. Δ drops very slowly as T increases from 0 to T_c. But near T_c, Δ drops precipitously to zero, as shown in Figure 3.10. A useful approximation to the theoretical behavior for $\Delta(T)$ is given by [86, 87]

$$\frac{\Delta(T)}{\Delta(0)} = \left[\cos\left(\frac{\pi t^2}{2}\right)\right]^{1/2}, \qquad (3.41)$$

where t is the reduced temperature T/T_c.

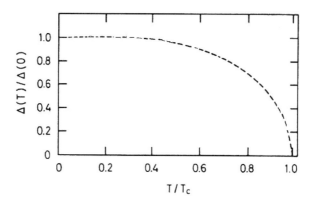

Figure 3.10: The predicted temperature dependence of the energy gap.

3.4 ELECTRICAL PROPERTIES, DC AND RF RESISTANCE

One can understand the zero dc resistance and the infinitesimally low rf resistance of a superconductor in terms of a simple two-fluid model; one fluid is a "superfluid" of paired electrons, and the other is a normal fluid of "free" electrons. At T_c, none of the electrons are paired. Below T_c, a fraction given by the Boltzmann factor, $\exp(-\Delta/k_B T)$ are *unpaired*. The number of normal electrons is

$$n_{\text{normal}} \propto \exp\left(-\frac{\Delta}{k_B T}\right). \tag{3.42}$$

At zero temperature all the electrons are paired. The conduction electrons which are paired carry a supercurrent with zero resistance. All the pairs behave coherently, in unison, to produce a supercurrent. The normal current and the super current flow in parallel. Since the supercurrent flows with zero resistance, it can carry the entire current, while the normal electrons remain inert.

Each pair carries the same net momentum imparted by the electric field. There is still an energy gap, provided that the energy gain of the Cooper pair from the electric field is less than 2Δ. To produce resistance, electrons have to be scattered, i.e., pairs have to be broken up. Cooper pairs only scatter if they gain more energy than the gap. As we mentioned, there are two types of scattering sites: impurities (including lattice imperfections) and phonons. An impurity is a fixed target and scattering will not increase the energy of the electrons in the pair. Therefore impurities will not scatter Cooper pairs. Lattice vibrations will also not scatter pairs as long as the thermal energy is smaller than the gap, i.e., $T < T_c$. To show resistance in a dc field, all existing pairs have to be stopped, and there are a very large number of pairs in the same quantum state. Therefore the dc resistance of a superconductor is exactly zero.

It is also illustrative to think about zero resistance in terms of the wave picture. In a normal metal, the electron velocity is high and the de Broglie wavelength turns out comparable to the lattice dimension. The electron waves can be scattered by obstacles of atomic dimensions, such as the disturbance in the perfect lattice created by interstitial impurities. In a superconductor, all the Cooper pairs are represented collectively by one wave. The macroscopic de Broglie wave is the main reason for the hallmark features of the superconducting state. When there is no current flow the CM of each Cooper pairs is at rest. When the superconductor carries current, all the pairs drift with one and the same velocity in the direction of the current. When the current ($\mathbf{j} = -n2e\Delta\mathbf{v}$) flows, it is carried by a very large number of Cooper pairs so that the drift velocity of the individual pairs is very low. Hence the wavelength of the Cooper pair assembly is very large, much larger than the dimensions of the lattice imperfections. Scattering by imperfections is impossible, leading to zero resistance.

When the energy gain of the pairs from the electric field exceeds Δ, superconductivity will breakdown. The maximum (critical) current density is given by

$$J_c = \frac{2en\Delta}{m_e v_F}. \tag{3.43}$$

In a later chapter we will relate this critical current to the critical magnetic field.

For rf currents, the surface resistance of a superconductor is nonzero, although it is very small compared to a normal metal. While the Cooper pairs move without friction, they do have inertia. Forces must be applied to make the time-varying current flow. (In a normal conductor, there is a similar effect arising from the mass of the electrons, but the resulting inductive impedance is overwhelmed by the resistance due to scattering.) Because of their inertia, the Cooper pairs do not screen the applied field perfectly. A time-varying electric field is present in the skin layer. It is induced by the time-varying magnetic surface field which penetrates into the "skin depth" (see below),

$$E_{\text{int}} \propto \frac{dH}{dt} \propto \omega H. \tag{3.44}$$

The time-varying electric field couples to the normal electrons to accelerate and decelerate them, leading to dissipation. The internal current is proportional to the internal electric field:

$$j_{\text{int}} \propto n_{\text{normal}} E_{\text{int}} \propto n_{\text{normal}} \omega H. \tag{3.45}$$

Since $P_{\text{diss}} \propto E_{\text{int}} j_{\text{int}}$ we find

$$P_{\text{diss}} \propto n_{\text{normal}} \omega^2 H^2. \tag{3.46}$$

As we saw in Chapter 2, the dissipated power can be couched in terms of a surface resistance:

$$P_{\text{diss}} = \frac{1}{2} R_s H^2. \tag{3.47}$$

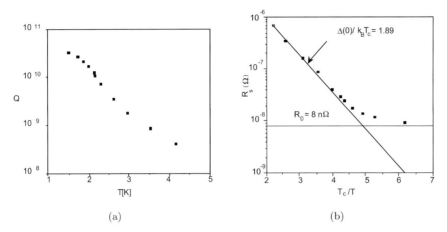

Figure 3.11: (a) Measured Q_0 versus T for a 1.5-GHz single-cell niobium cavity. (b) Same data as in (a) plotted as surface resistance versus inverse reduced temperature.

Therefore, since the number of normal electrons decreases exponentially, we have

$$R_s = A_s \omega^2 \exp\left(-\frac{\Delta(0)}{k_B T}\right). \quad (3.48)$$

This approximate relation is valid for $T < T_c/2$ when $\Delta(T)$ has reached the asymptotic value $\Delta(0)$, as shown in Figure 3.10. Equation 3.48 explains two of the prominent features of the superconducting state surface resistance:

1. R_s increases with the square of the rf frequency.

2. R_s decreases exponentially with temperature, as $\exp(-\Delta/T)$.

Figure 3.11(a) shows the measured Q_0 versus T for a 1.5 GHz single-cell cavity from $T = 4.2$ K to $T = 1.5$ K. When the data are replotted in Figure 3.11(b) as R_s versus T_c/T, it is clear that the surface resistance decreases exponentially as

$$\exp\left(-\frac{\Delta(0)}{k_B T_c} \frac{T_c}{T}\right). \quad (3.49)$$

The exponential temperature dependence, characteristic of the gap, has been well demonstrated. In general, A_s in Equation 3.48 depends on the material parameters, such as the Fermi velocity v_F, the London penetration depth $\lambda_L(0)$ (see Chapter 4), the coherence length ξ_0, and the mean free path of electrons.

Detailed expressions based on the BCS theory have been worked out to successfully calculate R_s, and computer programs exist to carry out the required numerical computations called for by the theory.

As we can see from Figure 3.11(b), below a certain temperature the experimentally observed surface resistance is higher than the BCS prediction and is accounted for by a residual resistance R_0:

$$R_\text{s} = R_\text{BCS}(T) + R_0. \tag{3.50}$$

We will discuss residual loss mechanisms in Chapter 9. A well-prepared niobium surface can reach a residual resistance of 10 to 20 nΩ. The record values are near 1 nΩ. The operating temperature of a superconducting cavity is usually chosen so that the BCS resistance is reduced to an economically tolerable value. For a 1.5 GHz cavity it is therefore necessary to cool the cavity to $T < 2$ K, at which temperature the BCS resistance is < 20 nΩ.

3.5 THERMAL CONDUCTIVITY IN THE SUPERCONDUCTING STATE

Finally we discuss the influence of superconductivity on the thermal conductivity. As already mentioned, at low temperatures the contribution of the phonons is negligible, because the number of phonons $\propto T^3$. The few remaining phonons are also scattered by electrons. Therefore the electronic contribution dominates near T_c. As T decreases below T_c, electrons freeze out into Cooper pairs which cannot transport thermal energy, as they cannot be scattered by lattice vibrations. Therefore, below T_c, the thermal conductivity drops sharply, as shown by the predictions [88] of the BCS theory in Figure 3.12 and compared to the experimental results [89] for a niobium sample of RRR = 195.

From the Wiedemann–Franz law, which relates the normal state thermal conductivity to the electrical conductivity, a convenient relationship can be derived between the RRR and the superconducting state thermal conductivity at 4.2 K.

$$\text{RRR} = 4k_\text{s}(\text{W/m-K}). \tag{3.51}$$

The formation of Cooper pairs has another important consequence on the thermal conductivity of niobium. As electrons freeze out they are no longer effective scatterers for the few remaining phonons, and the thermal conductivity actually rises. Figure 3.12 also shows the theoretically expected [90] increase of relative phonon conductivity below T_c. At sufficiently low temperature the phonon contribution to the thermal conductivity rises to a maximum determined by the mean free path of the phonons.

One of the important scattering sites for phonons are the boundaries of the crystal grains. For niobium, increasing the grain size can produce a peak in the thermal conductivity near 2 K [91], normally referred to as the "phonon peak" (Figure 3.12). Finally, as T approaches zero the phonon conductivity falls as T^3, as expected from the temperature dependence of the phonon density. An excellent review of the thermal conductivity of niobium can be found in [91].

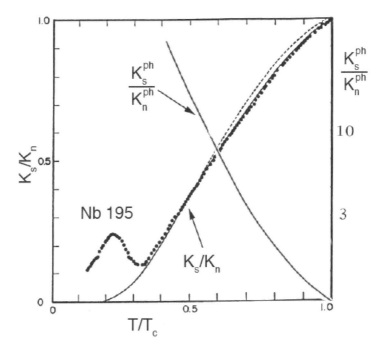

Figure 3.12: Measured thermal conductivity for a niobium sample with RRR = 195 [89] compared to the theoretical prediction [88]. Note the rapidly decreasing electronic component. The curve on the right shows the theoretical prediction for the thermal conductivity contribution from phonons when electron scattering is predominant. This is the cause of the phonon peak visible in the curve on the left.

CHAPTER 4

Electrodynamics of Normal and Superconductors

4.1 INTRODUCTION

In this chapter we start with a discussion of the surface resistance of normal conductors to introduce the skin effect [1] and the anomalous skin effect [92]. The treatment of the surface resistance of superconductors starts with the London two-fluid model which incorporates the Meissner effect and explains the f^2 rf frequency dependence. It also partially accounts for the mean free path dependence of the surface resistance. The London treatment introduces the idea of the superconducting penetration depth and its consequences for the attenuation of the magnetic field and the distribution of currents within a superconductor. Finally, we present some of the key results of the full-fledged BCS calculation for the rf surface resistance.

4.2 SKIN DEPTH AND SURFACE RESISTANCE OF NORMAL CONDUCTORS

The simple ideas of electrical conductivity discussed in the last chapter are useful in treating the ac conductivity of normal metals. Using Maxwell's equations it is possible to show ([1]) that for alternating fields, $\mathbf{E} = \mathbf{E}_0 \exp(i\omega t)$ with $\mathbf{j} = \sigma \mathbf{E}$, and $\omega \epsilon \ll \sigma$ (a good conductor):

$$\nabla^2 \mathbf{E} = i\mu_0 \sigma \omega \mathbf{E} \quad \text{or} \quad \nabla^2 \mathbf{E} = \tau_n^2 \mathbf{E} \quad \text{with} \quad \tau_n = \sqrt{i\omega\sigma\mu_0}. \tag{4.1}$$

Similar equations can be obtained for the current $\mathbf{j}$ and the magnetic field $\mathbf{H}$. The approximation $\omega \epsilon \ll \sigma$ is valid for microwave frequencies and for good conductors. It implies that the displacement current is negligible compared to the conduction current in metals.

For the simple planar example of a conductor filling half-space $x > 0$, as shown in Figure 4.1, assume that the electric field is in the z direction, and that there are no variations with y or z. A solution to the above differential equation is

$$E_z = E_0 \exp(-\tau_n x), \tag{4.2}$$

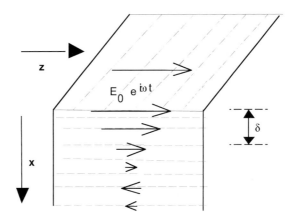

Figure 4.1: Penetration of rf electric field into normal conductor to a skin depth δ.

which can be written as

$$E_z = E_0 \exp\left(\frac{-x}{\delta}\right) \exp\left(\frac{-ix}{\delta}\right). \qquad (4.3)$$

AC fields penetrate a thickness δ (the skin depth)

$$\delta = \frac{1}{\sqrt{\pi f \mu_0 \sigma}}. \qquad (4.4)$$

Qualitatively speaking, the surface currents in the skin of the metal shield the electric field from the bulk of the metal. Note that the skin depth decreases with increasing frequency.

In a similar treatment to the above, the ac current density and magnetic fields are given by

$$j_z = j_0 \exp(-\tau_n x), \quad H_y = H_0 \exp(-\tau_n x). \qquad (4.5)$$

In analogy with dc resistance, the surface impedance Z_s is defined as the ratio of the surface electric field E_0 to the total current in the conductor, I. Since, in ac circuits, the current does not distribute itself uniformly over the conductor, we integrate to obtain the total current.

$$I = \int_0^\infty j_z(x)dx = \int_0^\infty j_0 \exp(-\tau_n x)dx = \frac{j_0}{\tau_n}. \qquad (4.6)$$

Using the relation $j_0 = \sigma E_0$, we get the surface impedance

$$Z_s = \frac{E_0}{j_0/\tau_n} = \frac{\tau_n}{\sigma} = \frac{\sqrt{i\omega\mu_0\sigma}}{\sigma} = R_s + iX_s. \qquad (4.7)$$

The impedance has an imaginary part because the surface field is not in phase with the total current in the conductor, due to the rate of change of magnetic flux in the conductor. The real part, R_s, gives the ac resistance, and the imaginary part, X_s gives the reactance.

The microwave surface resistance is thus

$$R_\text{s} = \sqrt{\frac{\pi f \mu_0}{\sigma}} = \frac{1}{\sigma \delta}. \tag{4.8}$$

4.3 THE ANOMALOUS SKIN EFFECT

We have seen that ac fields penetrate an amount δ, which decreases with increasing frequency. Therefore at very low temperature and/or high rf frequency, the skin depth may become shorter than the mean free path of an electron. In this case, electrons spend only part of the time between collisions in the field penetrated region. Qualitatively speaking, this makes electrons less effective carriers of rf current, reducing their effectiveness for shielding the electric field from the bulk of the metal and thereby leading to a higher rf surface resistance than can be expected from Equation 4.8. One might say that contrary to the dc case and contrary to intuition, the longer mean free path does not increase the rf conductivity. Therefore the behavior is called the anomalous skin effect.

More rigorously [92], the anomalous effect occurs because the relation $\mathbf{j}(0) = \sigma(\ell)\mathbf{E}(0)$ is valid only if $\mathbf{E}(\mathbf{R})$ varies slowly over a distance of the order of the mean free path. When the relation between current and fields becomes nonlocal, Ohm's law is no longer valid, and the current at a point depends on the electric field in a region around the point of size comparable to the mean free path. The current at a point is determined by the integrated effect of the field over a distance of order of ℓ

$$j(0) \propto \frac{\sigma}{\ell} \int \frac{\mathbf{R}(\mathbf{R}\cdot\mathbf{E})\, e^{-R/l} d\tau}{R^4}. \tag{4.9}$$

The results of the theory of the anomalous skin effect are characterized by a dimensionless parameter, α_s, which depends on the temperature-independent product, commonly referred to as the $\rho\ell$ product, of the mean free path, ℓ, and the resistivity, $\rho = 1/\sigma$,

$$\alpha_\text{s} = \frac{3}{4}\mu_0 \omega \left(\frac{1}{\rho\ell}\right)\ell^3. \tag{4.10}$$

α_s has a strong dependence on ℓ. The classical expression for surface resistance (ohm) applies when as $\alpha_\text{s} \leq 0.016$. At the other extreme, when as $\alpha_\text{s} \to \infty$

$$R_\text{n}(\ell \to \infty) = \left[\sqrt{3}\pi \left(\frac{\mu_0}{4\pi}\right)^2\right]^{1/3} \omega^{2/3} (\rho\ell)^{1/3} = 3.789 \times 10^{-5} \omega^{2/3} (\rho\ell)^{1/3}. \tag{4.11}$$

In this anomalous limit, the surface resistance is independent of the dc resistivity, or the temperature, because the $\rho\ell$ product is a material constant, independent of temperature. For intermediate values, i.e., as $\alpha_s \geq 3$, the normal state surface resistance is

$$R_n(\ell) = R(\infty)\left(1 + 1.157\alpha_s^{-0.2757}\right), \tag{4.12}$$

where $R(\infty)$ is the surface resistance in the extreme anomalous limit [93].

The anomalous limit is applicable to a very good conductor, such as copper, at microwave frequencies and at low temperatures (see Problem 6). The implications are significant. When a copper cavity is cooled to low temperature, even if the dc conductivity increases by a factor of 100, the largest improvement in surface resistance one can expect with respect to the room temperature surface resistance is only a factor of 6. From an economic point of view this meager improvement is not enough to pay for the refrigeration cost.

4.4 PERFECT CONDUCTORS

In the presence of an electric field, the electrons in an ideally perfect conductor are accelerated freely:

$$m\frac{\partial \mathbf{v}}{\partial t} = -e\mathbf{E}. \tag{4.13}$$

With the current density $\mathbf{j}_s = -n_s e \mathbf{v}$, the infinite conductivity equation can be written as

$$\frac{\partial \mathbf{j}_s}{\partial t} = \frac{n_s e^2}{m}\mathbf{E}. \tag{4.14}$$

The above equation is known as the first London equation. It should be contrasted with the conductivity equation for a normal metal

$$\mathbf{j}_n = \frac{ne^2\tau}{m}\mathbf{E}. \tag{4.15}$$

Using Maxwell's equation

$$\nabla \times \mathbf{E} = -\frac{\partial \mathbf{B}}{\partial t}, \tag{4.16}$$

the perfect conductivity equation gives

$$\frac{\partial}{\partial t}(\nabla \times \mathbf{j}_s) = \frac{n_s e^2}{m}\nabla \times \mathbf{E} = -\frac{n_s e^2}{m}\frac{\partial \mathbf{B}}{\partial t}, \tag{4.17}$$

or

$$\frac{\partial}{\partial t}\left(\nabla \times \mathbf{j}_s + \frac{n_s e^2}{1}\frac{\mathbf{B}}{m}\right) = 0. \tag{4.18}$$

Together with the other Maxwell equations, Equation 4.18 determines the magnetic field and current density inside a *perfect conductor*. The equation characterizes any medium that conducts electricity without dissipation, and it requires that the expression

$$\nabla \times \mathbf{j}_s + \frac{n_s e^2}{m}\mathbf{B} \tag{4.19}$$

PERFECT CONDUCTORS

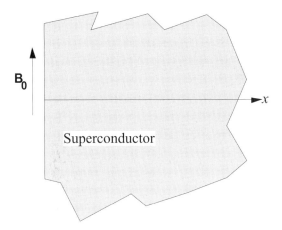

Figure 4.2: Semi-infinite superconducting plane extending in the z direction with parallel external magnetic field.

be independent of time. Further application of Maxwell's equation (see Problem 4) leads to

$$\nabla^2 \left(\frac{\partial \mathbf{B}}{\partial t} \right) = \frac{1}{\lambda_\mathrm{L}^2} \frac{\partial \mathbf{B}}{\partial t}, \quad \lambda_\mathrm{L}^2 = \frac{m}{n_\mathrm{s} e^2 \mu_0}. \tag{4.20}$$

To see what this implies, consider a plane boundary of a superconductor with a uniform magnetic flux density $\mathbf{B}_0$ applied parallel to this boundary (See Figure 4.2).

If the direction normal to the boundary is the x direction, the solution to the one-dimensional form of Equation 4.20 is

$$\frac{\partial \mathbf{B}(x)}{\partial t} = \frac{\partial \mathbf{B}_0}{\partial t} \exp\left(-\frac{x}{\lambda_\mathrm{L}} \right), \tag{4.21}$$

where $\mathbf{B}(x)$ is the flux density inside the metal. This means that $\partial \mathbf{B}(x)/\partial t$ decays exponentially as we penetrate into the superconductor. At a sufficient distance inside the superconductor, the flux density has a constant value and cannot change with time, independent of what is happening to the applied field outside. The flux density inside a perfectly conducting metal cannot change. If we start with zero flux density inside the superconductor metal and then increase the external field, the flux density inside the superconductor will stay the same. Perfect conductivity implies perfect shielding from external magnetic flux.

In a perfect conductor, $\partial \mathbf{B}/\partial t = 0$ inside. But $\mathbf{B} = $ constant is also allowed. Figure 4.3 describes what happens to the magnetic field $\mathbf{H}$ in two cases:

(a) when a perfect conductor is cooled to its perfectly conducting state in the absence of a field and then a field is applied;

(b) when a specimen is cooled to its perfectly conducting state in the presence of a magnetic field.

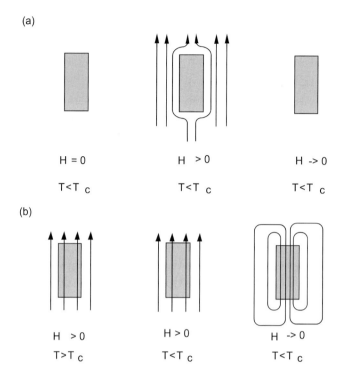

Figure 4.3: (a) Screening of external magnetic field by a perfect conductor. (b) Flux trapping in a perfect conductor.

The last step in each figure also shows what happens to the magnetic field lines when the externally applied field is reduced to zero. Notice the different final states. The transition from the normal to the perfectly conducting state is not reversible. The final state depends on the path of the transition. In the next section we will see that a superconductor does not behave like a perfect conductor because of the Meissner effect, a unique feature of superconductivity.

4.5 MEISSNER EFFECT

When a superconducting material is cooled through its transition temperature, T_c, in the presence of an external magnetic field, the magnetic flux is abruptly expelled from the volume of the superconductor. This is the Meissner effect which was discovered 22 years after the original discovery of superconductivity. The transition to the superconducting state is accompanied by the appearance of supercurrents on the surface which cancel the magnetic field in the interior. A dramatic demonstration of the Meissner effect is the "levitation" effect. If one takes a pellet of superconducting material that is above its transition temperature and places it on a permanent magnet pole, it will just sit there. As

MEISSNER EFFECT

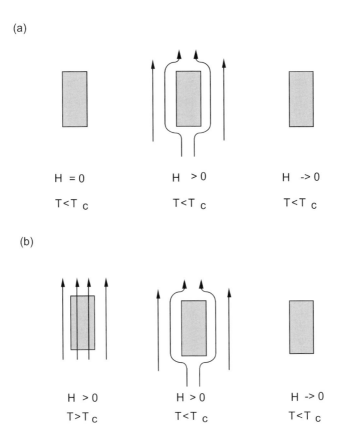

Figure 4.4: (a) Screening of external magnetic field by a superconductor. (b) Meissner effect (flux expulsion) in a superconductor.

the whole system is cooled down so that the material makes a transition into the superconducting state, the superconducting pellet will suddenly rise and float above the magnet. Because of the Meissner effect, the magnetic field is expelled from the volume of the superconductor. The resulting magnetic field configuration on the surface of the superconductor repels the magnetic field of the permanent magnet below, providing the force for the remarkable levitation.

A related aspect of the magnetic behavior of a superconductor is flux exclusion, as distinct from expulsion. An externally applied magnetic field (less than the critical field) cannot penetrate into the interior of a superconductor. It is excluded from the volume of the superconductor. A metal in the superconducting state therefore never allows magnetic flux density to exist in the interior.

Although the perfect exclusion (perfect shielding) aspect can be expected from the property of perfect conductivity, the expulsion aspect, the Meissner effect, cannot be explained by perfect conductivity alone. Later we will also see that, in the presence of impurities, serving as flux trapping sites, only partial flux expulsion may take place, resulting in an incomplete Meissner effect.

Perfect conductivity does not imply the Meissner effect, which demands that the flux density inside the superconductor be zero. In a superconductor, $\partial \mathbf{B}/\partial t = 0$ inside *and* $\mathbf{B} = 0$ inside. The transition of a superconductor from the normal to the superconducting state is therefore a reversible transition. The final state does not depend on the path of the transition. Figure 4.4 describes what happens to the magnetic field (H) in two cases:

(a) When the superconductor is cooled to its superconducting state in the absence of a field and then a field is applied, screening takes place.

(b) When a specimen is cooled to the superconducting state in the presence of a magnetic field, the Meissner effect expels the field.

Again, the last step in each figure also shows what happens to the magnetic field lines when the externally applied field is reduced to zero. Notice the same final states. The transition of a superconductor from the normal to the superconducting state is completely reversible. The final state does not depend on the path of the transition.

London realized that the Meissner effect requires that the quantity

$$\nabla \times \mathbf{j}_{\mathrm{s}} + \frac{n_{\mathrm{s}}e^2}{m}\mathbf{B} \tag{4.22}$$

be not only time-independent, but also have the value of zero.

$$\nabla \times \mathbf{j}_{\mathrm{s}} + \frac{n_{\mathrm{s}}e^2}{m}\mathbf{B} = 0. \tag{4.23}$$

Equation 4.23 is the second London equation, which incorporates the Meissner effect. It cannot be justified within classical physics. The magnetic properties of a superconductor cannot be accounted for by zero resistivity alone. Again, using the London equations and Maxwell's equations for static fields, one can show (see Problem 4) that

$$\nabla^2 \mathbf{H} = \frac{1}{\lambda_{\mathrm{L}}^2}\mathbf{H}, \quad \nabla^2 \mathbf{j}_{\mathrm{s}} = \frac{1}{\lambda_{\mathrm{L}}^2}\mathbf{j}_{\mathrm{s}}, \tag{4.24}$$

where λ_{L} is the London penetration depth.

For a simple case of a plane at right angles to the x axis, where the superconductor extends in the positive x direction, with magnetic field $\mathbf{H}$ pointing in the z direction at $x = 0$, the solution is

$$H_z = H_0 \exp\left(-\frac{x}{\lambda_{\mathrm{L}}}\right), \quad j_y = -\frac{1}{\lambda_{\mathrm{L}}}H_0 \exp\left(-\frac{x}{\lambda_{\mathrm{L}}}\right). \tag{4.25}$$

The London equations predict that the current density and magnetic field in a superconductor exist only within a layer of thickness λ_{L}. The effectiveness of the shielding supercurrents is therefore not perfect. In a small region near

SURFACE IMPEDANCE OF SUPERCONDUCTORS

the surface, there is some penetration of magnetic field. Another consequence is that current through a superconducting wire can only flow through a thin surface layer. No currents are allowed in the interior as they would generate magnetic fields inside the bulk.

The London equations are empirical restrictions on the ordinary equations of electromagnetism and are introduced so that the behavior of a superconductor deduced from these laws agrees with experiment — that is, both the Meissner effect and infinite conductivity. For a more precise theoretical treatment of the London equations based on the vector potential, see [3].

4.6 SURFACE IMPEDANCE OF SUPERCONDUCTORS IN THE TWO-FLUID MODEL

Consider, once again, the electric field and current flow to be in one dimension. In the two-fluid model, the normal component of the current is $j_n = \sigma_n E$ where

$$\sigma_n = \frac{n_n e^2 \tau}{m}, \tag{4.26}$$

and n_n is the number of unpaired electrons. We aim to arrive at parallel expressions for the superfluid component. For the case of rf currents, let

$$j_s = j_{s0} e^{i\omega t}. \tag{4.27}$$

We use the first London equation

$$\frac{\partial j_s}{\partial t} = \frac{1}{\mu_0 \lambda_L^2} E \tag{4.28}$$

to obtain

$$i\omega j_s = \frac{1}{\mu_0 \lambda_L^2} E \tag{4.29}$$

and arrive at a parallel expression to $j_n = \sigma_n E$ for the supercurrent component:

$$j_s = \frac{-i}{\omega \mu_0 \lambda_L^2} E = -i\sigma_s E. \tag{4.30}$$

Using the definition of λ_L, we get

$$\sigma_s = \frac{n_s e^2}{m\omega}. \tag{4.31}$$

Now the total two-fluid current becomes

$$j = j_n + j_s = (\sigma_n - i\sigma_s) E. \tag{4.32}$$

In analogy with the treatment for normal conductors we can show (Problem 10) that

$$\nabla^2 E = \tau_{\text{tot}}^2 E, \tag{4.33}$$

but in the two-fluid case we have

$$\tau_{\text{tot}} = \sqrt{\mu_0 \omega i \left(\sigma_{\text{n}} - i\sigma_{\text{s}}\right)}. \tag{4.34}$$

Consequently, the surface impedance of a superconductor is given by

$$Z_{\text{s}} = \sqrt{\frac{i\omega\mu_0}{\sigma_{\text{n}} - i\sigma_{\text{s}}}}. \tag{4.35}$$

To derive a simpler expression for Z_{s}, a very useful approximation is that $\sigma_{\text{n}} \ll \sigma_{\text{s}}$; i.e., the conductivity of the normal fluid is very much smaller than the "conductivity" of the superfluid, for two reasons. From

$$\sigma_{\text{n}} = \frac{n_{\text{n}} e^2 \tau}{m} \quad \text{and} \quad \sigma_{\text{s}} = \frac{n_{\text{s}} e^2}{m\omega} \tag{4.36}$$

we note that at $T \ll T_{\text{c}}$ the number of unpaired electrons, n_{n}, is very much smaller than the number of paired electrons, n_{s}. Secondly, for the normal conducting electrons, the relaxation time ($\approx 10^{-14}$ s) between collisions is very much smaller than the rf period ($\approx 10^{-9}$ s), i.e., $\tau \ll 1/\omega$. Therefore, with the usual definition of impedance

$$Z_{\text{s}} = R_{\text{s}} + iX_{\text{s}} \tag{4.37}$$

it is possible to show (Problem 11) that the real and imaginary parts of the impedance become

$$R_{\text{s}} = \frac{1}{2} \sigma_{\text{n}} \omega^2 \mu_0^2 \lambda_{\text{L}}^3, \tag{4.38}$$

and

$$X_{\text{s}} = \omega \mu_0 \lambda_{\text{L}}. \tag{4.39}$$

The magnitude of the surface resistance is very much less than the surface reactance, except near T_{c}. Once again we have shown the salient features of the surface resistance of a superconductor, namely,

$$R_{\text{s}} \propto \omega^2 \quad \text{and} \quad R_{\text{s}} \propto \sigma_{\text{n}} \propto n_{\text{n}} \propto \exp\left(-\frac{\Delta}{kT}\right). \tag{4.40}$$

Now we also see a very strong dependence of the surface resistance on another material parameter

$$\lambda_{\text{L}}^2 = \frac{m}{n_{\text{s}} e^2 \mu_0}. \tag{4.41}$$

This means that for a material with a low density of charge carriers, n_{s}, such as the new high-temperature cuprate superconductors, the penetration depth (and correspondingly the theoretical surface resistance) will be substantially larger, as we shall discuss in Chapter 14.

We also see from Equation 4.38 the remarkable result that the superconducting state surface resistance is proportional to the dc normal state conductivity

$$R_{\text{s}} \propto \tau \propto \ell, \tag{4.42}$$

SURFACE IMPEDANCE OF SUPERCONDUCTORS

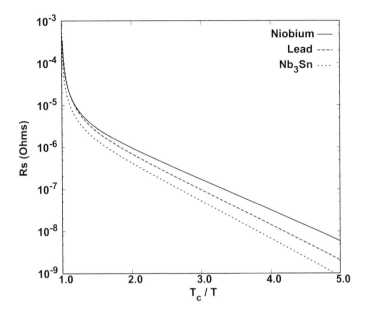

Figure 4.5: Theoretical surface resistance at 1.5 GHz of lead, niobium and Nb_3Sn as calculated from program [94]. The values given in Table 4.1 were used for the material parameters.

Table 4.1: Material parameters used for the caculations of Figure 4.5

Material parameter	Pb	Nb	Nb_3Sn
T_c [K]	7.19	9.20	18.00
Energy gap, Δ/kT_c	2.10	1.86	2.25
Penetration depth λ [Å]	280	360	600
Coherence length ξ [Å]	1110	640	60
Mean free path ℓ [Å]	10 000	500	10

which means that the superconducting state surface resistance *increases* with the mean free path. An intuitive (but limited) explanation for this unexpected behavior is that if the unpaired electrons have a higher conductivity, i.e. their mean free path is large, they draw a relatively higher current. Therefore keeping the normal fluid component more lossy lowers the superconducting state surface resistance!

This effect is observed as a lower Q_0 for a niobium cavity made with higher purity material. Note that we are discussing the theoretical part of the surface resistance, so that the change of R_s with mean free path is only apparent at temperatures and rf frequencies when the residual resistance part is negligible. Since most cavities are operated in a regime where the residual resistance dominates, the mean free path effect usually does not limit the operating Q_0.

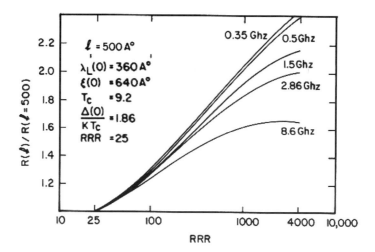

Figure 4.6: Relative increase of BCS surface resistance expected from increasing purity of niobium. The baseline surface resistance is for niobium with mean free path of 500 Å. Higher purity is characterized by increasing RRR. Curves are calculated from Halbritter's computer program [94] for several rf frequencies.

4.7 BCS TREATMENT OF SURFACE RESISTANCE

Based on the very successful BCS theory, expressions for the superconducting surface impedance have been worked out by Mattis and Bardeen [95] and by others. These expressions involve the material parameters λ_L, ξ_0, v_F, and ℓ. They are in a rather difficult form to obtain general formulas to work with. Separate computer programs have been written by Turneaure [96] and by Halbritter [94]. In Figure 4.5, we give some of the results from Halbritter's programs for niobium, lead and Nb_3Sn. Calculations from the theory agree well with experimentally measured R_s. The theory confirms the simplified form of the temperature dependence of Nb for $T < T_c/2$ and for frequencies much smaller than $2\Delta/h \approx 10^{12}$ Hz. A convenient expression and a good fit is

$$R_{\text{BCS}}(\text{ohm}) = 2 \times 10^{-4} \frac{1}{T} \left(\frac{f}{1.5}\right)^2 \exp\left(-\frac{17.67}{T}\right), \qquad (4.43)$$

where f is the rf frequency in GHz.

There is no correction in Equation 4.43 for the increase of surface resistance with RRR. To explore the effect of mean free path (and correspondingly the RRR) on R_s, we show in Figure 4.6 the surface resistance enhancement factor with RRR for various frequencies as calculated from the BCS theory with the computer program [94].

Experimental results [97] on 8.6-GHz cavities confirm this behavior of R_s as a function of mean free path, as shown in Figure 4.7. Another important case

BCS TREATMENT OF SURFACE RESISTANCE

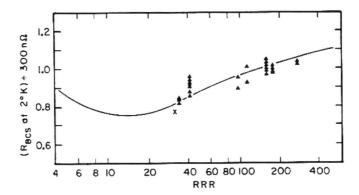

Figure 4.7: A comparison between 8.6-GHz measurements and predictions for the relative increase in BCS surface resistance due to increasing niobium purity.

of this effect is observed in 350-MHz cavities made from sputtered Nb on Cu [98]. Here the grain size and purity of the thin film are such the mean free path is very small (10 – 100 nm), and the Q_0 of a cavity at 4.2 K is a factor of two higher than a cavity made from high-RRR bulk Nb.

The relation $R_s \propto \sigma_n$ is only valid when $\xi_0/\ell \ll 1$. In this regime the London equation, which is a local equation, is valid and relates the current density at a point to the electric field at the same point. For these "London" superconductors the fields and currents vary slowly in space on the scale of the superconducting coherence length. However, when $\xi_0/\ell > 1$, i.e., for materials with a large coherence length or very short mean free path, non-local effects become important. The first London equation (Equation 4.28) breaks down, and the surface resistance reaches a minimum. For shorter mean free paths (i.e., very low RRR), or for larger coherence length, calculations [94] from the full BCS theory show that the surface resistance increases again as shown in Figure 4.7.

To correctly predict the frequency dependence for niobium from microwave frequencies to the gap frequency, $2\Delta/h$, it is necessary to take into account the anisotropy of the pairing energy induced by the anisotropy of the crystal lattice. The results of this full-fledged treatment [100] are given in Figure 4.8. Note how, at frequencies above 10 GHz, the frequency dependence starts to depart from the f^2 of the two-fluid model. This departure is confirmed by the experimental points [99].

The surface reactance can also be obtained from the Mattis–Bardeen formulation [95] and can be used to estimate the penetration depth for a superconductor. Near T_c, the temperature dependence of penetration depth agrees with that derived from the two-fluid model.

$$\lambda(T) = \lambda(0)[1 - (T/T_c)^4]^{-1/2}. \tag{4.44}$$

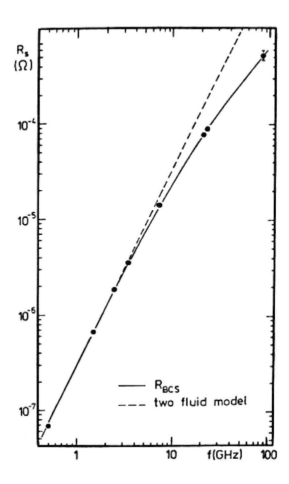

Figure 4.8: The two-fluid model predicts an f^2 frequency dependence for the superconducting state surface resistance. When the anisotropy of the pairing energy induced by the anisotropy of the crystal lattice is taken into account, the BCS theory predicts a departure from the f^2, as confirmed by the data [99]. (Courtesy of Wuppertal.)

The penetration depth is rather insensitive to temperature, except in the neighborhood of T_c, where it is very sensitive. Changes in $\Delta\lambda(T)$ can be measured by the changes in the resonant frequency of a cavity near T_c [101].

CHAPTER 5

Maximum Surface Fields

5.1 INTRODUCTION

The ultimate limit to the maximum achievable rf surface magnetic field is the rf critical field, $H_{\text{rf, crit}}$. Before we can discuss the theoretical maximum rf magnetic field that the surface of a superconducting cavity can withstand, we need to introduce some of the concepts related to the dc critical magnetic field. We will also discuss the surface energy associated with boundaries between superconducting and normal conducting regions. The concepts referred to here are developed in *Introduction to Superconductivity* by Tinkham [102]. In the last part of this chapter we present the results of efforts to confirm the theoretical predictions for the rf critical magnetic field of niobium and lead.

5.2 THE THERMODYNAMIC CRITICAL FIELD

When electrons condense into Cooper pairs, the resulting superconducting state becomes more highly ordered than the normal-conducting state. Since only those few electrons within $k_B T_c$ of the Fermi energy are involved, the entropy difference is small. As mentioned, the free energy in the superconducting state is lower than that in the normal-conducting state (Figure 5.1). Here free energy is

$$F = U_{\text{int}} - TS, \tag{5.1}$$

where U_{int} is the internal energy and S is the entropy. As discussed earlier, when an external dc magnetic field H_e is turned on, supercurrents flow in the penetration depth to cancel out the field in the interior. This raises the free energy of the superconducting state. When the external field rises to a value H_c so that the free energy of the superconducting state $F_s(H)$ becomes equal to the free energy of the normal state (F_n), the two phases are in equilibrium:

$$F_s(H) = F_n = F_s(H=0) + \mu_0 V_s \int_0^{H_c} H \, dH. \tag{5.2}$$

Here V_s is the volume of the superconductor. All the flux enters the superconductor at H_c, which is called the thermodynamic critical field. The second

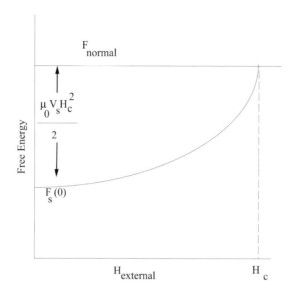

Figure 5.1: The superconducting state has a lower free energy than the normal state. In the presence of an external field, the free energy rises because of the work done to expel the magnetic field from the bulk of the superconductor volume. When the external field reaches H_c, the free energy in the superconducting state becomes equal to that in the normal state. There is a phase transition to the normal state and all the flux enters the superconductor.

term in Equation 5.2 is the work done on the superconductor to establish the screening currents:

$$F_\text{n} - F_\text{s}(0) = \frac{\mu_0 V_\text{s} H_\text{c}^2}{2}. \tag{5.3}$$

The BCS theory leads to an expression for the free energy of electrons in the superconducting state from which various thermodynamic properties, including H_c, can be derived. In particular, one can calculate the zero temperature thermodynamic critical field

$$\frac{\mu_0 H_\text{c}^2}{2} = \frac{3\gamma T_\text{c}^2}{4\pi^2}\left(\frac{\Delta(0)}{k_\text{B} T_\text{c}}\right)^2 = 0.236 \gamma T_\text{c}^2. \tag{5.4}$$

Here γ is the coefficient of the linear electronic specific heat in the normal state as discussed in connection with Equation 3.26. It is important to note that the thermodynamic critical field increases proportionately to T_c. Therefore the higher temperature superconductors, such as Nb_3Sn and the new high T_c cuprates, are interesting for accelerating cavities not only because of the possibility of their higher operating temperature, but also because of the promise of higher operating fields. Note that γ is related to the electronic density of states and thereby reflects the role of the electronic structure of the material.

POSITIVE SURFACE ENERGY SUPERCONDUCTORS (TYPE I) 93

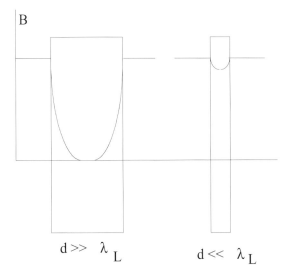

Figure 5.2: Attenuation of the magnetic flux density inside a thick superconducting slab compared to the same for a thin sheet.

Due to the temperature dependence of the energy gap $\Delta(T)$, the thermodynamic critical field falls to zero as the temperature is raised, and it reaches zero at T_c. The temperature dependence of the critical field is approximately (within 5%) given by

$$H_c(T) = H_c(0) \left[1 - \left(\frac{T}{T_c}\right)^2\right]. \tag{5.5}$$

The thermodynamic critical magnetic field is also related to the critical current density. As we mentioned in Chapter 4, the current flowing through a wire is confined to a surface layer λ_L. By equating the density of kinetic energy $1/2\, n_s m v_s^2$ to density of condensation energy $1/2\, \mu_0 H_c^2$, and using the definitions $j_s = -n_s e v$ for the supercurrent density and $\lambda_L^2 = m/n_s e^2 \mu_0$ for the penetration depth, we find that the maximum current density J_c is related to the critical flux density by $H_c = \lambda_L J_c$. Since the new high temperature superconductors have a large λ_L they also show a relatively lower J_c.

5.3 POSITIVE SURFACE ENERGY SUPERCONDUCTORS (TYPE I)

If we compare, as in Figure 5.2, the attenuation of the external magnetic flux density in a thick slab ($d \gg \lambda_L$) with the attenuation in a thin sheet ($d \ll \lambda_L$) we see that, because the distance is short, the internal flux density does not drop to zero in the thin sheet. As a result, less work is done to exclude the

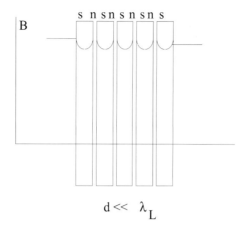

Figure 5.3: In the presence of an external magnetic flux density, it is energetically favorable for the flux to enter the superconductor in form of thin normal-conducting domains, except for considerations of the surface energy associated with the boundaries between the normal and superconducting domains.

magnetic flux and the free energy is lower for the thin sheet case. A natural question then arises: Is it energetically favorable for the magnetic flux to enter a thick material in very thin normal- conducting regions, as shown in Figure 5.3?

Such a situation does not always occur because it is takes some surface energy to form a superconducting/normal-conducting phase boundary. As we shall see below, there is an important class of superconductors (called Type II) for which this phase boundary energy is negative, so that at a certain field, called the lower critical field, H_{c1}, the flux does enter the material. There are superconductors for which this does not happen because they have a positive surface energy. These are called Type I superconductors. Since the density of the superconducting electrons changes appreciably only within the coherence distance ξ_0, and not discontinuously with position, the boundary between a superconducting and a normal-conducting region is not sharp. There is no spatially abrupt change from normal to superconducting behavior.

When the density of superconducting electrons is suppressed over the distance ξ_0, the free energy per unit area is raised by

$$\frac{\mu_0}{2} H_c^2 \xi_0. \tag{5.6}$$

In the presence of the external magnetic field H_e, the magnetic flux density penetrates a distance λ_L. By admitting the applied field H_e over λ_L, the energy per unit area is lowered by

$$-\frac{\mu_0}{2} H_e^2 \lambda_L. \tag{5.7}$$

The net boundary energy per unit area is

$$\frac{\mu_0}{2} \left(\xi_0 H_c^2 - \lambda_L H_e^2 \right). \tag{5.8}$$

POSITIVE SURFACE ENERGY SUPERCONDUCTORS (TYPE I)

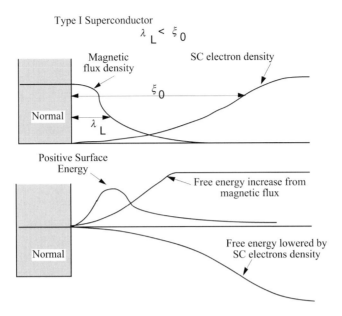

Figure 5.4: Positive surface energy for a Type I superconductor.

In general, the material properties of a superconductor are such that ξ_0 and λ_L are not the same, so that the two contributions do not cancel near the boundary. If $\xi_0 > \lambda_L$ (i.e., the density of superconducting electrons near the boundary is low) then there is a positive surface energy at the boundary (See Figure 5.4 and Figure 5.5) If $\xi_0 < \lambda_L$, the boundary energy is negative so that the total energy can be lowered by having superconducting/normal-conducting boundaries. When the boundary energy is positive, the Meissner state is the lower energy state. As mentioned earlier, positive surface energy superconductors are called Type I, and negative surface energy superconductors are called Type II. The Ginzburg–Landau parameter,

$$\kappa_{\text{GL}} = \frac{\lambda_L}{\xi_0}, \tag{5.9}$$

differentiates a Type I superconductor from a Type II superconductor. The relationship between ξ_0 and λ_L determines the response of a superconductor to an external magnetic field. In the more refined treatment, the GL parameter that separates the two regions obeys

$$\kappa_{\text{GL}} < \frac{1}{\sqrt{2}} \quad \text{Type I}, \quad \kappa_{\text{GL}} > \frac{1}{\sqrt{2}} \quad \text{Type II}. \tag{5.10}$$

Lead is a Type I superconductor with $\kappa_{\text{GL}} \approx 0.45$ and $H_c(0) = 800$ Oe. Niobium is a Type II superconductor with $\kappa_{\text{GL}} \approx 1$ and $H_c(0) = 2000$ Oe. More generally the coherence length and the penetration depth are both purity dependent and

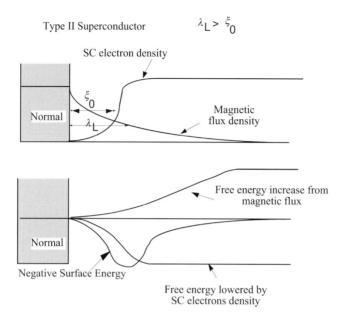

Figure 5.5: Negative surface energy for a type II superconductor.

temperature-dependent. As a result, κ_{GL} also depends on temperature and purity.

5.4 NEGATIVE SURFACE ENERGY SUPERCONDUCTORS (TYPE II)

Over the volume of the Type I superconductor, the free energy remains lower in the superconducting state than in the normal-conducting state up to H_c. But if $\xi_0 < \lambda_L$ (i.e., the superconductor has negative surface energy), it is energetically favorable for magnetic flux to enter and for interphase boundaries to appear below H_c. To minimize the total free energy, the boundary area tends to a *maximum*. Above a lower critical field, H_{c1}, the superconductor breaks up into finely divided normal and superconducting zones in a periodic lattice arrangement of flux vortices, as shown in Figure 5.6. The flux lines nucleate at the surface of the superconductor. Each vortex has a single flux quantum associated with it and a supercurrent circulating around it. The effective radius of the normal core is the coherence length. The core is a normal region because the increase in kinetic energy of a Cooper pair due to the circulating current exceeds the binding energy of the pair for distances smaller than the coherence length. The region between the vortices remain superconducting. DC currents can flow without loss through the superconducting regions. Therefore a Type II superconductor can carry dc currents without resistance, all the way up to H_{c2}.

NEGATIVE SURFACE ENERGY SUPERCONDUCTORS (TYPE II)   97

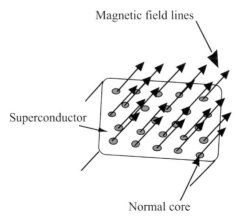

Figure 5.6: When the external magnetic field exceeds H_{c1} for a type II superconductor, an array of flux vortices enter the superconductor in the form of single-flux quanta. The cores are normal, but the region between the cores remains superconducting and lossless for dc current flow. The separation of the vortices is a few thousand nanometers.

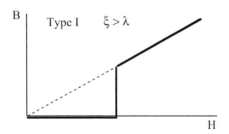

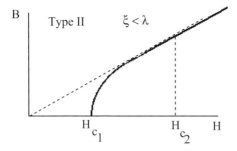

Figure 5.7: The magnetic behavior of a Type I superconductor compared with the same for a Type II superconductor. In a type I material all the flux enters the superconductor at the critical field H_c. In a type II material, flux starts to enter at H_{c1}. Between H_{c1} and H_{c2}, the superconductor is in a mixed phase. Above H_{c2}, the bulk becomes normal conducting.

As the external field continues to rise, the magnetic flux continues to enter the superconductor and the vortex density increases until an upper critical field H_{c2} when the normal cores of flux lines overlap. At this stage, the transition to the normal state is complete. Superconducting regions persist to fields higher than H_c, up to $H_{c2} = \sqrt{2}\kappa_{GL}H_c$. The magnetic behavior of a Type II superconductor is contrasted with that of a Type I superconductor in Figure 5.7. In certain materials, such as Nb-Ti, the upper critical field can assume very large values (e.g., 12 T) which makes them useful for high-field magnets needed for high-energy proton accelerators. But a high H_{c2} material provides no advantage for rf applications because the rf critical magnetic field is much less, as we will soon see.

The Ginzburg–Landau (GL) phenomenological theory links the electrodynamic properties with the thermodynamic properties of a superconductor [103]. In this theory, the density of superconducting electrons is made proportional to the magnitude of the wave function squared — i.e., $|\Psi(r)|^2$. Using a thermodynamic treatment, GL described the superconducting system in the presence of a magnetic field. They estimated H_{c1}, H_{c2}, and H_c in terms of the penetration depth, the coherence distance, and the flux quantum

$$\Phi_0 = \frac{h}{2e} = 2.07 \times 10^{-15} \text{T-m}^2, \tag{5.11}$$

where h is Planck's constant and e is the electron charge. The total magnetic flux associated with a circulating current is equal to a whole number of single flux quanta. In terms of the microscopic properties, they obtain

$$H_{c2} = \frac{\Phi_0}{2\pi\mu_0\xi_0^2}, H_c = \frac{H_{c2}}{\sqrt{2}\kappa_{GL}}. \tag{5.12}$$

There is no comparably simple expression for H_{c1}, but a convenient expression at large κ_{GL} is

$$H_{c1} \propto \frac{H_c}{\sqrt{2}\kappa_{GL}} \ln(\kappa_{GL}) = \frac{\Phi_0}{4\pi\mu_0\lambda_L^2} \ln(\kappa_{GL}). \tag{5.13}$$

Both H_{c1} and H_{c2} follow an approximate parabolic temperature dependence of Equation 5.5 as shown in Figure 5.8.

Bulk niobium is a Type II superconductor, and the values for the various dc critical fields at zero temperature are, in Oe

$$H_{c1} = 1700, \quad H_c = 2000, \quad H_{c2} = 2400. \tag{5.14}$$

More generally, H_{c2} of bulk niobium depends on the RRR and the state of anneal, as well as the degree of recrystallization [104]. The upper critical field of sputtered niobium films is much higher than that for bulk niobium. The films are Type II superconductors with $\kappa_{GL} = 3.5$–12, $H_{c2} = 2.5$–3.5 T [105].

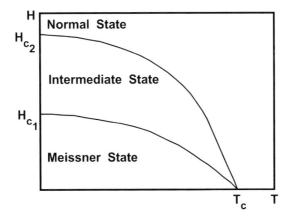

Figure 5.8: The three phases of a Type II superconductor are separated by the curves $H_{c1}(T)$ and $H_{c2}(T)$ which meet at T_c.

5.5 THE RF CRITICAL MAGNETIC FIELD

The above discussion describes the *equilibrium* condition for dc fields. In the absence of an external magnetic field, the phase transition at T_c takes place without a discontinuity in the entropy, and is referred to as a second-order phase transition. In the presence of an external magnetic field, the phase transition is of first order at $T_c(H)$, and is combined with a latent heat due to the discontinuity in entropy. The first-order phase transition takes place at nucleation centers. Because of the entropy discontinuity and required nucleation centers, there is a possibility for a "superheated" superconducting state to persist metastably at $H > H_c$, as well as for a "subcooled" normal-conducting state to persist metastably at $H < H_c$. We use the word "superheating" because the field exceeds the criteria for the phase transition.

Surface energy considerations lead to an estimate for superheated critical field values. In the process of phase transition, a boundary must be nucleated. In a Type I superconductor, the positive surface energy suggests that, in dc fields, the Meissner state could persist metastably beyond the thermodynamic critical field, up to the superheating field, H_{sh}. At this field, the surface energy per unit area vanishes:

$$\frac{\mu_0}{2}\left(H_c^2 \xi - H_{\text{sh}}^2 \lambda\right) = 0, \quad H_{\text{sh}} = \frac{1}{\sqrt{\kappa_{\text{GL}}}} H_c. \tag{5.15}$$

At H_{sh}, the conditions for metastability also disappear.

For Type II superconductors, it is also possible for the Meissner state to persist metabstably above H_{c1}. The dependence of the superheating critical

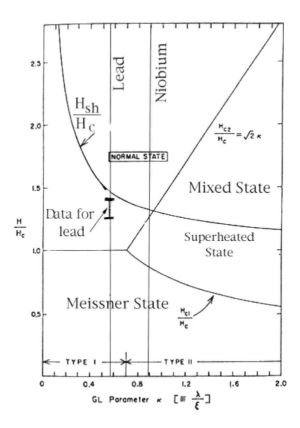

Figure 5.9: Phase diagram showing the Meissner, normal, mixed, and superheated states. The normalized critical fields H_{c1}, H_c, H_{c2} and H_{sh} are shown as functions of the GL parameter. The range of data for the rf critical field of lead is also shown.

field on κ_{GL} has been calculated [106] by solving the GL equations to show that

$$\begin{aligned}
H_{sh} &\approx \frac{0.89}{\sqrt{\kappa_{GL}}} H_c \quad \text{for} \quad \kappa \ll 1, \\
H_{sh} &\approx 1.2 H_c \quad \text{for} \quad \kappa \approx 1, \\
H_{sh} &\approx 0.75 H_c \quad \text{for} \quad \kappa \gg 1.
\end{aligned} \quad (5.16)$$

The phase diagram, including all the critical fields, is shown in Figure 5.9. In rf conditions, the fields change rapidly, within nanoseconds. The time it takes to nucleate fluxoids ($\approx 10^{-6}$ s [107]) is very long compared to the rf period ($\approx 10^{-9}$ s). Therefore there is a stronger tendency for the metastable superconducting state to persist up to H_{sh}. It is theoretically possible for the superconducting state to persist above H_c, but only up to H_{sh}. Therefore we expect that

$$H_{rf,\,crit} = H_{sh}. \quad (5.17)$$

THE RF CRITICAL MAGNETIC FIELD

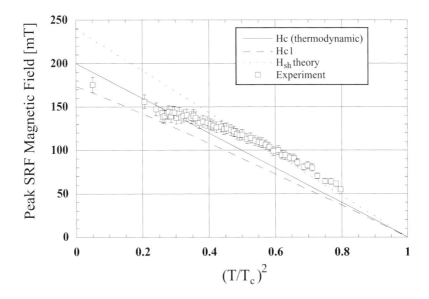

Figure 5.10: Measurement of the rf critical field for niobium and comparison to H_c, H_{c1} and H_{sh}.

In experiments [108], the rf critical field has been shown to be equal to H_{sh} for the Type I superconductors, In, Sn, and Pb. Experimental results for lead are compared to H_c and H_{sh} in Figure 5.9. Note that $H_{rf} > H_c$. As we mentioned in Chapter 1, the major difficulty in trying to reach the high value of the theoretical rf critical field in a superconducting cavity is the phenomenon of thermal breakdown of superconductivity at imperfections, which cause superconductivity to quench at low fields. The prospects of verifying the theory are improved near T_c where the theoretical fields are lower due to the inherent temperature dependence of the critical magnetic fields.

Figure 5.10 shows the results of rf critical magnetic field measurements [109] for niobium over the temperature range $0.5T_c < T < T_c$. We see that the critical rf magnetic field for niobium is greater than both H_{c1} and H_c. The rf critical magnetic field is close to H_{sh}, as expected from theory. The results of Figure 5.10 were obtained by using pulsed rf fields. In these experiments, the conditions for exceeding the thermal breakdown field were made more favorable by reducing the cavity fill time with a strongly coupled input probe and by using a high power pulsed rf source to reach high fields. As a result, the pulsed rf fields had a rise time shorter than the growth time of normal conducting regions.

What do these numbers imply for the ultimate performance that may be expected for niobium cavities? For a typical cavity geometry, $H_c/E_{acc} = 42$ Oe/MV/m leads to $E_{acc} = 57$ MV/m as the fundamental limit to the accelerating field for niobium.

It is important to note that the rf critical field does *not* depend on H_{c2}, so that high field magnet materials such as Nb-Ti do not offer higher operating fields for superconducting cavities. Indeed for rf superconductivity, it is essential to always operate in the Meissner state. It is even possible that the precipitates which serve as pinning centers for the dc flux lines may be harmful to rf performance.

5.6 MAXIMUM SURFACE ELECTRIC FIELD

At the theoretical magnetic field limit, a well-designed accelerating structure would have to support an rf surface electric field of 120 MV/m. This value has already been exceeded on niobium surfaces. In a specially designed "mushroom" cavity a surface rf electric field of 145 MV/m was reached in cw operation [110]. In another cavity [20], which was a radio frequency quadrupole like structure, 220 MV/m was reached in pulsed operation. These are the experimentally reached highest values; we know of no fundamental limit to the maximum rf electric field. However, as we mentioned in Chapter 1 (and will return to in Chapter 12 and Chapter 13), at high electric fields, field emission is a major practical limitation to the maximum achievable surface field.

Part II
Performance of Superconducting Cavities

CHAPTER 6

Cavity Fabrication and Preparation

6.1 INTRODUCTION

Niobium cavities can be constructed using three different methods: machining out of a solid piece, forming from sheet niobium, and depositing a niobium film on a preformed copper cavity substrate. For cavities machined out of a solid piece, the material and labor costs are excessive. The 11-cell, S-band cavity (Figure 1.9) used in the Cornell 12-GeV synchrotron test is an example of a cavity machined out of a slab of solid niobium. With the success of modern forming, welding and sputtering techniques, machining is rarely used.

As we wish to cover the most commonly used fabrication procedures, we will describe only the forming, welding, and surface preparation of sheet niobium cavities. Currently "deep drawing" and "spinning" of half-cells is common. Several laboratories are looking into alternate methods of multicell cavity forming to lower costs — for example, hydroforming from a tube [111, 112] and spinning multicells from a single sheet [113].

We will discuss the niobium films in Chapter 14. Note that niobium deposition onto copper requires a completed copper cavity as a substrate. The copper cavity is made in essentially the same way as the sheet niobium cavity except for surface preparation before the film deposition.

6.2 NIOBIUM

The metallurgy of niobium has been thoroughly studied because of its many uses in the manufacture of stainless steel alloys and superalloys for aerospace applications [114]. Because of its low neutron capture cross section, it is also used as cladding material for nuclear fuel elements; because of its inertness at ambient temperatures, it is used in the chemical industry. Niobium ore is found in Brazil, Canada, and Australia. After extraction from the ore, the most common method of consolidation and refinement for commercial high-purity sources is electron-beam melting in a furnace, such as the one schematically depicted in Figure 6.1. The beam is arranged to hit both the feedstock as well as the top of the ingot which is contained in a water-cooled copper sleeve. Impurities are boiled out of the pool and pumped away. As melting progresses

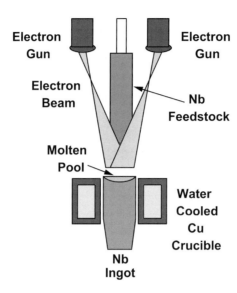

Figure 6.1: Schematic representation of an electron-beam furnace being used to purify niobium. Impurities are boiled out of the niobium and pumped away. The typical furnace vacuum is 10^{-4} to 10^{-5} torr. A better vacuum is desired to produce niobium with a high residual resistivity ratio (RRR).

the solidified ingot is continuously withdrawn through the sleeve. The rate of withdrawal has to be carefully coordinated with the melt rate of the raw material to ensure thorough melting and proper outgassing. This procedure drives out all volatile impurities. After a satisfactory ingot is cast, it is forged into a thick rectangular slab, annealed, and rolled in stages to the final sheet thickness. A final annealing is necessary for recrystallization. The temperature and time must be carefully controlled to obtain the desired grain size (see next section). These parameters are very sensitive to the RRR of the niobium. Niobium of purity 99.85% (by weight) is readily available from industrial sources in many forms: ingot, sheet, foil, rod, tubing and wire [115, 116, 117].

For the case of superconducting cavities, the purity of niobium purchased is important, in terms of bulk impurity content as well as inclusions from manufacturing steps, such as rolling. As we will discuss in Chapter 11, inclusions on the rf surface play the role of normal conducting nucleation sites for thermal breakdown. Dissolved impurities serve as scattering sites for the electrons not condensed into Cooper pairs, as we discussed in Chapter 3. These impurities lower the thermal conductivity and thereby limit the maximum tolerable surface magnetic field before the onset of thermal breakdown.

Among the metallic impurities, tantalum is found in the highest concentration (typically 500 ppm by weight) since all naturally occurring ores contain some tantalum. Presently the 500–1000 ppm wt impurity level is not perceived to be a problem since tantalum is a substitutional impurity which does not

NIOBIUM

Table 6.1: Expected residual resistivity ratio contribution for niobium for 1 ppm wt of impurities[a]

Element	RRR	Element	RRR
H	2640	Zr	102 000–239 000
N	4230	Hf	200 000
C	4380	W	262 000–721 000
O	5580	Mo	717 000
Ti	53 700	Ta	1 140 000

[a] The ideal RRR due to phonon scattering is 35 000. To obtain the RRR one must add the resistance contributions for each impurity element in parallel to the resistance contribution from phonons.

substantially affect the electronic properties. But tantalum-free niobium can be obtained by electro-deposition. Next in abundance after tantalum are the higher-temperature refractory elements, such as tungsten, zirconium, hafnium, and titanium, but these are usually found at the level of 10–50 ppm wt.

Among the light, interstitially dissolved impurities, oxygen is dominant due to the high affinity that niobium has for oxygen above 200 °C. Interstitial impurities such as oxygen are more dangerous than substitutional impurities such as tantalum. For example, oxygen decreases the T_c of niobium by 1 °-K/atomic %, whereas tantalum decreases T_c by 0.1 °K/atomic %.

The other intersitials are carbon, nitrogen, and hydrogen. The electron scattering effectiveness of the various impurities are shown in Table 6.1 in terms of the RRR of niobium for 1 ppm wt of each impurity [118]. Note that the interstitial impurities (O, N, C, and H) are particularly detrimental to the mobility of electrons. As we discussed in Chapter 3, by the Wiedemann–Franz law these impurities also have the strongest effect on the thermal conductivity. To obtain the RRR, one must add the resistance contributions for each impurity element in parallel to the resistance contribution from phonons. The contribution of the phonons is always present, so that the highest theoretical RRR for niobium is 35,000 [77]. The highest RRR ever achieved in a niobium sample was 33,000 [119].

The most convenient method to obtain high purity niobium for superconducting cavities is to remove the interstitials during the electron-beam melting stages of the ingot. The outgassing of niobium at high temperatures has been studied extensively [77], so that the reactions, degassing rates, and equilibrium impurity contents of oxygen, carbon, nitrogen, and hydrogen can be estimated from the starting content, melt rate, and partial pressures of residual gases in the vacuum furnace (see Problem 34). Results from such studies show that after the first few melts, the equilibrium contribution from the residual gas in the furnace dominates the final interstitial impurity content [120]. Multiple melts and progressive improvements in the furnace chamber vacuum have led to a steady increase in the RRR of commercial niobium over the last decade from

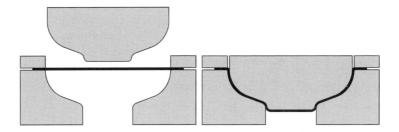

Figure 6.2: Deep drawing a niobium disk into a half cell.

30, typical of reactor grade niobium, to 300 [121]. Recently, niobium sheet of RRR = 500–700 became available from Russian sources [122].

6.3 FORMING SHEET NIOBIUM

Niobium can be formed and fabricated by practically all metallurgical and engineering techniques. The primary methods used today for fabricating superconducting cavities are deep drawing [123] and spinning [124].

6.3.1 Deep Drawing

Deep drawing is a forming process whereby a blank is pressed into shape using a set of dies as depicted in Figure 6.2. Figure 6.3 shows the dies used to deep draw half-cells for a 1.5-GHz cavity from 3-mm-thick sheet niobium. These dies are machined out of 7075-T6 aluminum alloy because of its high yield strength, ease of machining, and low cost. Dies made from traditional materials, such as steel or tungsten carbide, tend to friction weld, or gall, to the niobium, so aluminum- or copper-based alloys (e.g., AMPCO or beryllium copper) are preferred [125]. The starting blank for deep drawing is usually a niobium disk from a sheet. It is bolted across the female die with a hold-down plate. With appropriate torque on the bolts, the outer edge of the niobium is constrained without tearing at the clamped edges. Clean motor oil is painted onto the niobium for lubrication. The male die is placed in position and the assembly is squeezed in a hydraulic press (see Figure 6.4). For the 1.3-GHz cavity half-cell shown, 100 tons (8.9×10^5 N) of force are applied. To get the curvature required at the iris, the nose of the cup is then coined (Figure 6.5) with a coining ring and the male die (25 tons applied).

Like all sheet-forming methods, deep drawing is sensitive to the niobium's mechanical properties. For this one-step deep drawing process (no intermediate anneals), the sheet material must be suitably forgiving. In particular, small and uniform grain size (ASTM ≥ 4 = grain size < 50 μm) are essential as shown in Figure 6.6. (Each grain is a crystal of niobium.) The sheet niobium must be properly annealed to achieve complete recrystallization, and to remove lattice

FORMING SHEET NIOBIUM

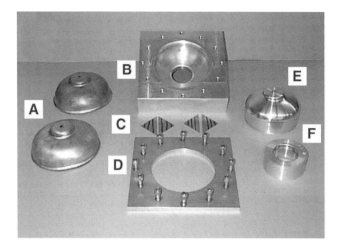

Figure 6.3: Dies and other hardware used for deep drawing niobium half-cells. All the dies are made out of 7075-T6 aluminum alloy. (A) Deep drawn half-cells ready to be trimmed and welded. (B) Female die. (C) Niobium shims placed outside the bolt circle of the female die. These are cut from the same sheet as the disk to be deep drawn. (D) Hold-down plate used to constrain the niobium edge. (E) Male die. (F) Coining ring used with (E) to obtain the correct curvature at the half-cell's iris.

defects, but without growing large grains. If the grains are too large, an "orange peeling" effect occurs.[1] If the material is incompletely recrystallized, it tears during deep drawing. If the grain sizes are not uniform (Figure 6.7) tearing will also occur.

In practice, the sheet niobium from industry has a yield strength in the neighborhood of 140 MPa (20×10^3 psi) for reactor grade niobium (RRR $\approx$ 30) and about 70 MPa (10×10^3 psi) for high-purity (RRR $\approx$ 250) niobium. The yield strength is very sensitive to purity, state of work and state of anneal. Above the yield point, niobium will flow plastically. Typical stress–strain curves for niobium are shown in Figure 6.8 [126]. (The yield strength is typically 50 MPa (7.3×10^3 psi) at 0.2% extension.) At cryogenic temperatures the yield strength increases by an order of magnitude, providing a greater safety margin against cavity collapse during accidental overpressure in the helium vessel. If the cavity is postpurified (see Section 6.6 below) by titanium at 1400 °C, the yield strength falls substantially. Plastic deformation is observed to start at about 5 MPa (725 psi). As shown in Figure 6.9, the maximum elongation will vary from 10 to 50% [126].

The ultimate tensile strength of niobium is about 200 MPa (29×10^3 psi) for RRR $\approx$ 30 and about (20×10^3 psi) 140 MPa for RRR $\approx$ 250 [127]. Above this stress, the material fractures. Niobium also has a low degree of work hardening,

---

[1] The surface takes on the texture of an orange peel.

Figure 6.4: Hydraulic press after a half-cell was deep drawn in the sandwich of aluminum male and female dies.

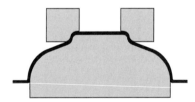

Figure 6.5: Coining the nose of the untrimmed cup to obtain the desired curvature.

which is advantageous for mechanical forming. In almost all cases when proper mechanical properties of the sheet are achieved, cavity parts can be deep drawn or spun to final shape without intermediate annealing. If work hardening occurs, niobium may be annealed and recrystallized by heating in a vacuum. The best annealing temperature depends on the RRR, e.g., 800-850 °C for RRR = 250. Care must be taken to protect the niobium with titanium foil to avoid lowering the RRR in the furnace. (See Section 6.6.)

FORMING SHEET NIOBIUM 111

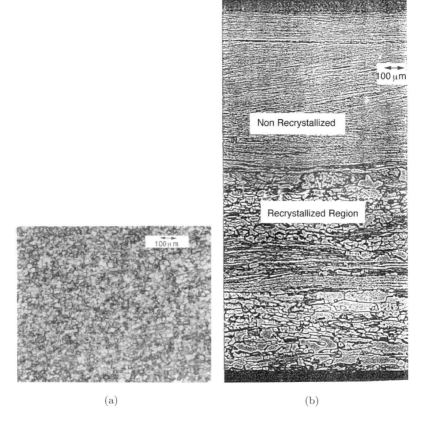

(a) (b)

Figure 6.6: Grain size variations in sheet niobium. (a) Uniform and fine grain material, suitable for deep drawing. (b) Incompletely recrystallised niobium, tears on deep drawing. Good drawing properties were subsequently recovered by reannealing.

6.3.2 Spinning

It is also common to spin half-cells instead of deep drawing them. When making large half-cells (i.e., < 500 MHz), spinning is often preferred to eliminate the need for the high-tonnage hydraulic press. A half-cell is made by spinning the sheet and slowly pushing it to form the desired shape on a mandrel (see Figure 6.10). Among the challenges in this process is achieving uniform thickness. Several laboratories are investigating the possibility of spinning complete cells or even multicells out of a single niobium sheet. This procedure has the potential to produce a cavity that has no seams which would reduce welding costs and eliminate the possibility of weld errors.

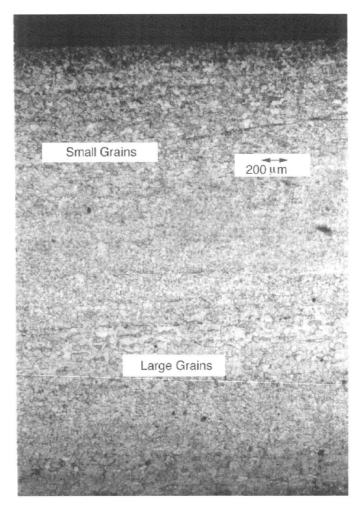

Figure 6.7: Nonuniform grains, unsuitable for deep drawing and not possible to recover by annealing.

FORMING SHEET NIOBIUM 113

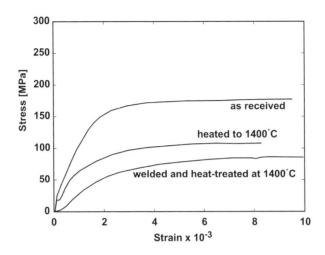

Figure 6.8: Stress–strain curves for niobium at room temperature for three samples: as received with no treatment, a heat treatment to 1400 °C, and electron-beam welded then heat-treated to 1400 °C. (Courtesy of CEBAF.)

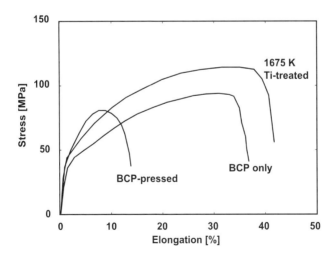

Figure 6.9: Elongation measurements on niobium for three different preparations; chemically polished (BCP), chemically polished and hydraulically pressed, and postpurified by a heat treatment with titanium. (Courtesy of CEBAF.)

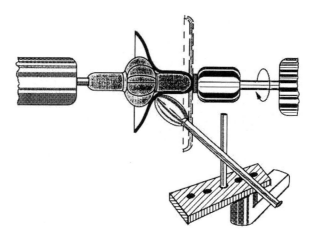

Figure 6.10: Spinning rig for an L-band cavity. By using a collapsible mandrel, this apparatus can spin a single cell cavity from a single sheet of niobium, a fabrication process under development. (Courtesy of INFN, Legnaro.)

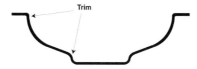

Figure 6.11: Locations where the half-cell needs to be trimmed.

6.4 TRIMMING

After deep drawing or spinning, the half-cells are trimmed to the final size for electron-beam welding (Figure 6.11). Trimming is done either on a lathe or a CNC (computer numerical control) milling machine. When trimming the formed parts, the high reactivity with oxygen must be taken into account and good cooling with lubricants must be provided (e.g., Accu-Lube[2]). As with all the machining operations, precautions must be taken to avoid scratching the intended rf surface. In a typical lathe-trimming operation, the half-cell is held sandwiched in a fixture similar to the deep drawing dies. For milling, the half-cell is held in place by a vacuum chuck.

Beam tubes are either purchased as tubes or rolled from sheet and electron-beam welded. Flanges for the beam tubes are machined. Because they are in regions of low magnetic field, the flanges can be made from reactor grade niobium instead of the high-purity material needed for the cells. If the desired magnetic field at the iris exceeds 200 Oe at the design gradient, the beam tube

---

[2] Accu-Lube, P.O. Box 823, Kant, WI 98035.

ELECTRON-BEAM WELDING

Figure 6.12: A single-cell niobium L-band cavity setup for electron-beam welding. The leaded glass door to the welder is open to permit this photograph. The electron gun is located directly above the cavity equator.

should also be made from high-RRR niobium.

The formed niobium parts are typically degreased in soap and water. Since the sheet material may have ferrous surface impurities from the rollers used to form the sheets, the parts are soaked in hot 50% sulfuric acid overnight. This acid solution removes the iron inclusions without attacking the niobium. The niobium parts then receive a light chemical etch [125] using a buffered chemical polish (BCP) of 1 part hydrofluoric acid, 1 part nitric acid, and 2 parts phosphoric acid[3] to remove approximately 5 microns from the surface. This 1:1:2 BCP removes about 2 microns per minute when the acid is at 15 °C. At room temperature, the etch rate is about 2.5 microns per minute. After this preliminary chemical treatment, it is advisable to check that there are no remaining iron inclusions by soaking the parts in clean water overnight and carefully inspecting the parts for rust spots. If iron is found, the sulfuric acid treatment and inspection must be repeated.

6.5 ELECTRON-BEAM WELDING

All of the parts are electron-beam welded together in a vacuum chamber as shown in Figure 6.12. The welds necessary for a typical cavity are shown schematically in Figure 6.13. The pressure in the chamber should be less than 10^{-5} torr. It is advisable to qualify the chamber vacuum by measuring the RRR of the weld region. The weld parameters are chosen to achieve full penetration "butt" welds with a smooth underbead by using a defocused electron beam.

---

[3]The 1:1:2 is acid mix refers to volume out of the bottle. In the United States, the concentrations in the bottles are 49% for HF, 69.5% for HNO_3, and 85% for H_3PO_4.

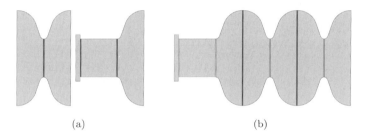

(a) (b)

Figure 6.13: Electron-beam welding of a cavity. (a) Iris welds (inside). (b) Equator welds (outside). The smaller-radius "iris" welds are done first. In addition to joining half-cells, this includes welding together the flanges and beam tubes to the end cells. The equator welds are done last.

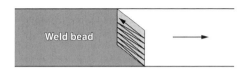

Figure 6.14: Rhombic raster pattern for the beam during an electron-beam weld. This rastering produces a well-defined and reproducibly defocused beam needed for trouble free electron-beam welding of niobium.

For example, to achieve full penetration in 1.6-mm-thick niobium, a typical set of electron-beam welding parameters are $V = 50$ kV and $I = 37$ mA at a 46-cm/min weld speed. It is important not to use a focused beam, as this will produce significant weld spatter, resulting in a myriad of weakly attached niobium beads and balls. The vapor column from a focussed beam may also leave voids in the molten niobium weld seam [128].

One technique to produce a well-defined and reproducibly defocused beam is to raster a slightly defocused beam to cover a pattern in the shape of a rhombus, commonly referred to as the "rhombic raster weld"[125] (Figure 6.14). Superconducting joints of high quality can be made in this manner. Figure 6.15 shows an example of the smooth "underbead" of a full penetration weld in 1.5-mm-thick niobium. The underbead is the melted zone on the side of the niobium opposite the side where the electron beam is incident. The joint at the cell's equator is a region of high surface current and can show strong heating if the weld is done poorly. On the rare occasion that there is a weld defect, it can be located by cryogenic thermometry during the acceptance test and fixed by grinding down the spot on the interior (see Chapter 11).

Usually, the welding is done in several steps. In the first step, all welds except the equator welds are done. In modern electron-beam welding, the iris welds are very smooth, but it is still good practice to perform the extra step

ELECTRON-BEAM WELDING

Figure 6.15: The smooth underbead of a full penetration electron-beam weld in 1.5-mm-thick niobium.

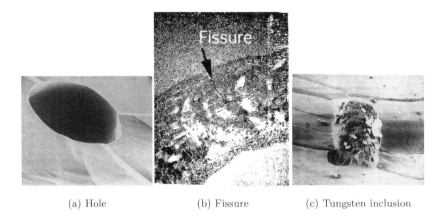

(a) Hole   (b) Fissure   (c) Tungsten inclusion

Figure 6.16: Scanning electron micrographs of defects found in TIG welds. It is very difficult to avoid contamination in the TIG weld, so electron-beam welding is preferred. (Courtesy of Wuppertal and CERN.)

of grinding the iris weld[4] to ensure a smooth inner surface in this high-electric-field region and to avoid geometric field enhancement. After the grinding, the cavity pieces are cleaned by etching with 1:1:2 BCP to remove about 20 μm from the surface. The equators are then electron-beam welded and the cavity is complete. The welds are inspected from the inside with a borescope or a well-lit angle mirror inserted into the cavity. It is important to find a smooth underbead. The typical width of the underbead (Figure 6.15) is about 0.4 cm.

Tungsten inert gas (TIG) welding of niobium is not recommended because the experience to date has not been favorable. Even when great precautions were taken to remove water vapor and oxygen from the surrounding argon atmosphere, the welds were frequently found to be contaminated, leading to fissures. Occasionally the welding electrode tip would touch the weld zone, leaving tungsten inclusions. Figure 6.16 shows three examples of such defects in TIG welds found by researchers at CERN [129, 130]. Also the overall heat deposition in a TIG weld is much greater than for electron-beam welding, leading to more shrinkage and deformation of niobium cavity parts.

---

[4]The grinding is done with a ScotchBrite™ abrasive wheel.

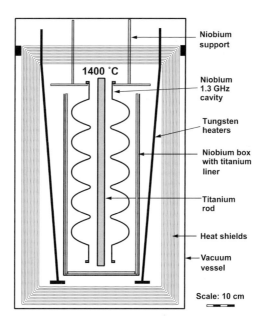

Figure 6.17: Schematic representation of a niobium cavity being "post-purified" in a high vacuum furnace using titanium as a getter for the oxygen.

6.6 POSTPURIFICATION

As we will discuss in Chapter 11, if higher fields are desired, the thermal conductivity of the niobium in the completed cavity may be improved by a "postpurification" step. The purity of the niobium is increased by solid state gettering [131] of the interstitial oxygen using yttrium [120] or titanium [132] at high temperature. The foreign metal is vapor deposited on the niobium surface and in the same step, the high temperature decreases the diffusion time of the oxygen in niobium. Over a few hours, oxygen collects in the deposited layer. If yttrium is used, the best temperature is 1200–1250 °C, as both the vapor pressure of yttrium and the diffusion rate of oxygen in niobium are sufficiently high. If titanium is used, temperatures at 1350–1400 °C, are required because of the lower vapor pressure. Typically the RRR improves a factor of two in a few hours primarily due to the removal of interstitial oxygen. With longer times, titanium can also remove nitrogen and carbon [133].

Figure 6.17 shows the arrangement for post purifying a complete cavity in a high-vacuum furnace. The cavity must be clean to avoid diffusing impurities into the bulk. Chemically etching away about 100 μm is generally sufficient cleaning. Only a light etch (≈ 5 μm) is required if the cavity was previously cleaned.

In the setup of Figure 6.17, a large quantity of yttrium or titanium is required and the cost of yttrium can become quite high; therefore, titanium is

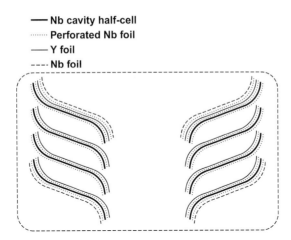

Figure 6.18: Outline of the setup for purifying the niobium in half-cells of a cavity. This assembly would be fired at 1200–1250 °C in a UHV furnace.

preferred. Also, because of its lower vapor pressure, a titanium liner will keep the furnace cleaner during heat treatment at lower temperatures (e.g., 900 °C) as is sometimes required for removal of hydrogen (Chapter 9). The solid state gettering process can also be used at the half-cell stage. Figure 6.18 shows a setup for purifying half-cells. This layered package would go in the furnace. In this case, it is important to verify that the subsequent electron-beam welding will preserve the higher RRR achieved. This process leaves several microns of titanium or yttrium that has to be chemically removed. Titanium diffuses deeply into niobium grain boundaries, requiring removal of an additional 80 μm of the inside rf surface. Yttrium does not diffuse as deeply into niobium so that a 10 to 20 μm removal is enough. The exterior surface must also be etched because the deposited layer hinders the thermal transport [134] (see Chapter 11). Measurements of the thermal resistance show that if titanium is used, a 30 μm removal restores the heat transfer coefficient. When the cavity is postpurified by Ti at 1400 °C, the yield strength falls substantially [126] (Figure 6.8).

6.7 TUNING

If the cavity has more than one cell, the cells need to be tuned relative to each other so that the accelerating field is the same for each cell. Figure 6.19 shows an apparatus used to tune L-band cavities. Each cell is tuned by squeezing or stretching it while gripping the irises at either end. Small variations between cells are sufficient to alter the field profile substantially. The topic of obtaining the proper tune for multicell cavities will be dealt with thoroughly in Chapter 7.

Figure 6.19: A device used to tune the cells of a multicell cavity. Holding the irises on both ends of the cell, the cell is stretched or squeezed. (Courtesy of DESY.)

6.8 SURFACE PREPARATION

To achieve the optimum rf performance, the surface of the cavity must be as close as possible to ideal. As we will discuss in Chapters 11 through 13, microscopic contaminants can severely limit the performance, either by magnetic heating or by electron field emission. The high level of cleanliness required is comparable to that required in the semiconductor industry where clean rooms of Class 10-100 are routine. The 10 in Class 10 refers to the number of particles of size 0.5 μm or larger in one cubic foot of air [135].

6.8.1 Chemical Treatment

For a newly fabricated cavity, a clean rf surface is achieved by chemically etching away a surface layer (≈ 100 μm). This removes the mechanically damaged layer as well as any evaporated niobium scale deposited on the surface during welding. Only a light etch (≈ 5 μm) is necessary if the cavity has previously received the (≈ 100 μm) etch. After etching, the cavity must be thoroughly rinsed with ultraclean water, taking precautions so that no contaminants come in contact with the clean rf surface. A potential danger during chemical treatment of niobium is the "Q disease" which is known to be caused by hydrogen contamination [136] (Chapter 9). Proper precautions must be taken to avoid polluting the niobium with hydrogen. Experience has shown that hydrogen contamination can be avoided by using the appropriate acid solution (1:1:2 BCP)

SURFACE PREPARATION

and keeping the acid temperature low <18 °C [137].

Chemical etching of niobium parts (i.e., the cavity pieces in Figure 6.13) is done by immersing them completely in the acid bath for the required time. Small-scale cavities and structures may also be etched in this manner if the volume of acid required is manageable. Figure 6.20 shows the etching of a 1.5-GHz five-cell by "dunking." Gentle agitation is often used.

When the rf surfaces of larger structures such as a 9-cell 1.3-GHz cavity or a single-cell 500-MHz cavity require etching, a dunk chemistry is very awkward because of the large volume of acid required. In these cases, the cavity is filled with acid using a "closed" system. Figure 6.21 shows such a system. With the operator in a safe, remote area, valves can be opened and closed to fill the cavity with cooled acid, drain the acid, and rinse the cavity with ultrapure water.[5]

Some laboratories use electropolishing instead of chemical etching [138]. Niobium is the positive electrode (anode) and the cathode is made from aluminum. The electrolyte consists of H_2SO_4 and 40% HF in a ratio of 85:10 by volume. A constant flow rate of 60 liters/minute is used. The voltage is 25 volts and the current density is 50 mA/cm^2. The temperature is 30 °C. The cavity is rotated at 0.4 to 1 rpm. The thickness of the removed niobium is estimated from the total current and cross-checked with an ultrasonic thickness gauge. The advantage of electropolishing is a smooth, mirror-like surface finish with no sharp steps at the grain boundaries. A disadvantage is that the etch rate is slower than chemical etching. There is also surface contamination by sulfur which must be removed by ultrasonic rinsing in H_2O_2. Finally, there is more hydrogen contamination due to hydrogen evolved at the negative electrode; therefore, the cathode must be enveloped by a porous teflon membrane to keep the bubbles away from the niobium. Figure 6.22 depicts an electropolishing setup [139]. It is necessary to remove the hydrogen by heating in a vacuum furnace at about 800 °C after electropolishing. During the furnace degassing step, it is advisable to surround the outside of the cavity with titanium foil to preserve the RRR.

6.8.2 Rinsing

After acid etching or electropolishing, the cavity is placed in a closed loop with the ultrapure water system for several hours. The resistivity of the water should be close to theoretically pure (>18 MΩ-cm) and the water inlet should be filtered to eliminate particles >0.3 µm. Water is recirculated for several hours through the cavity in series with the water purification system to continuously and thoroughly remove any chemical and particulate residue from the niobium surface. With ultraclean water inside the cavity, the plumbing attached to the cavity is pinched off and the cavity is carried into the dust-free clean room where the water is drained. The cavity surface thus only comes in contact with filtered air after the final ultrapure water rinse.

Many laboratories have found that the rf surface can be made even cleaner if chemistry is followed by high-pressure rinsing of the cavity [140] with ultrapure

---
[5]The rinse water is sent to a neutralization tank.

(a)

(b)

Figure 6.20: A five-cell L-band cavity being etched by immersion in an acid bath. (a) Cavity being transferred from the acid bath to the water bath. (b) The chemistry is done in a ventilated chamber with a remotely operated crane.

CLEAN ASSEMBLY

Figure 6.21: The closed acid chemistry system at Cornell. This system permits an operator to remotely actuate valves in order to perform acid etching of the inner surface of larger cavities. A 5-cell 1.3-GHz cavity is shown. The acid is stored in a chilled reservoir before being admitted to the cavity. Valves permit the cavity to be safely filled and drained first with acid and then with ultrapure water.

water (Figure 6.23). At TJNAF, for example [141], 1000 psi (69 bar) water is sprayed through 12 stainless steel nozzles each having a 0.3-mm-diameter orifice. The potent jets of water are scanned across all parts of the rf surface to dislodge and sweep away microscopic contaminants that have adhered to the surface. These contaminants may have been swimming in the acid bath or have been airborne and landed on the surface after the cavity was removed from the acid.

6.9 CLEAN ASSEMBLY

The newly cleaned surface, still wet from ultrapure water, is exposed only to filtered air in a Class 10-100 clean room. The laboratory workers in the vicinity need to wear special particulate-free clothing and follow strict protocols to reduce particulate generation.

The surface must be dried before the cavity is evacuated and cooled down

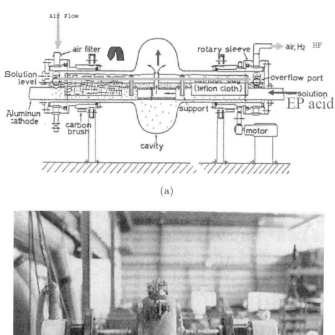

Figure 6.22: Electropolishing equipment at KEK. (a) Schematic. (b) Single-cell cavity setup for electropolishing.(Courtesy of KEK.)

for rf tests. There are several different approaches to this last step. At Cornell, the surface is dried by blowing warmed, filtered nitrogen gas through the cavity. At Jefferson Lab, the surface is rinsed with ultrapure methanol or ethanol to displace the water, and the surface then dries in Class 10 air. Once the surface is dry, all the cavity's ports are sealed, and the cavity is attached cleanly to its test stand and evacuated. At KEK, the wet cavity is assembled and dried by applying heat while pumping. A clean room assembly at KEK is shown in Figure 6.24 [142].

Care must be taken to ensure that the vacuum system is thoroughly clean

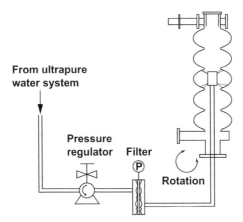

Figure 6.23: A schematic depiction of the high-pressure water-rinsing system at Jefferson Lab (formerly CEBAF). The cavity is slowly articulated so that the 12 jets of water reach all the interior surface to dislodge contaminants. (Courtesy of CEBAF.)

and dust-free. The cavity itself is evacuated slowly to avoid turbulent flow and reduce the risk of contaminants from the vacuum system reaching the cavity. By the same token, it is essential to maintain dust free conditions while attaching input and output couplers to the cavity. Figure 6.25 shows the use of a portable clean room assembled around a cavity for installing a high power input coupler. Similar precautions are necessary when the superconducting structure is installed in an accelerator.

6.10 SUMMARY

As we have presented, there are many steps to fabricating a high-performance niobium cavity. For a quick reference, below is a list to summarize the important steps and pertinent issues.

- Acquire sheet material,

    Check mechanical properties: grain size, yield strength

    Check RRR

    Inspect for scratches and inclusions, grind if necessary

- Form half-cells

    Deep drawing or spinning

- Trim

- Degrease

Figure 6.24: A cavity pair being assembled in a clean room at KEK. (Courtesy of KEK.)

- Light etch (≈ 5 μm)
- De-rust
- Inspect for scratches, defects or rust
- Electron-beam weld iris
- Grind iris
- Light etch (≈ 5 μm)
- Electron-beam weld equator
- Inspect weld
- Etch complete cavity (≈ 100 μm)
- Postpurify in furnace (for highest RRR)

SUMMARY

Figure 6.25: A portable clean room is used to attach cleanly the high power input coupler to a cavity at KEK. (Courtesy of KEK.)

- Etch inside (≈ 100 μm) and outside (≈ 30 μm)
- Tune to correct frequency and field flatness
- Final chemistry (≈ 5 μm)
- Rinsing
    Closed loop rinsing
    High-pressure rinsing
- Dry in clean room
- Assemble end flanges and couplers in clean room
- Evacuate

CHAPTER 7

Multicell Field "Flatness" Tuning

7.1 INTRODUCTION

A multicell cavity is a structure with multiple resonators (cells) coupled together. As with any set of coupled oscillators, there are multiple modes of excitation of the full structure for a given cell excitation mode (e.g., TM_{010}). For a $\beta \simeq 1$ multicell cavity the preferred accelerating mode provides an equal kick in each cell. This is accomplished by making the fields in neighboring cells π radians out of phase with each other and requiring the particle to cross a cell in one-half an rf period (see Figure 7.1).

For a given amount of stored energy in the cavity, having equal fields in each cell provides two advantages: The net accelerating voltage is maximized, and the peak surface EM fields are minimized. This "flat" field profile is only achieved when the cells are properly tuned relative to each other. Cell-to-cell tuning is usually needed only after initial fabrication and after significant etching or deformation due to heat treatment.

In this chapter a method for tuning a multicell cavity is developed in detail. The interested reader might compare our method with that of Smith [143], Tajima [52], or Sekutowicz [144]. Our development will go through the following steps.

1. Model the cells as capacitively coupled LC oscillators.

2. Examine the modes and accommodate for the beam tubes.

3. Model an ill-tuned cavity by introducing a frequency error in each cell of the LC circuit model.

4. Use a first-order perturbation method to determine the effect of the frequency errors on the voltage in each cell.

5. Show how to make measurements on a multicell structure to characterize the cavity's field flatness.

6. Carefully invert step 4 and use measurements to predict frequency corrections for each cell.

7. Give examples of tuning a multicell cavity.

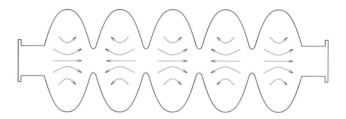

Figure 7.1: Sketch of the electric field lines of the π-mode of a 5-cell accelerating cavity.

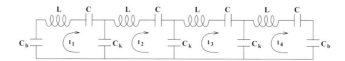

Figure 7.2: Equivalent circuit for a 4-cell cavity with beam tubes.

7.2 CIRCUIT MODEL

For the analysis, we will use the circuit model shown in Figure 7.2. This model is general enough for our purposes. Unlike a real cavity, the model has only one resonant mode for each cell. Since we are interested in exciting only the cells' TM_{010} mode, this presents no disadvantage.

In this model, L and C are the characteristic inductance and capacitance for each cell. The cell-to-cell coupling is done capacitively via C_k. This is appropriate for the shapes we used for $\beta \simeq 1$ cavities. For low-β structures the coupling is usually inductive [145]. The beam tubes are modeled by a capacitance C_b. Since $Q \gg 1$ for these cavities (even at room temperature) the resistance in our circuit is taken to be zero.

Apply Kirchhoff's voltage summation rule to each current loop I_j. Recall that impedances of an inductor and a capacitor are $i\omega L$ and $1/(i\omega C)$. The following coupled equations for an N-cell cavity result:

$$\left(\frac{1}{i\omega C_b} + i\omega L\right) I_1 + \left(\frac{1}{i\omega C}\right) I_1 + \left(\frac{1}{i\omega C_k}\right)(I_1 - I_2) = 0$$
$$\frac{1}{i\omega C_k}(I_j - I_{j-1}) + \left(i\omega L + \frac{1}{i\omega C}\right) I_j + \left(\frac{1}{i\omega C_k}\right)(I_j - I_{j+1}) = 0, \quad (7.1)$$
$$1 < j < N$$
$$\left(\frac{1}{i\omega C_k}\right)(I_N - I_{N-1}) + \left(i\omega L + \frac{1}{i\omega C}\right) I_N + \left(\frac{1}{i\omega C_b}\right) I_N = 0.$$

If we multiply through by $i\omega C$ and define $\omega_0^2 = 1/LC$, $k = C/C_k$, $\gamma = C/C_b$, and

$$\Omega = \frac{\omega^2}{\omega_0^2}, \quad (7.2)$$

CIRCUIT MODEL

then the system of equations becomes

$$\begin{aligned}
(1+k+\gamma)I_1 - kI_2 &= \Omega I_1, \\
-kI_{j-1} + (1+2k)I_j - kI_{j+1} &= \Omega I_j, \quad 1 < j < N \\
-kI_{N-1} + (1+k+\gamma)I_N &= \Omega I_N.
\end{aligned} \qquad (7.3)$$

This can be expressed as the tridiagonal matrix equation

$$\begin{pmatrix} 1+k+\gamma & -k & 0 & \cdots & 0 \\ -k & 1+2k & -k & & \vdots \\ 0 & & \ddots & & 0 \\ \vdots & & -k & 1+2k & -k \\ 0 & \cdots & 0 & -k & 1+k+\gamma \end{pmatrix} \mathbf{v} = \Omega \mathbf{v}, \qquad (7.4)$$

where the notation is used as a reminder that for a cavity, the cell voltages v_i are used instead of the currents, I_j, in the circuits. In the following, the subscript on a vector quantity will refer to the cell number. The superscript will designate the mode in the passband.

7.2.1 Compensating for Beam Tubes

We know that a "flat" π mode is desired so that the Nth normalized eigenvector must be

$$\mathbf{v}^{(N)} = \frac{1}{\sqrt{N}} \begin{pmatrix} 1 \\ -1 \\ 1 \\ \vdots \end{pmatrix}. \qquad (7.5)$$

This restriction is enough to determine the beam tube parameter γ.

To see this, apply (7.5) in the general matrix equation (7.4). The first two equations are

$$1 + 2k + \gamma = \Omega^{(N)} \qquad (7.6)$$
$$-1 - 4k = -\Omega^{(N)}. \qquad (7.7)$$

Here $\Omega^{(N)}$ denotes the eigenvalue corresponding to $\mathbf{v}^{(N)}$. Solving this system of two equations, we find that $\gamma = 2k$.

The general matrix equation becomes

$$\mathbf{A}\mathbf{v} = \Omega \mathbf{v}, \qquad (7.8)$$

where

$$\mathbf{A} \equiv \begin{pmatrix} 1+3k & -k & 0 & \cdots & 0 \\ -k & 1+2k & -k & & \vdots \\ 0 & & \ddots & & 0 \\ \vdots & & -k & 1+2k & -k \\ 0 & \cdots & 0 & -k & 1+3k \end{pmatrix}. \qquad (7.9)$$

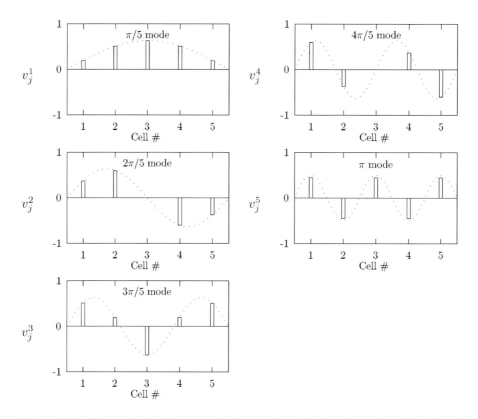

Figure 7.3: Computed orthonormal eigenmodes of a 5-cell cavity. The dotted line shows the equivalent modes of a cavity with $N \to \infty$.

7.2.2 Eigenvectors

The solution to (7.8) is found to be

$$v_j^{(m)} = B^{(m)} \sin\left[m\pi\left(\frac{2j-1}{2N}\right)\right] \quad (7.10)$$

for the cell number $j = 1, \ldots N$ in mode $m = 1 \ldots N$. $B^{(m)}$ is a normalizing coefficient that can be shown to be

$$B^{(m)} = \sqrt{(2 - \delta_{mN})/N}, \quad (7.11)$$

where δ_{mN} is the Kroneker delta.[1] Figure 7.3 shows the eigenvectors for a five-cell cavity.

---

[1] $\delta_{mm} = 1$ and $\delta_{mn} = 0$ when $m \neq n$.

MODELING AN OUT-OF-TUNE CAVITY

7.2.3 Eigenvalues

To find the eigenvalues, substitute (7.10) into the jth equation of (7.8).

$$(1 + 2k)v_j - k(v_{j-1} + v_{j+1}) = \Omega v_j. \tag{7.12}$$

Note the useful relation

$$\begin{aligned}
v_{j-1}^{(m)} + v_{j+1}^{(m)} &= B^{(m)} \sin\left[\left(m\pi \frac{(2j-1)}{2N}\right) - \frac{m\pi}{N}\right] \\
&\quad + B^{(m)} \sin\left[\left(m\pi \frac{(2j-1)}{2N}\right) + \frac{m\pi}{N}\right] \\
&= 2v_j^{(m)} \cos\left(\frac{m\pi}{N}\right).
\end{aligned} \tag{7.13}$$

Use the above equation in (7.12) and solve for $\Omega^{(m)}$ to give the eigenvalue of mode m:

$$\Omega^{(m)} = \left(\frac{f_m}{f_0}\right)^2 = 1 + 2k\left[1 - \cos\left(\frac{m\pi}{N}\right)\right]. \tag{7.14}$$

7.2.4 Dispersion Diagram

It can be seen from Equation 7.14 that the mode spacing will increase with stronger cell-to-cell coupling, k, and will decrease as the number of cells, N, increases. Figure 7.4 shows the relative positions of the modes for a 5-cell structure. If we assume that each cell is excited in the TM_{010} mode, then this plot shows the frequencies in the "TM_{010} passband". As $N \to \infty$, all points on this curve are filled in.

The term "passband" comes from the analysis of a periodic structure where there is a continuous band of propagating modes in the neighborhood of a particular cell mode. In the case of a periodic structure composed of cavity cells, we would find that the mode designation $j\pi/N$ describes the phase advance per cell of the waveguide mode. In a cavity that has equal and opposing traveling waves to form a standing wave, the phase advance between neighboring cells can take on only the values 0 or π.

7.3 MODELING AN OUT-OF-TUNE CAVITY

Let us now consider a perturbation that takes the cavity out of tune. Let each of the cell capacitances, C, vary a bit by letting

$$C_j = \frac{C}{1 + e_j}, \tag{7.15}$$

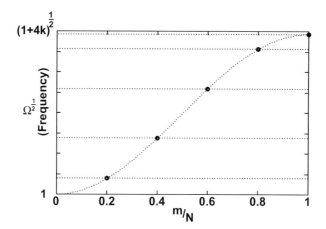

Figure 7.4: Modes in a "passband" of a 5-cell cavity. The dotted line shows the modes of a cavity as $N \to \infty$.

where $|e_j| \ll 1$. The jth equation in the current loop analysis becomes

$$\left(\frac{1}{i\omega C_k}\right)(I_j - I_{j-1}) + \left[i\omega L + \frac{1}{i\omega C}(1 + e_j)\right]I_j + \left(\frac{1}{i\omega C_k}\right)(I_j - I_{j+1}) = 0. \tag{7.16}$$

The cell perturbations then produce

$$\begin{pmatrix} 1+3k+e_1 & -k & 0 & \cdots & & 0 \\ -k & 1+2k+e_2 & -k & & & \vdots \\ 0 & & \ddots & & & 0 \\ \vdots & & -k & 1+2k+e_{N-1} & -k \\ 0 & \cdots & & 0 & -k & 1+3k+e_N \end{pmatrix} \mathbf{v}' = \Omega' \mathbf{v}'. \tag{7.17}$$

We can express this as a perturbation on the original matrix as

$$(\mathsf{A} + \mathsf{E})\mathbf{v}' = \Omega'\mathbf{v}', \tag{7.18}$$

where $\mathsf{E}_{ij} = \delta_{ij} e_j$.

Next we apply a first order perturbation technique to determine how these cell errors affect the cell voltages in the π mode. Before doing that, we will digress briefly to review nondegenerate perturbation theory.

7.4 REFRESHER ON PERTURBATION TECHNIQUES

Given a matrix equation

$$\mathsf{M}\mathbf{x}^{(m)} = \lambda^{(m)} \mathbf{x}^{(m)} \tag{7.19}$$

REFRESHER ON PERTURBATION TECHNIQUES

with a known set of eigenvalues $\lambda^{(m)}$ and orthonormal eigenvectors $\mathbf{x}^{(m)}$, we want to estimate the solutions to the system perturbed by $\mathbf{G}$:

$$(\mathbf{M} + \alpha \mathbf{G})\mathbf{x}'^{(m)} = \lambda'^{(m)}\mathbf{x}'^{(m)}. \tag{7.20}$$

The parameter α ranges from 0 to 1 to allow us to turn on the perturbation slowly. Also it is useful because the power of α in the calculation serves to keep track of the order of the perturbation.

Expand $\mathbf{x}'^{(m)}$ in terms of the complete set of unperturbed states $\mathbf{x}^{(m)}$:

$$(\mathbf{M} + \alpha \mathbf{G})\left(\sum_n C_{nm}\mathbf{x}^{(n)}\right) = \lambda'^{(m)}\left(\sum_n C_{nm}\mathbf{x}^{(n)}\right). \tag{7.21}$$

Apply (7.19):

$$\sum_n C_{nm}\lambda^{(n)}\mathbf{x}^{(n)} + \sum_n \alpha C_{nm}\mathbf{G}\mathbf{x}^{(n)} = \lambda'^{(m)}\left(\sum_n C_{nm}\mathbf{x}^{(n)}\right). \tag{7.22}$$

Multiply by $\mathbf{x}^{(r)}$ and use the fact that the unperturbed eigenvectors are orthonormal, $\mathbf{x}^{(r)} \cdot \mathbf{x}^{(n)} = \delta_{rn}$.

$$C_{rm}\lambda^{(r)} + \sum_n \alpha C_{nm}\left(\mathbf{x}^{(r)}\right)^{\mathrm{T}}\mathbf{G}\mathbf{x}^{(n)} = \lambda'^{(m)}C_{rm}. \tag{7.23}$$

In this notation, $\left(\mathbf{x}^{(r)}\right)^{\mathrm{T}}$ is the transpose of $\mathbf{x}^{(r)}$.

An exact expression for the coefficients is then

$$C_{rm}\left(\lambda'^{(m)} - \lambda^{(r)}\right) = \sum_n \alpha C_{nm} <r|\mathbf{G}|n>, \tag{7.24}$$

where

$$<r|\mathbf{G}|n> \equiv \left(\mathbf{x}^{(r)}\right)^{\mathrm{T}}\mathbf{G}\mathbf{x}^{(n)}. \tag{7.25}$$

If the perturbation $\mathbf{G}$ is not large, then $C_{mm} \simeq 1$ and $C_{nm} \ll 1$ where $n \neq m$. We assume that the eigenvalues are nondegenerate, that is, $\lambda^n \neq \lambda^m$ where $n \neq m$.

$$\lambda'^{(m)} - \lambda^{(m)} = \sum_n \frac{\alpha C_{nm}}{C_{mm}} <m|\mathbf{G}|n> \quad \text{for } r = m \tag{7.26}$$

$$\frac{C_{rm}}{C_{mm}} = \frac{1}{\lambda'^{(m)} - \lambda^{(r)}} \sum_n \frac{\alpha C_{nm}}{C_{mm}} <r|\mathbf{G}|n> \quad \text{for } r \neq m. \tag{7.27}$$

The coefficient ratios are computed in this way. If needed, the normalization may be imposed afterward.

Zero order:

$$\mathbf{x}'^{(m)} \to \mathbf{x}^{(m)} \quad C_{nm} \to \delta_{nm} \quad \lambda'^{(m)} \to \lambda^{(m)}. \tag{7.28}$$

First order: Use the zero order values in the right-hand side of (7.26) and (7.27).

$$\lambda'^{(m)} = \lambda^{(m)} + \alpha <m|\mathsf{G}|m>, \qquad r=m \qquad (7.29)$$

$$C_{mm} = 1 \qquad (7.30)$$

$$C_{rm} = \frac{\alpha}{\lambda^{(m)} - \lambda^{(r)}} <r|\mathsf{G}|m>, \qquad r \neq m. \qquad (7.31)$$

Using the first-order perturbation expressions, we can write

$$\mathbf{x}'^{(m)} = C_{mm}\mathbf{x}^{(m)} + \sum_{n \neq m} C_{nm}\mathbf{x}^{(n)}$$

$$= \mathbf{x}^{(m)} + \sum_{n \neq m} \frac{\alpha}{\lambda^{(m)} - \lambda^{(n)}} <m|\mathsf{G}|n> \mathbf{x}^{(n)}. \qquad (7.32)$$

Second order: use the first-order values in the right-hand side of (7.26) and (7.27). The process continues to arbitrary order.

7.5 APPLYING THE PERTURBATION

Using the first-order perturbation approach on the perturbed eigensystem in (7.18) we find

$$\delta\Omega^{(m)} = \Omega'^{(m)} - \Omega^{(m)} \qquad (7.33)$$

$$= <m|\mathsf{E}|m> = \sum_{j,k} v_j^{(m)} \delta_{jk} e_j v_k^{(m)} \qquad (7.34)$$

$$= \sum_j v_j^{(m)} e_j v_j^{(m)}. \qquad (7.35)$$

For predicting the amplitude errors we find

$$\delta\mathbf{v}^{(m)} \equiv \mathbf{v}'^{(m)} - \mathbf{v}^{(m)} \qquad (7.36)$$

$$= \sum_{n \neq m} \frac{1}{\Omega^{(m)} - \Omega^{(n)}} \sum_j v_j^{(m)} e_j v_j^{(n)} \mathbf{v}^{(n)}. \qquad (7.37)$$

When we measure the cavity's field flatness, we indirectly measure the perturbations in field, $\delta\mathbf{v}_{(m)}$, not the cell errors in capacitance given by $\mathbf{e}$, the vector of components e_j. It would be very convenient to have a transformation matrix H so that we could write

$$\delta\mathbf{v}^{(m)} = \mathsf{H}\mathbf{e}. \qquad (7.38)$$

This simple expression could then be solved to get $\mathbf{e}$.

By rearranging the summation in (7.37) we find this transformation matrix to be

$$H_{lk}^{(m)} = \sum_{j \neq m} \frac{v_l^{(j)} v_k^{(j)} v_k^{(m)}}{\Omega^{(m)} - \Omega^{(j)}} = \sum_{j \neq m} \frac{v_l^{(j)} v_k^{(j)} v_k^{(m)}}{2k\left[\cos\left(j\pi/N\right) - \cos\left(m\pi/N\right)\right]}. \qquad (7.39)$$

"BEAD PULLING" TO MEASURE THE FIELD PROFILE

Since it is the $m = N$ mode or the π mode that we are trying to tune, it is worth writing that expression explicitly:

$$H_{lk}^{(N)} = \sum_{j \neq N} \frac{v_l^{(j)} v_k^{(j)} v_k^{(N)}}{\Omega^{(N)} - \Omega^{(j)}} = \sum_{j \neq N} \frac{v_l^{(j)} v_k^{(j)} v_k^{(N)}}{2k \left[\cos\left(j\pi/N\right) + 1\right]}. \tag{7.40}$$

At this point, it is useful to see how one goes about measuring $\delta \mathbf{v}^{(m)}$.

7.6 "BEAD PULLING" TO MEASURE THE FIELD PROFILE

The field flatness of a cavity is measured by perturbing each cell in succession using a small metal bead while the frequency of the π-mode is measured. In practice the "bead" might be a tiny (compared to the wavelength) segment of metal tube on a fishing line strung through the cavity along the axis (Figure 7.5). As will be shown, because the same bead is used to perturb successive cells, there is no need to calibrate the bead. Once the tuning parameters are calculated, each cell is tuned by squeezing or stretching it. Figure 6.19 is a photograph of an apparatus that tunes the cell.

We can use first-order perturbation theory again to predict the effect of the bead on the already out-of-tune π mode.

$$\delta \Omega'^{(N)}_j = \left(v'^{(N)}_j\right)^2 e_{\text{bead}}. \tag{7.41}$$

Recall that the primed quantities designate the out-of-tune solution. In this case we are relying upon Nature providing the out-of-tune solution; that is, we will use the measured amplitudes of the π mode. From (7.2) and (7.41) we find

$$\delta f'^{(N)}_j = \frac{1}{2} \frac{\delta \Omega'^{(N)}_j}{\Omega'^{(N)}} f'^{(N)}, \tag{7.42}$$

$$= \frac{1}{2} \frac{f'^{(N)}}{\Omega'^{(N)}} \left(v'^{(N)}_j\right)^2 e_{\text{bead}}. \tag{7.43}$$

Each $\delta f'^{(N)}_j$ is measured directly when the bead is in the center of cell j.

To use Equation 7.38, we need to work with the unperturbed modes. Let us expand the amplitudes for small perturbations.

$$\left(v'^{(N)}_j\right)^2 = \left(v^{(N)}_j\right)^2 \left(1 + \frac{\delta v^{(N)}_j}{v^{(N)}_j}\right)^2 \simeq \left(v^{(N)}_j\right)^2 \left(1 + \frac{2\delta v^{(N)}_j}{v^{(N)}_j}\right). \tag{7.44}$$

Substitute this into (7.43) and solve for $\delta v^{(N)}_j$:

$$\delta v^{(N)}_j = \left\{ \frac{\delta f'^{(N)}_j}{\left(v^{(N)}_j\right)^2} \left[\frac{2\Omega'^{(N)}}{e_{\text{bead}} f'^{(N)}}\right] - 1 \right\} \frac{v^{(N)}_j}{2}. \tag{7.45}$$

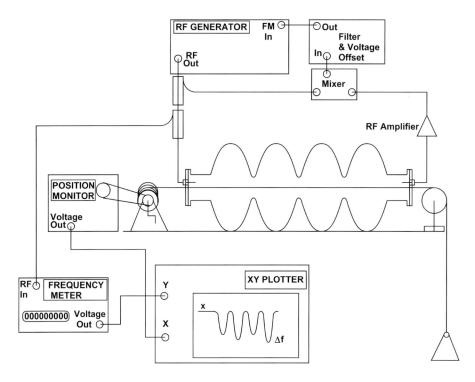

Figure 7.5: Schematic diagram of cavity-tuning apparatus. A metal bead is sent through the cavity and the cavity's frequency change is plotted as a function of position of the bead. Maximum frequency excursions are recorded as $\delta f'^{(N)}_j$.

If the term in brackets in Equation 7.45 were known, we could relate the frequency shifts observed when pulling a bead through the structure to the amplitude errors in each cell.

Consider what happens when the measured frequency shifts are averaged:

$$<\delta f'> = \frac{1}{N}\sum_j \delta f'^{(N)}_j = \frac{1}{2}\frac{f'^{(N)}}{\Omega'^{(N)}}\frac{e_{\text{bead}}}{N}\sum_j \left(v'^{(N)}_j\right)^2. \quad (7.46)$$

Since the $\mathbf{v}'$ are normalized, we have

$$<\delta f'> = \frac{1}{2}\frac{f'^{(N)}}{\Omega'^{(N)}}\frac{e_{\text{bead}}}{N}. \quad (7.47)$$

Using this, the unperturbed eigenvector for the Nth mode (7.5), and (7.45), we find an expression that involves only the unperturbed mode and the measured frequency shifts:

$$\delta v^{(N)}_j = \left[\frac{\delta f'^{(N)}_j}{<\delta f'>} - 1\right]\frac{v^{(N)}_j}{2}. \quad (7.48)$$

"BEAD PULLING" TO MEASURE THE FIELD PROFILE

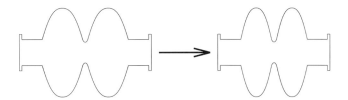

Figure 7.6: Decreasing the frequency of both cells of a 2-cell cavity.

We are almost ready to use

$$\delta \mathbf{v}^{(N)} = \mathbf{H}\mathbf{e}$$

to solve for $\mathbf{e}$ except for one difficulty: $\mathbf{H}$ cannot be inverted. It is singular. This makes sense because $\mathbf{H}$ transforms the absolute capacitance errors into amplitude changes. Information is lost in the transformation.

To illustrate this, consider a two-cell cavity. If the perturbed cavity has $e_1 = e_2$, the cells are still in tune with respect to each other (see Figure 7.6). The relative cell amplitudes are the same as the unperturbed cavity, and $\delta \mathbf{v}^{(N)} = 0$ for both cases. The error vector $\mathbf{e}$ is not unique. We cannot simply solve for it.

To make up for the lost information, we require an absolute frequency reference. Assume (temporarily) that one of the cells is "correct"; that is, it has $e_j = 0$. Arbitrarily assume it is the last cell and strip off the last row and column of $\mathbf{H}$ to produce a reduced matrix $\mathbf{H_r}$. Similarly, we construct vectors of length $N-1$, $\mathbf{e_r}$ and $\delta \mathbf{v_r}$, to write

$$\mathbf{e_r} = (\mathbf{H_r})^{-1} \delta \mathbf{v_r}. \qquad (7.49)$$

Using $\mathbf{e_r}$ directly to correct the cells in the cavity will change the π-mode frequency, f^π, to match the frequency of the last cell. To produce no net change in f^π, we construct a corrected vector $\mathbf{e_c}$ with an average cell correction of zero:

$$(\mathbf{e_c})_j = -\left[(\mathbf{e_r})_j - <\mathbf{e_r}>\right] \quad \text{for } 1, \ldots, (N-1) \qquad (7.50)$$
$$(\mathbf{e_c})_N = <\mathbf{e_r}>, \qquad (7.51)$$

where

$$<\mathbf{e_r}> = \frac{1}{N} \sum_{j=1}^{N-1} (\mathbf{e_r})_j. \qquad (7.52)$$

Note we have divided by N and not $N-1$.

Recall that the components of $\mathbf{e_c}$ describe errors in capacitances. We can convert these to π-mode frequency corrections for each cell via

$$\delta \mathbf{f_c} = \mathbf{e_c} \frac{f^\pi}{2N} \qquad (7.53)$$

to bring all the cells in tune without changing f^π.

If, in addition to the cells being out of tune with respect to each other, f^π is not equal to the desired value, f^π_{desired}, each cell can be altered by the same amount, so that f^π is altered but the field flatness is unaffected. For a "flat" cavity, each cell has an equal contribution to determining the π-mode frequency. The frequency correction for each of the cells should be

$$\delta f^\pi = \frac{f^\pi_{\text{desired}} - f^\pi_{\text{measured}}}{N}. \tag{7.54}$$

The net result is that each cell in succession needs to be stretched or squeezed. Cell j needs to be altered so that f^π changes by $(\delta f_c)_j + \delta f^\pi$.

7.7 CONSTRUCTING THE MODEL FROM MEASUREMENTS

Constructing H in Equation 7.40 requires knowledge of all the $\mathbf{v}^{(m)}$ and k. The $\mathbf{v}^{(m)}$ are given explicitly in (7.10). To a reasonable degree, k can be found by measuring the frequencies of two modes in the passband. For this we make the approximation that the perturbed eigenvalues are the same as the unperturbed.

Recall from (7.14) that

$$\Omega^{(m)} = \left(\frac{f^{(m)}}{f_0}\right)^2 = 1 + 2k\left[1 - \cos\left(\frac{m\pi}{N}\right)\right]$$

for a mode m.

Measure the frequency of two modes, $f^{(m)}$ and $f^{(p)}$. Combine the above equation for both modes to obtain

$$k = \frac{\frac{1}{2}\left[\left(f^{(p)}\right)^2 - \left(f^{(m)}\right)^2\right]}{\left(f^{(m)}\right)^2\left[1 - \cos(p\pi/N)\right] - \left(f^{(p)}\right)^2\left[1 - \cos(m\pi/N)\right]}. \tag{7.55}$$

If we measure $f^{(N)}$ and $f^{(1)}$, this becomes

$$k = \frac{\frac{1}{2}\left[\left(f^{(N)}\right)^2 - \left(f^{(1)}\right)^2\right]}{2\left(f^{(1)}\right)^2 - \left(f^{(N)}\right)^2\left[1 - \cos(\pi/N)\right]}. \tag{7.56}$$

Knowing k, we can compute $\Omega^{(m)}$ for the unperturbed case using (7.14).

7.8 TWO-CELL WORKED EXAMPLE

To see how the machinery works, let us go through the near trivial example of tuning a 2-cell cavity such as the one shown in Figure 7.7.

Suppose that the target frequency for the π mode is 1300.000 MHz. The resonant frequencies of the two modes of the TM_{010} passband are measured to be

TWO-CELL WORKED EXAMPLE

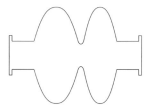

Figure 7.7: Two cell cavity in need of tuning.

$$\begin{aligned} f_1 &= 1\,290.427 \text{ MHz} \quad (\pi/2 \text{ mode}), \\ f_2 &= 1\,302.165 \text{ MHz} \quad (\pi \text{ mode}). \end{aligned}$$

Pulling a bead and monitoring f_2, we measure maximum frequency excursions of

$$\begin{aligned} \delta f_1'^{(2)} &= -78 \text{ kHz} \quad (\text{cell 1}), \\ \delta f_2'^{(2)} &= -74 \text{ kHz} \quad (\text{cell 2}). \end{aligned}$$

We use the measured resonances to approximate the cell-to-cell coupling constant:

$$k = \frac{\frac{1}{2}\left[\left(f^{(2)}\right)^2 - \left(f^{(1)}\right)^2\right]}{2\left(f^{(1)}\right)^2 - \left(f^{(2)}\right)^2 [1 - \cos(\pi/2)]} = 9.307684 \times 10^{-3}.$$

We determine the eigenvectors to an ideal cavity using (7.10):

$$\mathbf{v}^{(1)} = \tfrac{1}{\sqrt{2}}\begin{pmatrix} 1 \\ 1 \end{pmatrix}, \qquad \mathbf{v}^{(2)} = \tfrac{1}{\sqrt{2}}\begin{pmatrix} 1 \\ -1 \end{pmatrix}.$$

We compute $\mathsf{H}^{(N)}$ from (7.40). Only $H_{11}^{(2)}$ is needed since $\mathsf{H_r}$ is only of dimension $(N-1) \times (N-1)$ and $N = 2$.

$$\begin{aligned} H_{11}^{(2)} &= \frac{v_1^{(1)} v_1^{(1)} v_1^{(2)}}{2k [\cos(\pi/2) + 1]} \\ &= \frac{(1/\sqrt{2})(1/\sqrt{2})(1/\sqrt{2})}{2k} \\ &= 18.99255. \end{aligned}$$

With this element we obtain the 1×1 matrix:

$$\mathsf{H_r} = (\,18.99255\,).$$

We use (7.48) to compute cell amplitude errors from the measured frequency shifts. Only the first $(N-1)$ errors are needed.

$$\begin{aligned} \delta v_1 &= \left(\frac{\delta f_1'^{(2)}}{<\delta f'>} - 1\right)\frac{v_1^{(2)}}{2} \\ &= \left(\frac{-78}{-76} - 1\right)\frac{1}{2\sqrt{2}} \\ &= 9.30 \times 10^{-3}. \end{aligned}$$

We obtain the one-element vector:
$$\delta \mathbf{v_r} = (9.30 \times 10^{-3}).$$

We get $\mathbf{e_r}$ from
$$\begin{aligned}\mathbf{e_r} &= (\mathbf{H_r})^{-1} \delta \mathbf{v_r} \\ &= (18.99255)^{-1} \times (9.30 \times 10^{-3}) = 4.90 \times 10^{-4}.\end{aligned}$$

We build a frequency error vector that does not change the π-mode frequency:
$$<\mathbf{e_r}> = \frac{1}{N} \sum_{j=1}^{N-1} (e_r)_j = \frac{4.90 \times 10^{-4}}{2} = 2.45 \times 10^{-4},$$

$$\begin{aligned}(e_c)_1 &= -[(e_r)_1 - <\mathbf{e_r}>] \\ &= -\left(4.90 \times 10^{-4} - 2.45 \times 10^{-4}\right) \\ &= -2.45 \times 10^{-4}, \\ (e_c)_2 &= <\mathbf{e_r}> \\ &= 2.45 \times 10^{-4}.\end{aligned}$$

We convert these capacitance corrections to frequency corrections for the π mode.
$$\delta \mathbf{f_c} = \mathbf{e_c} \frac{f^\pi}{2N} = \frac{1302.165 \text{ MHz}}{4} \begin{pmatrix} -2.45 \\ +2.45 \end{pmatrix} \times 10^{-4} = \begin{pmatrix} -80 \\ +80 \end{pmatrix} \text{ kHz}.$$

We must correct for the fact that the untuned π mode had the wrong frequency. We impose an additional frequency correction to compensate:
$$\begin{aligned}\delta f^\pi &= \frac{f^\pi_{desired} - f^\pi_{measured}}{N} = \frac{(1300.000 - 1302.165)}{2} \\ &= -1.0825 \text{ MHz}.\end{aligned}$$

The end result is that the frequencies of both cells will be lowered. The first cell's frequency is lowered more than the second:
$$\delta \mathbf{f}_{applied} = \delta \mathbf{f_c} + \delta f^\pi = \begin{pmatrix} -1162.5 \\ -1002.5 \end{pmatrix} \text{ kHz}.$$

To lower the frequency of a cell, the cell is squeezed (shortened along its z axis).

7.9 FIVE-CELL CAVITY EXAMPLE

Tuning a multicell cavity is usually an iterative process. Our first-order technique will make reasonable guesses, but a few iterations are often needed to

FIVE-CELL CAVITY EXAMPLE   143

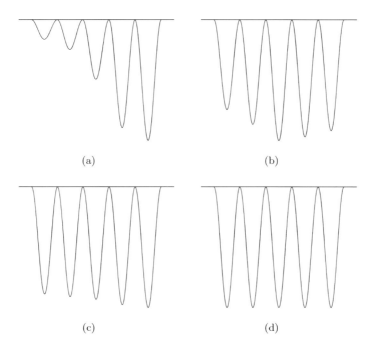

Figure 7.8: Bead pull results on a 1.3-GHz 5-cell cavity. The change of frequency of the π mode is plotted as the perturbing bead is pulled through the axis of the structure. The peak of each hump is proportional to the stored energy in the cell. (a) After initial fabrication. (b) After first tuning. (c) After second tuning. (d) After third tuning.

converge upon a reasonable field flatness. To illustrate, the tuning of a 1.3-GHz 5-cell cavity [145] is presented in Figure 7.8. In this example, the cycle of pulling a bead, doing the calculations, and mechanically deforming the cells was performed three times. Had this structure undergone heat treatment to purify the niobium, it is likely that more iterations would have been needed, because the niobium would be softer. Softer cavities are harder to tune one cell at a time because while deforming one cell, the neighboring cells tend to deform a little as well.

CHAPTER 8

Cavity Testing

8.1 INTRODUCTION

The principles and techniques discussed in this chapter are used to evaluate the overall performance of a cavity. This is required for an acceptance test or to find out the effects of a newly invented surface treatment on the performance of a superconducting cavity. The methods discussed here also serve as a technique for the diagnosis of any anomalous behavior of a superconducting cavity (e.g., multipacting or field emission). The most generally used diagnostic technique is to measure the response of the cavity to rf fields. The most important figure of merit in evaluating the rf performance is the unloaded quality factor (Q_0) of the cavity as a function of the peak electric field. This gives the ubiquitous Q_0 versus E curve.

The rf techniques only give information on the average behavior of the cavity; that is, the power losses are averaged over the entire cavity rf surface. To resolve the details of the local behavior, one uses detectors positioned at various points of the cavity. Usually, anomalous power dissipation is localized in assorted hot spots around the cavity. The spatial temperature distribution can be very effectively mapped using an array of thermometers attached to the outer surface of the cavity. Thermometry has emerged as a powerful tool in investigating the behavior of superconducting cavities.

8.2 RF MEASUREMENTS

To excite a resonant mode of the cavity, we need to connect the cavity to an rf source. In practice, a setup such as that shown in Figure 8.1 might be used.

The power from the rf source is carried to the cavity via a coaxial cable in a TEM mode. The center conductor protrudes into the beam tube of the cavity which forms the outer conductor. The strength of the input coupler is adjusted by changing the penetration of the center conductor. The fixed output coupler is also called a *transmitted power probe* since it picks up power transmitted through the cavity. As will be discussed in Chapter 12, this transmitted power probe can also serve as an electron pickup probe. In this arrangement with coaxial couplers, the TEM mode that propagates in the input probe weakly couples

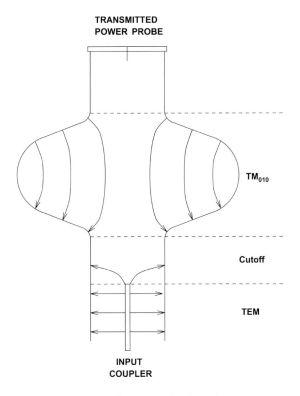

Figure 8.1: Cavity with rf probes.

to the cavity fields via an intermediate non-propagating mode. The coupling strength is a measure of the overlap between the cavity fields and the fields that leak into the cutoff region from the input coupler. Since the cavity fields for the fundamental (accelerating) mode fall off exponentially with distance in the cutoff region, the coupling strength increases exponentially with the coupler's insertion distance. As will be seen, it is useful to be able to adjust the input coupler to make the cavity act as a matched load.

8.2.1 Undriven Cavity

If we consider the behavior of the cavity operating at its fundamental resonant frequency after the rf drive is switched off, we can obtain some general insight into the cavity behavior. After the rf is turned off, the total power being lost will be the sum of the power dissipated in the cavity walls (ohmic losses) and the power that leaks out each coupler:

$$P_{\text{tot}} = P_{\text{c}} + P_{\text{e}} + P_{\text{t}}. \tag{8.1}$$

Here P_{e} designates the power leaking back out the input coupler (currently the rf drive is off) and P_{t} is the power coming out the transmitted power coupler.

RF MEASUREMENTS

Analogous to the intrinsic quality factor Q_0 introduced in Chapter 2;

$$Q_0 = \frac{\omega U}{P_c}, \tag{8.2}$$

we can define a "loaded" quality factor, Q_L, as

$$Q_L \equiv \frac{\omega U}{P_{tot}} \tag{8.3}$$

to characterize a cavity with couplers.

When the cavity is left to "ring down," the stored energy satisfies

$$\frac{dU}{dt} = -P_{tot} = -\frac{\omega U}{Q_L}. \tag{8.4}$$

If the losses are quadratic (i.e., no field emission or other anomalous losses) the solution to this simple differential equation is

$$U = U_0 \exp\left(\frac{-\omega t}{Q_L}\right), \tag{8.5}$$

where U_0 is the stored energy at $t = 0$. The energy in the cavity thus decays exponentially with a time constant

$$\tau_L = \frac{Q_L}{\omega}. \tag{8.6}$$

By measuring this decay time, we can determine the loaded quality factor of the cavity.

Equation 8.6 is useful for determining the loaded Q of a cavity connected to input and output rf lines. It takes into account the leakage through the couplers. We are more interested in determining the intrinsic properties of the cavity, not the measurement probes; that is, we want the Q_0 of the cavity without the contribution of the couplers. Substituting (8.1) into the inverse of (8.3) suggests that we can assign a quality factor to each loss mechanism such that

$$\frac{P_{tot}}{\omega U} = \frac{P_c + P_e + P_t}{\omega U} \tag{8.7}$$

$$\frac{1}{Q_L} = \frac{1}{Q_0} + \frac{1}{Q_e} + \frac{1}{Q_t}, \tag{8.8}$$

where we have defined "external" quality factors for the couplers

$$Q_e \equiv \frac{\omega U}{P_e} \tag{8.9}$$

and

$$Q_t \equiv \frac{\omega U}{P_t}. \tag{8.10}$$

We can rewrite (8.8) by defining the "coupling parameters" as

$$\beta_\mathrm{e} \equiv \frac{Q_0}{Q_\mathrm{e}} \qquad (8.11)$$

and

$$\beta_\mathrm{t} \equiv \frac{Q_0}{Q_\mathrm{t}}, \qquad (8.12)$$

so that we can write

$$\frac{1}{Q_\mathrm{L}} = \frac{1}{Q_0}\left(1 + \beta_\mathrm{e} + \beta_\mathrm{t}\right). \qquad (8.13)$$

If we can determine the coupling parameters and the loaded Q, then Q_0 can be calculated. The term *coupling strength* is appropriate for the β's since we can rewrite them as

$$\beta_\mathrm{e} = \frac{P_\mathrm{e}}{P_\mathrm{c}} \qquad (8.14)$$

and

$$\beta_\mathrm{t} = \frac{P_\mathrm{t}}{P_\mathrm{c}}. \qquad (8.15)$$

We see that the β's tell us how strongly the couplers interact with the cavity. If β is large, it implies that the power leaking out of the coupler is large with respect to the power being dissipated in the cavity walls. Note that even though the external Q of a coupler is dependent only on the cavity and coupler geometry, the β of the coupler is also dependent on the Q_0 of the cavity.

8.2.2 Driven Cavity with One Coupler

Generally we make the fixed transmitted power probe coupling very weak (i.e., $\beta_\mathrm{t} \ll 1$) so that we can safely neglect the influence of the output coupler in evaluating the rf performance. That being the case, we now consider a cavity driven on resonance with one coupler. Due to its prevalence in the literature, when only considering one coupler we will drop the subscript on β_e and call the input coupling parameter just β.

When the rf is on, the cavity–coupler system is a driven nonlinear harmonic oscillator. The "nonlinear" aspect is necessary because we include the possibility of anomalous losses. We would like to be able to predict the cavity's steady-state response and also develop a differential equation for the transient response.

Even though a real cavity has many excitation modes, we are only concerned with exciting the single fundamental cavity mode via a single propagating mode in our transmission line. A thorough and very readable analysis of a waveguide coupled to a cavity considering all the modes is performed by Slater [146].

Since even a normal conducting cavity has a high Q_0 (i.e., > 100), the analysis can be simplified considerably by using the quasi-static approximation which is exploiting the fact that the time constant of the cavity response is long relative to the period of the rf. To model the driven cavity, we will construct

RF MEASUREMENTS

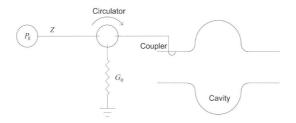

Figure 8.2: Schematic of cavity with generator and circulator.

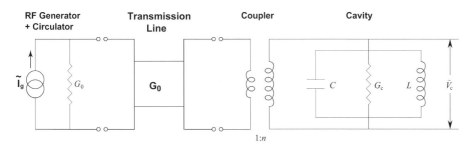

Figure 8.3: Equivalent circuit for a cavity with one coupler being driven by an rf generator.

an equivalent circuit containing lumped elements and analyze it as if the cavity were a linear device.

The system we want to model is shown schematically in Figure 8.2. Between the rf generator and the cavity is an isolator — that is, a circulator connected to a load. This ensures that signals coming from the cavity are terminated in a matched load. In addition to simplifying our analysis, the isolator has the benefit of protecting the generator from reflections coming from the cavity–coupler system.

The equivalent circuit for this system is shown in Figure 8.3. The rf generator and isolator combination is modeled by an ideal current source and a shunt admittance G_0. A lossless transmission line with a characteristic admittance G_0 connects the generator with the coupler. The cavity's input coupler is modeled with a transformer, allowing us the possibility of matching the transmission line to cavity with an arbitrary impedance.

We want to match up the circuit parameters with the cavity quantities discussed earlier. The stored energy in the circuit resonator is

$$U = \frac{CV_c^2}{2}. \tag{8.16}$$

Isolated from the external world, its dissipated power is

$$P_c = \frac{G_c V_c^2}{2}. \tag{8.17}$$

From these and (8.2) we can determine that

$$Q_0 = \frac{\omega_0 U}{P_c} = \sqrt{\frac{C}{L}}\frac{1}{G_c}. \tag{8.18}$$

When the coupler and transmission line is connected and the current source is off, the cavity sees the additional admittance G_0/n^2 of the transmission line through the transformer. This contributes to the total power losses a term

$$P_e = \frac{G_0}{n^2}\frac{V_c^2}{2} \tag{8.19}$$

describing the losses through the coupler. By using this expression and (8.16) in Equation 8.9, we can then see that

$$Q_e = n^2\sqrt{\frac{C}{L}}\frac{1}{G_0}. \tag{8.20}$$

Again we can see that the external Q is independent of the cavity losses G_c. From the ratio of Q_0 to Q_e, the external coupling is then found to be

$$\beta = \frac{G_0}{n^2 G_c}. \tag{8.21}$$

Although we will not apply this quantity until Chapter 17, for completeness we mention here that the shunt impedance is modeled by

$$R_a = \frac{2}{G_c}. \tag{8.22}$$

We can compute the power getting into the cavity by computing the reflected power and using conservation of energy. Seen from the transmission line, the reflection coefficient for the cavity is

$$\Gamma = \frac{1 - Y_c'/G_0}{1 + Y_c'/G_0}, \tag{8.23}$$

where Y_c' is the input admittance of the cavity–coupler system. It is primed to remind us that it is seen through the transformer. Summing the parallel admittances of the cavity elements in Figure 8.3 and multiplying by n^2 because of the transformer, we find that

$$\frac{Y_c'}{G_0} = \frac{n^2 G_c}{G_0} + i\frac{n^2}{G_0}\sqrt{\frac{C}{L}}\left(\frac{\omega}{\omega_0} - \frac{\omega_0}{\omega}\right), \tag{8.24}$$

where $\omega_0 = 1/\sqrt{LC}$ and ω is the angular frequency of the generator. In terms of the cavity parameters, this expression becomes

$$\frac{Y_c'}{G_0} = \frac{1}{\beta} + iQ_e\left(\frac{\omega}{\omega_0} - \frac{\omega_0}{\omega}\right), \tag{8.25}$$

RF MEASUREMENTS

giving a relative conductance

$$\frac{G'_c}{G_0} = \frac{1}{\beta} \tag{8.26}$$

and a relative susceptance of

$$\frac{B'_c}{G_0} = Q_e \left(\frac{\omega}{\omega_0} - \frac{\omega_0}{\omega} \right). \tag{8.27}$$

The reflected power is $P_f|\Gamma|^2$, where P_f is the power traveling forward along the transmission line toward the cavity. Through conservation of energy, the power flowing into the cavity is thus

$$P_{\text{in}} = P_f \left(1 - |\Gamma|^2\right). \tag{8.28}$$

If we explicitly separate Y'_c into its component conductance and susceptance as $Y'_c = G'_c + iB'_c$ and apply Equation 8.23, then (8.28) can be written as

$$P_{\text{in}} = P_f \frac{4(G'_c/G_0)}{[1 + (G'_c/G_0)]^2 + (B'_c/G_0)^2}. \tag{8.29}$$

Using Equation 8.25 in the above yields

$$P_{\text{in}} = \frac{4P_f/(Q_e Q_0)}{(1/Q_0 + 1/Q_e)^2 + (\omega/\omega_0 - \omega_0/\omega)^2} \tag{8.30}$$

after some rearranging.

Transient behavior is often treated by Fourier analysis. This can be done for this model in the usual manner and will give the solution for a particular choice of a drive signal. Unfortunately, the Fourier method requires that the cavity is a linear oscillator. A cavity loaded with electron field emission that has its Q_0 dependent upon the strength of the electric field could not be analyzed using this method. We instead seek to derive a differential equation that would allow us to solve the general case of the high-Q nonlinear oscillator.

Borrowing the method used by Slater [146], let us recognize that a damped oscillator can be considered to have a complex frequency. The positive imaginary component gives rise to damping. If the imaginary component of the frequency is negative, the oscillator would have its amplitude increasing over time.

Let us take $\omega = \omega_1 + i\omega_2$. The amplitude of oscillation in the cavity will change as $\exp(-\omega_2 t)$ and the stored energy will go as the square of that, $\exp(-2\omega_2 t)$. With this rate of change, we can write down the differential equation to describe the time evolution of the stored energy:

$$\frac{dU}{dt} = -2\omega_2 U. \tag{8.31}$$

Making the oscillation frequency complex will make an important change in the cavity admittance. If we assume that the losses are small and that we are driving the cavity very close to resonance, there will be a negligible effect on B'_c/G_0 but the relative conductance in (8.26) becomes

$$G'_c/G_0 = 1/\beta - 2\omega_2 Q_e/\omega_0. \tag{8.32}$$

From (8.31) we can solve for ω_2 and substitute it into the above equation to give

$$G'_c/G_0 = \frac{1}{\beta} + \frac{Q_e}{\omega_0 U}\frac{dU}{dt}. \tag{8.33}$$

This additional term is the correction needed to the admittance to account for the cavity being out of equilibrium.

Let us recompute the power flowing into the cavity using this more general form of the admittance. When driving the cavity on resonance, we find

$$P_{\text{in}} = \frac{\frac{4P_f}{Q_e}\left(\frac{1}{Q_0} + \frac{1}{\omega_0 U}\frac{dU}{dt}\right)}{\left(\frac{1}{Q_e} + \frac{1}{Q_0} + \frac{1}{\omega_0 U}\frac{dU}{dt}\right)^2}. \tag{8.34}$$

By conservation of energy, we know that the power flowing into the cavity goes into changing the level of the stored energy, dU/dt, or is dissipated in the cavity's loss mechanisms, $\omega_0 U/Q_0$, so we can write

$$P_{\text{in}} = \frac{dU}{dt} + \frac{\omega_0 U}{Q_0}. \tag{8.35}$$

By dividing through by $\omega_0 U$, we can rearrange this expression to match the form of one part of Equation 8.34:

$$\frac{P_{\text{in}}}{\omega_0 U} = \frac{1}{Q_0} + \frac{1}{\omega_0 U}\frac{dU}{dt}. \tag{8.36}$$

Applying this in (8.34) and solving for P_{in} gives a rather simple expression:

$$P_{\text{in}} = \sqrt{\frac{4P_f \omega_0 U}{Q_e}} - \frac{\omega_0 U}{Q_e}. \tag{8.37}$$

If we now eliminate P_{in} between (8.35) and (8.37), we find

$$\frac{dU}{dt} = \sqrt{\frac{4P_f \omega_0 U}{Q_e}} - \frac{U}{\tau_L}. \tag{8.38}$$

If there is no rf drive, the first term is zero and this reduces as expected to (8.4).

Equation 8.38 is complete but not intuitively helpful since it has U appearing in both terms on the right-hand side. Using

$$\frac{dU}{dt} = 2\sqrt{U}\frac{d\sqrt{U}}{dt}, \tag{8.39}$$

RF MEASUREMENTS

Equation 8.38 can be written as

$$\frac{d\sqrt{U}}{dt} = \frac{1}{2\tau_L}\left(\sqrt{U_0} - \sqrt{U}\right), \tag{8.40}$$

where U_0, defined as

$$U_0 \equiv \frac{4\tau_L{}^2 \omega P_f}{Q_e}, \tag{8.41}$$

is the equilibrium value of the cavity stored energy for the given level of rf drive. Since the amplitude of oscillation is proportional to $\sqrt{U}$, Equation 8.40 states that the rate at which the amplitude changes is proportional to how far the amplitude is from its steady-state value.

We have been focusing on the stored energy and the power getting into the cavity, but the reflected power — or more generally the reverse traveling power — also deserves close attention because it provides valuable information about the input coupling strength. By applying (8.25) in (8.23), we see that in steady state, the reflection coefficient is

$$\Gamma(\omega) = \frac{\beta - 1 - iQ_0\delta}{\beta + 1 + iQ_0\delta}, \tag{8.42}$$

where

$$\delta = \frac{\omega}{\omega_0} - \frac{\omega_0}{\omega}. \tag{8.43}$$

When driving the cavity exactly on resonance, this reduces to

$$\Gamma = \frac{\beta - 1}{\beta + 1}. \tag{8.44}$$

The circuit model we are using makes a definite choice for the phase of this reflection coefficient,[1] but in general, the phase of this coefficient will be determined by the type of input coupler used and the choice of reference plane position.

Since Equation 8.44 is only valid in steady state, we would like to find a general expression for the reflected power using the methods applied earlier — that is, $\omega \to \omega_1 + i\omega_2$. Most of the work is already done since for a cavity being driven on resonance we can find the reflected power through conservation of energy using the results obtained earlier. The reflected power is equal to the rf generator power less the power that enters the cavity. Using Equation 8.37, we obtain

$$P_r = P_f - P_{in} = P_f - \sqrt{\frac{4P_f\omega_0 U}{Q_e}} + \frac{\omega_0 U}{Q_e}. \tag{8.45}$$

With a little insight, we can recognize the right hand side as the square of a binomial:

$$P_r = \left(\sqrt{\frac{\omega_0 U}{Q_e}} - \sqrt{P_f}\right)^2 = \left(\sqrt{P_e} - \sqrt{P_f}\right)^2. \tag{8.46}$$

---

[1] The choice of phase used here is the "detuned short" position since the cavity–coupler system acts like a short circuit (−1) far from resonance.

This expression suggests that we can picture the reflected signal as a superposition of two signals. One is the direct reflection of the incident power signal and the other is the signal emitted from the cavity through the coupler.

8.3 CAVITY BEHAVIOR EXAMPLES

8.3.1 Steady State

Let us examine the most common solutions for the stored energy and the reflected power Equations 8.40 and 8.46. The first case is when P_f is a constant and the cavity is in steady state — that is,

$$\frac{dU}{dt} = 0. \tag{8.47}$$

Then the stored energy, as already stated, is constant: $U = U_0$.

Making use of the relationships between β, Q_e, Q_0, and Q_L, we can rewrite U_0 in terms of β explicitly:

$$U_0 = \frac{4\beta P_\text{f}}{(1+\beta)^2} \frac{Q_0}{\omega}. \tag{8.48}$$

In this way we see that U_0 is maximized when $\beta = 1$ and the cavity acts like a perfectly matched load. We can get the reflected power directly from the reflection coefficient (8.44) to find that

$$P_\text{r} = \left(\frac{\beta-1}{\beta+1}\right)^2 P_\text{f}. \tag{8.49}$$

We can use this expression to gain a method of finding the coupling strength by measuring P_r/P_f in steady state. Taking the square root of both sides we obtain

$$\pm\sqrt{\frac{P_\text{r}}{P_\text{f}}} = \frac{\beta-1}{\beta+1}. \tag{8.50}$$

Because β is always nonnegative, we must use the positive root when $\beta > 1$ and the negative root when $\beta < 1$. Solving for β and yields

$$\beta = \frac{1 \pm \sqrt{P_\text{r}/P_\text{f}}}{1 \mp \sqrt{P_\text{r}/P_\text{f}}}, \tag{8.51}$$

where the upper sign is used for $\beta > 1$ (overcoupled), and the lower sign is used for $\beta < 1$ (undercoupled). In Section 8.3.3, we will see how to determine if the coupler is over or under-coupled.

These relationships are essential for the experimental determination of β from the forward and reflected power measurements in the steady state.

CAVITY BEHAVIOR EXAMPLES

Table 8.1: Instantaneous value of emitted power for various values of β

β	P_e
0	0
$\frac{1}{3}$	P_r in steady state
1	P_f
∞	$4P_f$

8.3.2 Switch RF Off

When the rf drive, P_f, is abruptly turned off at $t = 0$, the instantaneous equilibrium value of the stored energy becomes zero ($U_0 = 0$) so that the differential equation (8.40) becomes simply

$$\frac{d\sqrt{U}}{dt} = -\frac{\sqrt{U}}{2\tau_L}. \tag{8.52}$$

This is easily solved, yielding the result

$$\sqrt{U(t)} = \sqrt{U_i} \exp\left(-\frac{t}{2\tau_L}\right). \tag{8.53}$$

This states that the field decays with a time constant $2\tau_L$. Squaring the above expression, we find our previous result:

$$U(t) = U(0) \exp\left(\frac{-t}{\tau_L}\right). \tag{8.54}$$

Now the reverse traveling power (8.46) consists only of the emitted power. If we let the cavity come to equilibrium before shutting off the rf, we can use (8.48) to write

$$P_r = P_e = \frac{\omega U(t)}{Q_e} = \frac{4\beta^2}{(1+\beta)^2} P_f \exp\left(\frac{-t}{\tau_L}\right). \tag{8.55}$$

Here P_f is the value of the rf drive just before it was turned off.

There are a few values of β that give some interesting results for the emitted power. These are shown in Table 8.1.

At time $t = 0$, the exponential in Equation 8.55 is unity, and this expression can be solved for β to yield the very useful expression

$$\beta = \frac{1}{2\sqrt{\frac{P_f}{P_e}} - 1}. \tag{8.56}$$

By abruptly shutting off the drive rf and measuring the ratio of the former drive rf to the instantaneous emitted power, we can find β rather simply. Note that this expression does not require any prior knowledge of whether the cavity is over- or undercoupled, unlike the steady state method in (8.51). Often, both the steady-state method and the instantaneous method are used as complementary measurements.

8.3.3 Switch RF On

Now let us start with an empty cavity and turn P_f on at $t = 0$. We expect U to rise until it reaches the equilibrium value U_0. The exact solution is found from the differential equation (8.40). To simplify the appearance of the expression, let us write and solve it in terms of the electric field in the cavity. Since $E \propto \sqrt{U}$ we can write

$$\frac{dE}{dt} = \frac{1}{2\tau_L}(E_0 - E). \tag{8.57}$$

With the boundary condition that $E(0) = 0$, we can readily find that the solution is

$$E(t) = E_0\left[1 - \exp\left(\frac{-t}{2\tau_L}\right)\right]. \tag{8.58}$$

This is just like the case of the decaying field in the previous section, except that the field is "decaying" to a level that is nonzero.

If we square both sides of the above expression and interpret E^2 as U, we obtain

$$U(t) = U_0[1 - \exp\left(\frac{-t}{2\tau_L}\right)]^2 \tag{8.59}$$

describing how the stored energy will rise when the rf is abruptly turned on.

To see how the reflection changes as the stored energy rises, we apply this solution in (8.46) to find

$$P_r = \left\{1 - \frac{2\beta}{1+\beta}\left[1 - \exp\left(-\frac{t}{2\tau_L}\right)\right]\right\}^2 P_f. \tag{8.60}$$

If $\beta > 1$, P_r will pass "through" zero and reach its nonzero steady-state value. When we consider again that the reverse traveling signal is composed of two opposing signals, as U rises, we are seeing the emitted signal, $\sqrt{P_e}$, canceling and then dominating the direct reflection ($\sqrt{P_f}$), resulting in a phase change of π radians for the reverse traveling signal. This fact becomes important when one is concerned about standing wave patterns in the drive system. For example, in the facility for testing 1.3-GHz cavities at Cornell, there is an rf window in the rf feed line (see Figure 8.4). The length of the input line is selected to position the rf window at voltage minimum in the standing wave that results as the cavity is filling. In addition to low-power, unity-coupled operation, this test apparatus is designed to be operated with $\beta > 10^4$ and a peak pulsed P_f of 2 MW. If the rf drive pulse is long enough, the reverse wave changes phase by 180°. The voltage minimum at the rf window becomes a maximum. To avoid this problem, the coupling and rf pulse length are adjusted so that the drive rf is turned off just as the reverse power goes to zero during the cavity fill.

8.4 RECTANGULAR PULSES

In practice, measurements of β are usually done by driving the cavity with a rectangular pulse. The pulse length is selected to be long enough so that it

FREQUENCY DOMAIN MEASUREMENTS 157

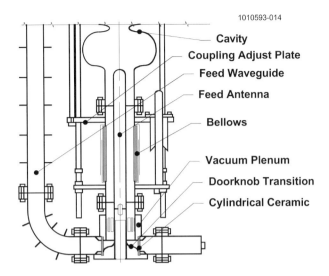

Figure 8.4: The 1.3-GHz high peak power test stand at Cornell. The input antenna length is selected so that when filling the cavity, an electric field minimum occurs at the ceramic window. When greatly overcoupled and applying high-power pulses, we make sure the input pulse is off before the reverse power changes phase and shifts the standing wave pattern.

drives the cavity to near equilibrium. Figure 8.5 shows what would be measured with an rf crystal detector and an oscilloscope for three different input coupling strengths. The stored energy is measured by measuring the transmitted power, P_t, via a weakly coupled transmitted power probe (see Figure 8.1). Although this is not measured directly, the reverse wave amplitude, V_r, is also shown in Figure 8.5 to illustrate the change of phase in the overcoupled case.

With the crisp turn on and off of the rf, the shape of the reflected power measurement alone is sufficient to permit the measurement of β even if the amplitude has no absolute calibration (but it does require a proper zero). From (8.60) we see that the first peak is proportional to P_f, the plateau is proportional to P_r (assuming the pulse is long enough), and the second peak is proportional to P_e. The ratios P_r/P_f and P_e/P_f needed for finding β in (8.51) and (8.56) can be measured directly from the oscilloscope trace.

8.5 FREQUENCY DOMAIN MEASUREMENTS

The finite cavity Q causes the resonance to be broadened in the frequency domain. We have already seen in (8.30) that the power getting into the cavity is a function of the drive rf. In steady state (and with one coupler), all the

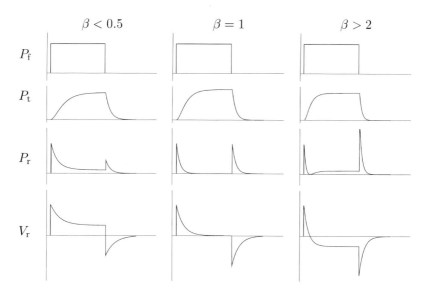

Figure 8.5: Rectangular drive pulses and their effects on the cavity for three different values of coupling. V_r is the reverse wave amplitude $\propto \sqrt{P_r}$.

power getting into the cavity is dissipated in cavity loss mechanisms:

$$P_{\text{in}}(\omega) = P_c(\omega) = \frac{\omega_0 U(\omega)}{Q_0}. \tag{8.61}$$

The above expression tells us that the stored energy has the same frequency dependence as P_{in}. Solving for the stored energy we find

$$U = \frac{4P_f/\omega_0}{(1/Q_L)^2 + (\omega/\omega_0 - \omega_0/\omega)^2}. \tag{8.62}$$

which is the familiar Breit–Wigner function (see Figure 8.6) describing an oscillator's response to a harmonic driving force. As we expect, when $\omega = \omega_0$, the stored energy is maximized. The frequency dependence is one way of measuring Q_L. Note that U is reduced to half its peak value when both squared terms in the denominator are equal. Letting $\Delta\omega = \omega - \omega_0$, this point occurs when

$$\Delta\omega = \pm\omega_0/(2Q_L) = 1/(2\tau_L). \tag{8.63}$$

The full width at half-maximum (FWHM) of the resonance is $1/\tau_L$. Another way of expressing this is that $Q_L = \omega/(2\Delta\omega)$. (Here $\Delta\omega$ is half the resonance width.)

In principle, this resonance could be mapped out using an adjustable frequency source and an rf detector sampling a small amount of the cavity's energy — for example, the transmitted power from a weakly coupled probe. In practice, the width of the resonance is measured by either a network analyzer or a spectrum analyzer.

FREQUENCY DOMAIN MEASUREMENTS 159

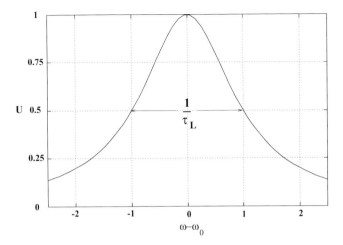

Figure 8.6: Classical shape of a resonance.

The network analyzer plays both the roles of the rf source and the rf detector. It can sweep through the desired frequency range and detect the response. The spectrum analyzer plays only the role of the detector, but it is able to resolve all the frequency components of its input. The cavity would be driven on resonance, and its transmitted power signal would be fed to the spectrum analyzer. Both devices would present a response versus frequency plot like that shown in Figure 8.6.

For a typical niobium superconducting cavity operating at a frequency of 1.3 GHz and a temperature of 1.5 K, $Q_0 = 1 \times 10^{10}$. With $\beta = 1$, $2\Delta\omega$ is only 2 Hz. This is far too narrow to measure.

In normal conducting cavities, on the other hand, $\Delta\omega$ can be measured since it is a few kilohertz; $\Delta\omega$ and ω_0 can then be used to deduce Q_L.

In order to properly interpret the loaded Q that is measured in the frequency domain, we need to measure the input coupling β. This can be done using the same network analyzer used to measure the Q_L. A simple way to do this is to measure P_r and P_f exactly on resonance and apply Equation 8.51 giving β in steady state. Recall that the choice of sign in (8.51) requires knowing whether the cavity is under- or overcoupled. One can tell this readily by sweeping the frequency and examining a polar plot of the reflection coefficient. Figure 8.7 shows the polar plots at three different couplings. The loops shown are mapped out by sweeping the frequency near the resonance. The angle of the plotted point is the phase of the reflected signal relative to the drive signal. The radius of the point is the magnitude of the ratio of the reflected power to the drive power. Far off resonance we would expect the radius to be equal to one if there were no cable attenuations. A point at the origin represents zero reflection.

One need only to observe the position of the loop plotted: If the loop encompasses the origin, the cavity is overcoupled; if it does not, the cavity is

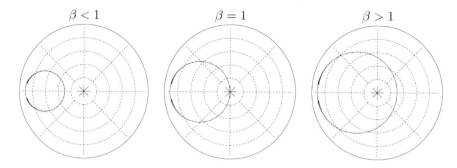

Figure 8.7: Polar plots of the reflection from a cavity under test for three different couplings as the frequency is swept across a span much greater than the FWHM. These plots are obtained by using Equation 8.42.

undercoupled. If it passes through the origin (the reflection goes to zero on resonance), the cavity is unity coupled.

8.6 RF EQUIPMENT AND ELECTRONICS

Because of the sharp resonance of a superconducting cavity, a phase-locked loop is required to keep the rf source operating at the peak of the cavity resonance. A schematic diagram of the rf equipment attached to a cavity is shown in Figure 8.8. Three types of power measurements are made: the forward power (P_f), the reverse power (P_r), and the transmitted power (P_t). By phase-lock we mean that the transmitted power and a sampling of the drive rf are fed into an rf mixer. The mixer generates a voltage proportional to the phase difference between the two input signals. The output of the mixer is then used to determine the frequency of the voltage-controlled oscillator (VCO). A phase shift is introduced into one of the input signals to the mixer so that the mixer generates zero voltage when the cavity is operating on resonance.

Steady-state power measurements are made using a power meter. As needed, the transient values are measured using rf detectors that convert the rf power to a voltage sent to an oscilloscope. Other equipment such as a spectrum analyzer might also be used for transient measurements.

A single-cell cavity is shown schematically in the context of its vertical test arrangement in Figure 8.9 [147]. As we will discuss further in Chapter 12, x-ray detectors may be used instead of thermometers to gain information about localized sites of cavity losses.

8.7 MEASURING Q_0 VERSUS E

To evaluate a cavity's performance, we measure the Q_0 as a function of the cavity's field level. The first measurement to be carried out is the input coupling

MEASURING Q_0 VERSUS E

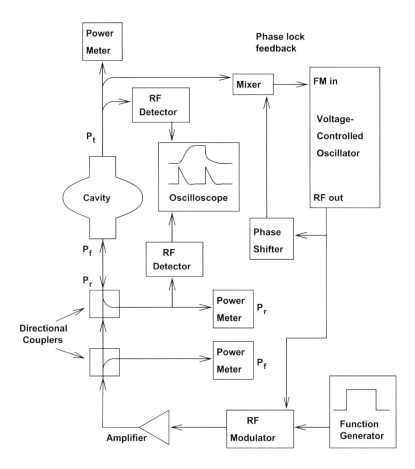

Figure 8.8: Schematic of the test equipment for rf measurements on a superconducting cavity.

β. As we illustrated before, we can measure this from the reverse power in steady state using (8.51) or in pulsed operation using (8.56). Unfortunately, the steady-state amplitude method requires knowing ahead of time whether the cavity is under- or overcoupled. The transient reflection will contain this information. If the reflection goes to zero before reaching a nonzero steady-state value, $\beta > 1$ (see Figure 8.5).

Often the step of determining β is made easy by adjusting the input coupler penetration so that the reflected power is zero, the cavity is matched, and $\beta = 1$.

Once β is known, the decay time, τ_L, is obtained by measuring of the exponential fall of the stored energy immediately after the input power is switched off. For this measurement either the reverse power or the transmitted power could be used since both couplers play equivalent roles in the absence of the rf drive.

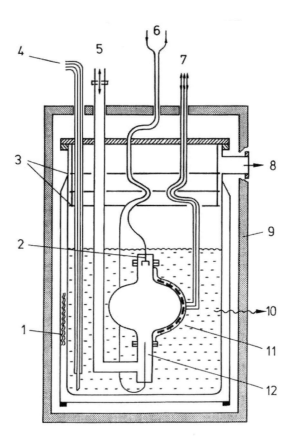

Figure 8.9: Test arrangement for a superconducting cavity. 1, superinsulation or liquid nitrogen shield; 2, rf output coupler and electron pick up probe; 3, thermal radiation shields; 4, liquid helium supply; 5, to UHV pumps and controls; 6, to rf drive and controls; 7, to diagnostic data acquisition system; 8, to helium gas pump system; 9, shielding against x-rays; 10, to radiation monitor; 11, surface thermometers or x-ray detectors; 12, rf input coupler. (Courtesy of Wuppertal.)

Once β and τ_L are known, we can obtain Q_0 via

$$Q_0 = \omega_0 \tau_L (1 + \beta). \tag{8.64}$$

As long as the losses in the cavity walls are proportional to the energy in the cavity, we expect Q_0 to remain constant as the field level in the cavity is raised. This is the case for ohmic losses. However, losses such as field emission will generally increase nonlinearly with the stored energy and hence will decrease the Q_0. Therefore it is important to measure the decay time at a field level that is low enough so that Q_0 is independent of the field level. Otherwise, the decay will not be a simple exponential, and the time constant cannot be measured this easily.

MEASURING Q_0 VERSUS E

We determine the field level in the cavity from the stored energy. Knowing the forward power and the coupling, we first determine the dissipated power in the cavity in steady state:

$$P_c = \frac{4\beta}{(1+\beta)^2} P_f. \tag{8.65}$$

The stored energy is then found from the definition of Q_0, giving

$$U_0 = \frac{P_c Q_0}{\omega}. \tag{8.66}$$

Or, as a short cut, we can express the stored energy directly in terms of the measured quantities:

$$U_0 = \frac{4\beta}{1+\beta} P_f \tau_L. \tag{8.67}$$

For any cavity geometry, the square of the peak surface electric field is proportional to the stored energy:

$$E_{pk} = \kappa_e \sqrt{U}. \tag{8.68}$$

Generally, for realistically shaped cavities, κ_e must be determined using computer codes as discussed in Chapter 2.

At the low-field level, before any nonohmic loss mechanisms are significant, it is also useful to simultaneously measure the transmitted power, P_t, and thereby calibrate the output probe. From

$$P_t = \frac{\omega_0 U}{Q_t} = \frac{U}{\tau_t} \tag{8.69}$$

we determine the calibration constant, τ_t. Since Q_t is held fixed during the experiment, hereafter a measurement of P_t will give E_{pk} through

$$E_{pk} = \kappa_e \sqrt{U/\tau_t}. \tag{8.70}$$

Recall that the transmitted power probe is coupled to the cavity very weakly to the cavity to minimize its effect on the measurements. Typically, $Q_t \approx 10^{13}$. Thus we can ignore its effect on the stored energy in the cavity.

Once this calibrated low field point is found, the rest of the Q_0 versus E curve can be mapped. The field level of the remaining points can be determined solely from a measurement of P_t and the proportionality constants determined above. Knowing U_0, Q_0 is determined by finding β from the reflected power, measuring the incident power, and applying

$$Q_0 = \frac{\omega_0 U_0 (1+\beta)^2}{4\beta P_f}. \tag{8.71}$$

8.8 STRONGLY COUPLED INPUT

If the input coupler is fixed so that Q_e is much lower than Q_0, the cavity losses play only a small role in determining the value of U. Making the rf measurements described above is then of little help in determining Q_0. As we shall see in Chapter 17, this situation may arise when a cavity is installed into an accelerator beam line. The input coupling is usually chosen to optimize the power delivery to the beam. In this case, Q_e is typically 10^6 and can even be as low as 10^5 for high beam current applications. In such a case, Q_e for the input coupler can be measured directly from the resonance bandwidth. Since $\beta \gg 1$, $Q_e \simeq Q_L$, the steady state stored energy in Equation 8.48 becomes

$$U_0 = \frac{4 Q_e P_f}{\omega}. \tag{8.72}$$

This means that the stored energy (and the cavity fields) can be determined directly from the forward power.

To determine Q_0 during the cold test, in the presence of the heavily coupled input probe, one can measure the dissipated power directly through the cryogenic losses. The stored energy is measured as described above. The helium boil-off is measured and the power losses are deduced (1 W of dissipated power = 1.4 liquid liters per hour of boil-off = 1.1 m^3/hr of helium gas at STP). It is helpful to calibrate the detector that keeps track of the helium boil-off against a resistive heater placed in the liquid helium bath with the rf power to the cavity turned off.

8.9 TEMPERATURE MAPPING

The rf measurements discussed so far give information only on the average behavior of the cavity, since we measure the total rf losses. To resolve the local distribution of energy losses from various mechanisms, one uses temperature mapping as a powerful diagnostic technique. As we will discuss in subsequent chapters, thermometry-based diagnostics have played a key role in improving the understanding of residual resistance, thermal breakdown, multipacting, and field emission.

In the first demonstration [148] a rotating chain of carbon resistors positioned a few millimeters away from the cavity wall successfully determined the location of a thermal breakdown in a niobium cavity. Since this pioneering demonstration, the technique has gone through many generations of improvement. The bakelite insulation of the resistors (56 or 100 ohm, 1/8 Watt, Allen–Bradley) is ground off to increase thermometer sensitivity, the resistors are held in contact with the cavity wall using beryllium-copper spring fingers, and a computer-based data acquisition system is used to read-out the many resistors and to control the angular position of the rotating arm that holds the thermometers. (See Figure 8.10 [147].)

TEMPERATURE MAPPING

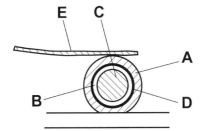

A: copper tube housing
B: bakelite insulation
C: carbon-filled body of resistor
D: gap filled with conduction silver
E: copper beryllium spring

(a) Cross section of a carbon thermometer.

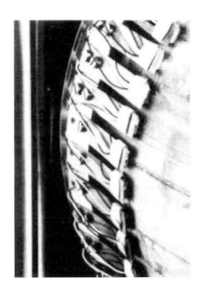

(b) Rotating arm on a 350-MHz cavity.   (c) Close-up of rotating arm.

Figure 8.10: Rotating thermometry system used at CERN. (Courtesy of CERN.)

In a clever advance, the sensitivity of the technique was increased by maintaining the helium bath in a subcooled state. In this state the bath temperature is held slightly above the superfluid helium temperature (2.17 K) by pressurizing to about one atmosphere. Bubbles are suppressed so that the cooling of the cavity wall by nucleate boiling is substantially reduced. For a given rf heat input, the temperature of the cavity wall is enhanced, making it possible to detect the heating due to defects, multipacting or field emission at lower field levels than is possible in nucleate boiling helium. This technique is particularly suitable for low-frequency cavities (e.g., $f < 500$ MHz) and for accelerating field levels < 10 MV/m because the operating temperature is usually near 4.2 K. Figure 8.11 shows a typical temperature map with line heating patterns due to impacting electrons from field emission [149].

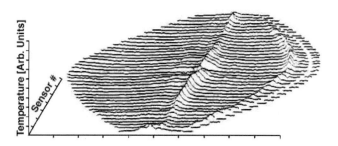

Figure 8.11: Temperature map acquired using the CERN rotating arm thermometry system. The stripes of heating are due to the impact of field emitted electrons. (Courtesy of CERN.)

For higher-field studies and especially for higher rf frequency (e.g., $f > 1$ GHz), operation in superfluid helium is essential. In subcooled helium the low heat transfer coefficient of the bath as well as the higher operating temperature can force the rf surface temperature to increase unstably at high fields due to BCS losses. As an added benefit, spatial resolution of superfluid thermometry is increased by a factor of 2.5, so that the FWHM of thermometer signals is reduced from centimeters to a few millimeters. However, because of the excellent cooling of the superfluid helium in which the thermometers are immersed, the efficiency of the thermometers[2] is greatly reduced, from near 100% of the wall temperature to about 20–30%. In addition, the thermometers have to be held fixed by the use of a thermally conducting grease, such as Apiezon N. Consequently, a large number of thermometers are required to cover the entire cavity (e.g., 700 for a single-cell 1.5-GHz cavity). One benefit of this extensive coverage is that the time to acquire a temperature map is reduced to a fraction of a second from the usual 30 minutes required for the rotating system. Therefore this technique is a good research tool, but it is not practical for diagnostics on multicell structures. Movable thermometers for use in superfluid have also been developed, but their sensitivities are only a few percent. These have been used successfully for multicell structure performance diagnostics.

To achieve higher-sensitivity thermometers in superfluid helium, it is important to bring the exposed heat-sensitive carbon element in close proximity to the cavity wall, as shown in Figure 8.12 [150]. Only a thin electrically insulating coating of GE varnish is used. It is equally important to isolate the rest of the thermometer from the superfluid by special epoxy-based sealants that are impervious to superfluid. The intimate contact between the cavity wall and the sensor is improved by using a beryllium–copper spring-loaded contact pin.

Figure 8.13 shows a system developed at Cornell [150] for 1.5-GHz single-cell cavities. It consists of 756 thermometers which, at 1.6 K, can detect a

---

[2]The thermometer efficiency is defined as the ratio of the detected temperature increase to the actual wall temperature rise.

TEMPERATURE MAPPING

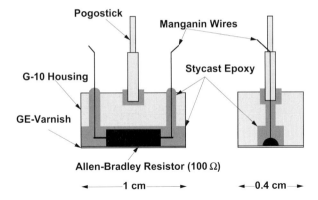

Figure 8.12: The "pogo-stick"-type thermometer used at Cornell.

Figure 8.13: Fixed thermometry system used at Cornell.

temperature rise of 0.2 mK with a scan time of 0.1 s. Increasing the scan time increases the sensitivity; in a 2.5-s scan, signals of 60 μK above ambient can be resolved.

8.9.1 A Cavity Test Using Thermometry

To illustrate some of the information derived from a cavity test, we turn to the results of a typical experiment. The performance of this cavity was mediocre, mainly because it did not receive a fresh chemical etch prior to this test. The cavity was just rinsed with methanol, even though it was exposed to room air for many months since its rf test.

The Q_0 versus E_{pk} curve (Figure 8.14) was obtained by the method outlined previously. In addition, temperature maps were taken at regular intervals, to

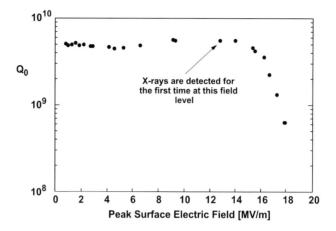

Figure 8.14: Illustrative Q_0 versus $E_{\rm pk}$ curve obtained during the test of a mediocre cavity.

monitor the heat distribution as a function of field.

At low fields, Q_0 is about 5×10^9. Using the fact that the geometry factor for this cavity is $G = 270\ \Omega$, we find that the average surface resistance is

$$R_{\rm s} = \frac{G}{Q_0} = 54\ {\rm n}\Omega. \tag{8.73}$$

Since Q_0 is roughly constant at low fields, we expect to only find Joule heating below 12.5 MV/m in this example. The losses scale as $E_{\rm pk}^2$ just as the stored energy. This observation is indeed borne out by the temperature data. Figure 8.15 shows a typical temperature map taken at $E_{\rm pk} = 4.5$ MV/m. The map is a flattened view of the cavity. The irises are at the top and bottom of the map, and the equator of the cavity runs horizontally along the line marked by thermometer #10. Three sites of significant heating were detected and they display characteristics consistent with Joule heating, probably arising from magnetic field induced losses. This is especially likely for sites 1 and 2 which are near the equator, where the magnetic field is high.

A log–log plot of the heating at site 2 versus $E_{\rm pk}$ (Figure 8.16) yields a straight line with slope 2.33 ± 0.05, confirming nearly ohmic behavior. If the measured temperature rise were due purely to Joule heating, this slope should be exactly 2. Hence there are indications that additional, unexplained heating processes were active. In later chapters, we will discuss the application of thermometry to the study of other types of losses.

TEMPERATURE MAPPING

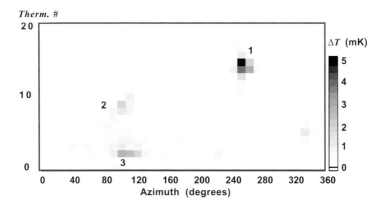

Figure 8.15: Typical temperature map showing heating sites.

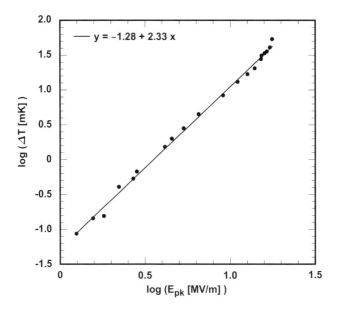

Figure 8.16: The electric field dependence of the heating at a particular site (number 2) on a 1.5-GHz cavity. The slope is very close to 2, indicating ohmic heating.

CHAPTER 9

Residual Resistance

9.1 INTRODUCTION

As we saw in Chapter 3 and Chapter 4, below a certain temperature the experimentally observed surface resistance is higher than the BCS prediction. This is accounted for by a residual (temperature-independent) resistance R_0:

$$R_\text{s} = R_\text{BCS}(T) + R_0. \tag{9.1}$$

The operating temperature of a superconducting cavity is usually chosen so that the BCS resistance is reduced to an economically tolerable value. For example, for a 1.5 GHz cavity, it is necessary to cool the cavity to below 2 K, at which temperature the BCS resistance becomes less than 20 nΩ. Residual losses can arise from several sources. For example, residues from chemical etching, foreign material inclusions, or condensed gases are common. Therefore, cleanliness during forming, welding, and surface preparation are essential. Successive reprocessing of the same cavity may show variations in R_0, indicating that we do not yet know how to control all the important variables.

9.2 TYPICAL RESIDUAL LOSSES

A well-prepared niobium surface can reach a residual resistance of 10 to 20 nΩ, corresponding to Q_0 values of about 2×10^{10}. The record low values are in the range of $1-2$ nΩ [151]. Figure 9.1 shows the distribution of R_0 values of 1-cell, 1.5-GHz cavities prepared by standard chemical treatment.

The temperature map of Figure 9.2 shows the distribution of sources of residual loss in a single-cell, 3–GHz cavity prepared by standard chemical treatment [152]. The temperature map was taken at 1.6 K and $H_\text{pk} = 1000$ Oe, just below the quench field of 1030 Oe. Since there was no detectable field emission in this test, the map is an excellent indicator of the residual loss sources. The low field Q_0 was 10^{10}, but dropped to 5×10^9 at 1000 Oe. This shows that residual losses can be field-dependent. At the equator, a region of high magnetic field, we see a high-resistance spot, which we are accustomed to calling a "defect". Such a defect is ultimately the cause of thermal breakdown. There are also several smaller resistive spots at the equator. At the upper iris, a region of high

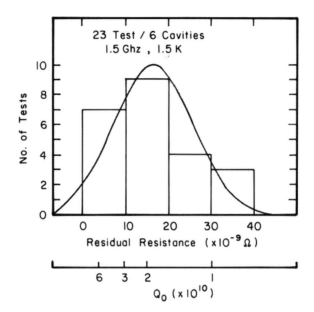

Figure 9.1: Statistical distribution of residual Q_0 and corresponding residual surface resistance for 23 tests on six single-cell, 1.5 GHz niobium cavities prepared by standard chemical treatment.

electric field, we see sources of dielectric loss. Over the remainder of the surface the residual losses are uniformly quite low. Knowing the sensitivity of the thermometers from separate calibrations, it is possible to estimate the local surface resistance to an accuracy of about 1 nΩ. The measured surface resistance in the low loss regions is found to be 4 n$\Omega \pm$ 1 nΩ, corresponding to a Q_0 of 7×10^{10}. At the same temperature (1.6 K), the BCS losses are also 4 nΩ. Therefore the residual loss over a large part of the cavity surface is measured by thermometry to be less than 1 nΩ, the accuracy limit of temperature measurement.

Although other basic causes for residual resistance have been suggested, such as direct phonon generation by rf fields [153], the uniformly low losses detected in this temperature map suggest that such basic mechanisms must contribute less than 1 nΩ at 3 GHz.

In some cases, the reason for residual losses may be extraneous to the superconducting surface — for example, a bad joint design, or excessive losses in a coupler. Joints at cavity flanges can contribute significantly if these are placed in a high field region. Indium wire is frequently used as the joining material. The superconducting transition temperature of indium is 3.4 K, so that the wire as well as the contact regions may add losses. To avoid joint losses in a velocity-of-light structure, the niobium beam tube length between the high field cell and the end flange should be made sufficiently long (see Problem 29) that the fields decay to acceptably low values.

TRAPPED MAGNETIC FLUX

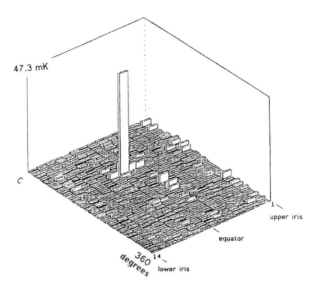

Figure 9.2: Temperature map on a single cell, 3-GHz cavity showing the distribution of residual losses. There was no field emission and the map is taken at a high field (1000 Oe) (Courtesy of Wuppertal.)

9.3 TRAPPED MAGNETIC FLUX

A well-understood and controllable source of residual loss is trapped dc magnetic flux from insufficient shielding of the earth's magnetic field, or other dc magnetic fields in the vicinity of the cavity. To get the highest Q_0 a superconducting cavity must be well-shielded from the earth's field. Under ideal conditions, if the external field is less than H_{c1}, the dc flux will be expelled from the bulk of a superconductor due to the Meissner effect. However, if there are lattice defects or other inhomogeneities in the material, the flux lines may be "pinned," and trapped within the material. It is suspected [154] that the oxide layer on the niobium surface serves as a source for flux pinning sites. The flux comes through the cavity wall in current vortices which contain single quanta of magnetic flux, as discussed in Chapter 5. The dc field decays in a distance λ_L from the center of a normal core of size $\approx \xi_0$. Experimental studies show that when niobium cavities are cooled down in the presence of a small dc magnetic field (e.g., less than 1 Oe) all the flux within the volume of the cavity is trapped [154].

We can estimate the rf losses due to 100% trapped flux. The trapped field H_{ext} over an area A breaks up into N fluxoids, each with a quantum of flux Φ_0 such that

$$AH_{\text{ext}} = N\Phi_0, \tag{9.2}$$

where Φ_0 is the flux quantum.

As shown in Figure 9.3, at the center of the normal core of these flux tubes, the superconducting order parameter vanishes. Over a distance ξ_0, the order pa-

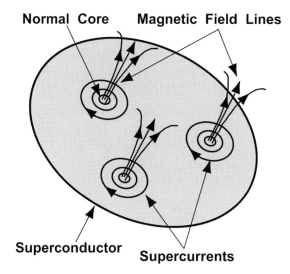

Figure 9.3: Field lines and supercurrents for magnetic flux trapped in a superconductor.

rameter reaches its maximum value. Therefore the contribution to the residual resistance from N fluxoids is the normal state resistance R_n times the fraction of the normal-conducting area.

$$R_\mathrm{mag} = N \frac{\pi \xi_0^2}{A} R_\mathrm{n} = \frac{H_\mathrm{ext} \pi \xi_0^2 \mu_0 R_\mathrm{n}}{\Phi_0}. \tag{9.3}$$

According to the theory of Type II superconductors, the upper critical field (discussed in Chapter 5) is given by

$$H_\mathrm{c2} = \frac{\Phi_0}{2\pi \mu_0 \xi_0^2};$$

therefore

$$R_\mathrm{mag} = \frac{H_\mathrm{ext}}{2 H_\mathrm{c2}} R_\mathrm{n}. \tag{9.4}$$

Using $H_\mathrm{c2} = 2400$ Oe, and the classical normal state surface resistance for niobium with $RRR = 300$, $R_\mathrm{n} \approx 1.5$ mΩ at 1 GHz,

$$R_\mathrm{mag} = 0.3(\mathrm{n}\Omega) H_\mathrm{ext}(\mathrm{mOe}) \sqrt{f(\mathrm{GHz})}. \tag{9.5}$$

The $\sqrt{f}$ dependence is justified because the normal core size is of order ξ_0 and comparable to the mean free path. Experimental results indicate that the coefficient in Equation 9.5 is about 1 mΩ. If a 1-GHz superconducting cavity is not shielded from the earth's magnetic field, the maximum Q_0 value could be limited to below 10^9.

9.4 RESIDUAL LOSSES FROM HYDRIDES

An interesting consequence of Equation 9.4 is that cavities made of a thin film of niobium on copper are much less sensitive to the ambient dc magnetic field because the upper critical field of sputtered, fine-grained niobium is between 15,000 and 35,000 Oe [155]. Therefore the Q_0 of a Nb/Cu cavity is up to a factor of ten less sensitive to the external dc magnetic field than a niobium cavity made from bulk sheet metal.

9.4 RESIDUAL LOSSES FROM HYDRIDES

An important residual loss mechanism arises when the hydrogen dissolved in bulk niobium precipitates as a lossy hydride at the rf surface [136]. This residual loss, also known as the "Q disease," is a subtle effect that depends on many factors besides hydrogen concentration. In particular, the rate of cool-down and the amount of other interstitial impurities present play important roles.

If the bulk hydrogen concentration in niobium exceeds 2 ppm by weight, there is a clear danger of hydride formation at the rf surface during cool-down, resulting in a high residual surface resistance. The effect can be severe enough to lower the Q_0 to 10^8 depending on the amount of hydrogen dissolved and the cool-down rate of the cavity [136], as shown in Figure 9.4.

As delivered, commercial niobium typically has less than 1 wt ppm of dissolved hydrogen because the material in its final form is usually annealed for recrystallization. But hydrogen concentration can increase during chemical etching, especially if the temperature of the acid etch is allowed to rise above 20 °C or if hydrogen bubbles are not allowed to escape freely. This is often the case because most cavities form closed vessels during acid etching. At 10 wt ppm, hydride precipitation and high residual loss are certain for a high-purity niobium cavity, even if fast cool-down is attempted.

According to the phase diagram of the Nb–H system [77], the required concentration of hydrogen to form the hydride phases is very high at room temperature (4.6×10^3 wt ppm for the θ phase and 7.5×10^3 wt ppm for the ϵ phase). Therefore these phases do not form. As the temperature is lowered, the hydrogen concentration needed to form the hydride phases decreases. Above 150 K the danger of hydride formation is still not very serious, because the concentration required is still relatively high. Hence a cavity can be cooled as slowly as desired to 150 K. As the temperature is lowered below 150 K, the hydrogen concentration required to form the hydride phases decreases to a dangerous level, so that islands of the hydride phase may form even when the concentration is as low as 2 wt ppm. The hydride precipitates at favorable nucleation sites. If these are at the surface they increase the residual loss. Below 150 K, the low critical hydride formation concentration poses a great danger of high residual loss. Also, the diffusion rate of hydrogen between 150 and 60 K remains quite significant, so that hydrogen can move to accumulate to critical concentrations at nucleation sites. Only when the temperature is reduced to below 60 K does the diffusion of hydrogen slow down enough that hydrogen can no longer accumulate at hydride centers.

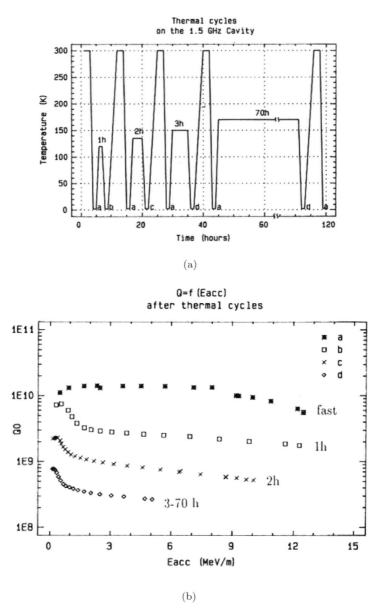

Figure 9.4: (a) Temperature cycling history of a 1.5-GHz niobium cavity. (b) Q_0 values measured after each cycle shown in (a). If the cavity is cooled rapidly there is no Q problem. The Q disease grows more serious with longer time spent in the dangerous temperature region. Not only does the Q_0 fall, but there is a marked degeneration of Q_0 with increasing field. (Courtesy of Saclay.)

Therefore when a cavity with a large bulk hydrogen concentration is cooled to liquid helium temperature, the length of time it is kept between 150 and 60 K will determine the extent of hydride formation and the accompanying residual losses. Figure 9.4 shows progressively degraded Q_0 values obtained when the cavity is held for significant length of time at $T = 120$–170 K. A characteristic signature of the Q disease is the marked increase of resistance with field. This could be due to the fact that one of the hydride phases is a superconductor at low field [136]. Another effect related to the Q disease is that when a cavity with a freshly etched surface is cooled down for the first time, the Q_0 degradation is not observed, even if the cool-down is slow through the dangerous region. We attribute this benefit to the lack of nucleation centers which become numerous after the hydride phase precipitates during the first thermal cycle.

The Q disease can be mitigated by rapid cooling of the cavity through the dangerous temperature regime. But the best remedy is to degas the hydrogen entirely by heating the niobium cavity in vacuum ($p < 10^{-6}$ torr) at 700 °C to 900 °C. Although the danger of oxygen pickup at these temperatures is minimal, it is advisable to protect the RRR of the niobium by surrounding the outside of the cavity with titanium wrap, as discussed in Chapter 6.

An interesting aspect of the Q disease is that niobium with high oxygen content (> 100 wt ppm) is not affected by the hydride-induced residual loss problem. The many interstitial oxygen sites serve as hydrogen traps, so that hydrogen cannot accumulate elsewhere in sufficient concentrations to form the hydride. Unfortunately, this implies that greater care must be taken to avoid the hydrogen disease as we continue to raise the RRR of niobium cavities to avoid thermal breakdown.

9.5 RESIDUAL LOSS FROM OXIDES

There have been many speculations that the natural oxide layer of niobium, typically 5–10 nm thick, is an important contributor to residual resistance. The loss tangent of Nb_2O_5 was estimated [156] to be $\tan(\delta) < 10^{-6}$ from a niobium cavity with a 40-nm oxide layer produced by electrolytic anodization. Systematic experiments [97] now place a lower limit of 1–2 nΩ for the oxide contribution. In this study, 8.6-GHz cavities were fired at 1400 °C to dissolve the oxide layer into the bulk. The dissolution of the layer was confirmed by companion Auger studies on similarly treated niobium samples. RF tests on the oxide free cavities showed a residual loss of 5–10 nΩ — i.e., comparable to typical niobium cavities with their natural oxide layer. Further, when the oxide layer was regrown by controlled exposure to clean oxygen (0.1 torr for 2–16 hours), the increase in rf loss was less than 1–2 nΩ. The fact that the oxide was indeed restored was confirmed by reabsorbing the oxide into the niobium at 300 °C and detecting the presence of the corresponding oxygen in the bulk through the BCS surface resistance at 4.2 K, which decreased by the amount expected from the lower electron mean free path. When the oxide free cavities were heated to 300 °C, they did not show any change in BCS resistance, confirming the absence of oxygen in the rf surface.

An interesting result that emerged from this 8.6-GHz study was the reduced sensitivity of oxide-free cavities to residual loss from trapped magnetic flux when the cavities were cooled in the presence of external fields of a few 100 mOe. The observed sensitivity was $0.3 - 0.6$ nΩ/mOe compared to the sensitivity of 3 nΩ/mOe for oxidized cavities. Presumably, the removal of the oxide by heat treatment to 1400 °C eliminated the flux trapping centers, so that the flux was able to freely move out of the bulk when the cavity was cooled below T_c.

CHAPTER 10

Multipacting

10.1 INTRODUCTION

Multipacting in rf structures is a resonant process in which a large number of electrons build up spontaneously absorbing rf power so that it becomes impossible to increase the cavity fields by raising the incident power. The electrons collide with structure walls, leading to a large temperature rise and eventually, in the case of superconducting cavities, to thermal breakdown. Although multipacting can be avoided in modern $\beta = v/c = 1$ cavities by selecting the proper cavity shape, it was a major performance limitation in the past. At present, multipacting still plagues many types of rf vacuum structures, such as low-β cavities, couplers, transmission lines, and rf windows.

Although our description of multipacting will be for a few select rf structures, the principles we discuss apply to the general problem of multipacting in other rf components as well. However, one should bear in mind that multipacting is a complex phenomenon, so that the general principles we will develop cannot be applied to every situation.

10.2 EXPERIMENTAL OBSERVATION OF MULTIPACTING IN CAVITIES

During the test of a superconducting cavity, such as the one shown in Figure 1.6 or the muffin-tin cavity (Figure 1.10), the onset of multipacting is usually recognized when the field level in the cavity remains fixed, as if a barrier were present, even as more rf power is supplied. In effect, the Q_0 of the cavity abruptly reduces at the multipacting threshold, as shown in Figure 10.1. When multipacting is encountered, the behavior of the forward, transmitted, and reflected power in a unity coupled cavity ($\beta = 1$) is depicted in Figure 10.2.

In many cases, it is found that such a multipacting barrier can be surmounted by processing. This is done by allowing multipacting to progress for several minutes, while slowly raising the rf power. Eventually, and sometimes abruptly, the Q_0 improves and the multipacting ceases. Often, further multipacting barriers appear at higher fields; they may process as well. Barriers which can be processed are called *soft* barriers, whereas others, which persist, are known as

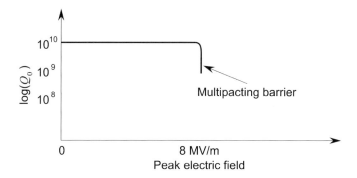

Figure 10.1: Q_0 versus E_{pk} curve for a superconducting cavity when multipacting is encountered.

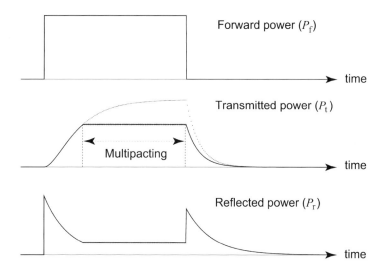

Figure 10.2: Power traces for a unity coupled cavity experiencing multipacting. The dotted curve in the case of the transmitted power indicates the levels that would be achieved if multipacting were not active.

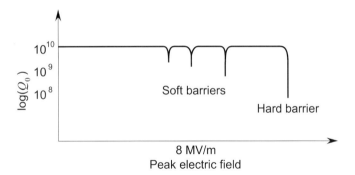

Figure 10.3: Q_0 versus E_{pk} curve for a superconducting cavity when several multipacting barriers are encountered while the fields are raised for the first time.

hard barriers. Thus, when raising the fields in a cavity for the first time, the Q_0 versus E_{pk} curve may look like Figure 10.3. Subsequently, soft multipacting barriers which have been processed do not reappear, provided that the cavity is kept under vacuum.

A processed soft barrier may reappear after the cavity is exposed to air, which indicates that multipacting is strongly dependent on the condition of the first few monolayers of the rf surface. Surface adsorbates can also strongly affect the multipacting behavior.

10.3 MULTIPACTING BASICS

The accepted mechanism for multipacting is as follows: An electron is emitted from one of the structure's surfaces. This may be precipitated by a cosmic ray, photoemission, or an impacting field emission electron. The emitted electron is accelerated by the rf fields and eventually impacts a wall again, thereby producing secondary electrons. The number of secondary electrons depends on the surface characteristics and on the impact energy of the primary. In turn, the secondaries are accelerated and, upon impact, produce another generation of electrons. The process then repeats. The electron current increases exponentially if the number of emitted electrons exceeds the number of impacting ones and if the trajectories satisfy specific resonance conditions which we will discuss shortly. The increase in current is limited only by the available rf power and by space-charge effects.

To determine the type of trajectories which can result in multipacting, consider an electron being emitted at position $\mathbf{x}_0$ on the cavity wall. The rf fields vary as
$$\mathbf{E}(\mathbf{x}, t) = \mathbf{E}(\mathbf{x}) \sin \omega_\mathrm{g} t$$
and
$$\mathbf{H}(\mathbf{x}, t) = \mathbf{H}(\mathbf{x}) \cos \omega_\mathrm{g} t.$$

Here ω_g is the drive frequency of the cavity (usually $\omega_g \approx \omega_0$) and we denote the phase $\omega_g t$ at the time of electron emission by φ_0. The electron is accelerated until it impacts with the wall at some other point $\mathbf{x}_1$ at a phase φ_1. Adopting the simplified view that the impact energy (K) alone determines the number of secondaries emitted, then $\delta(K)$ new electrons are emitted, where $\delta(K)$ is the secondary emission coefficient (SEC), which is material dependent. The secondaries follow another trajectory to an impact point $\mathbf{x}_2$ at phase φ_2, provided the phase φ_1 is such that $\mathbf{E}(\mathbf{x}_1, \varphi_1)$ points towards the surface; otherwise the electrons do not escape from the cavity wall. Following the impact at $\mathbf{x}_2$ the process repeats. Thus, the initial emission at $\mathbf{x}_0$ is associated with a series of impact points

$$\mathbf{x}_1, \mathbf{x}_2, \mathbf{x}_3, \ldots$$

and impact phases

$$\varphi_1, \varphi_2, \varphi_3, \ldots$$

The process ends at step k if $\mathbf{E}(\mathbf{x}_k, \varphi_k)$ prevents secondaries from escaping the wall. If this does not occur for any impact location, then the initial emission site $\mathbf{x}_0$ can *potentially* lead to multipacting.

Whether multipacting does indeed take place depends on the wall material, which determines the functional form of $\delta(K)$. The number of electrons N_e emitted after the kth impact is

$$N_e = N_0 \prod_{m=1}^{k} \delta(K_m), \qquad (10.1)$$

where N_0 is the number of electrons initially emitted at $\mathbf{x}_0$ and K_m is the kinetic energy at the mth impact. For multipacting we require that $N_e \to \infty$ as $k \to \infty$. Obviously this is satisfied if $\delta > 1$ for *all* impact sites, although other scenarios are conceivable. Knowledge of the SEC of the wall material is thus critical for predicting multipacting.

10.4 SECONDARY ELECTRON EMISSION

For most materials $\delta(K)$ exceeds unity in a range from a few tens of electron-volts to a few thousand electron-volts. Relativistic effects therefore have little influence on multipacting. The general form of $\delta(K)$ for many materials is similar to that shown in Figure 10.4.

Although $\delta(K)$ varies from material to material, the general shape of the function has a simple explanation. An impacting (primary) electron loses most of its energy by interacting with electrons in the wall, where the number of charges it interacts with is proportional to the impact energy. At low energy, the primary loses all its energy within a thin surface layer. The charges it interacts with have no trouble escaping from the material, and the number of secondaries is proportional to the energy of the primary. Since all the interactions occur at the surface, the SEC is very sensitive to adsorbates at the surface.

SECONDARY ELECTRON EMISSION

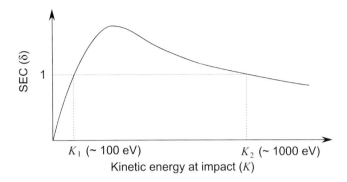

Figure 10.4: Generic dependence of the secondary emission coefficient (δ) on the impact kinetic energy (K).

Table 10.1: Secondary emission coefficient for a variety of materials

Material	δ_{max}	K_{max} (eV)	K_1 (eV)	K_2 (eV)
Ag	1.5	800	200	> 2000
Al	1.0	300	300	300
Au	1.4	800	150	50
C (diamond)	2.8	750		> 5000
C (graphite)	1.0	300	300	300
C (soot)	0.45	500	None	None
Cu	1.3	600	200	1500
Fe	1.3	400	120	1400
Nb	1.2	375	150	1050
Pb	1.1	500	250	1000
Ti	0.9	280	None	None
Al_2O_3 (layer)	2–9			
MgO (layer)	3–15			

Source: *CRC Handbook of Chemistry and Physics*, 65th edition.

The SEC increases with energy until the primary penetrates the wall so far that the secondaries produced at the greater depths no longer escape. Raising the impact energy any further increases the primary's penetration depth, reducing the fraction of secondaries that escape the wall. Hence $\delta(K)$ begins to decline. The peak in $\delta(K)$ usually occurs at a few hundred electron-volts, and the points at which $\delta(K)$ passes through one (if at all) are denoted the first and second crossovers (K_1 and K_2).

There is a large material dependent variation in $\delta(K)$. For example, aluminum oxides, which are often used in rf windows, have a high SEC, whereas pure aluminum has a relatively low SEC. Table 10.1 gives some SEC parameters for a number of materials, including the maximum SEC (δ_{max}), which occurs at an impact energy K_{max}.

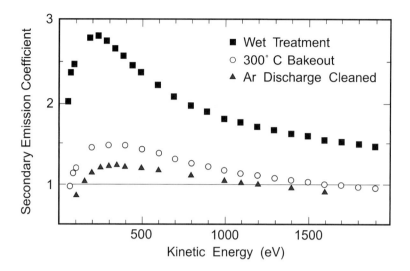

Figure 10.5: Secondary emission coefficient for niobium following various treatments [157].

Even for different samples of the same material $\delta(K)$ can vary significantly, and Table 10.1 only applies to very clean samples. Since the emission of charges occurs at the surface, sample preparation is critical in determining the SEC. This is illustrated in Figure 10.5, which shows $\delta(K)$ for niobium following several different treatments. We see that the wet treatment enhances secondary emission over the entire measured range of impact energies.

Adsorbates that can enhance secondary emission include hydrocarbons (from pump-oil vapors) and lubricants used in polyethylene [158, 159]. Polyethylene bags are therefore inappropriate for storage of rf components susceptible to multipacting.

Since many of the adsorbates are only a few monolayers thick, they can be desorbed by electrons impacting the surface. In many cases, continuous multipacting serves to lower $\delta(K)$, which explains why some multipacting barriers are soft.

10.5 COMMON MULTIPACTING SCENARIOS

A common situation arises if the charges return to their point of origin (or very near to it) after only a small number of impacts. This situation is referred to as n-point multipacting, where n is the number of impact sites along the electron trajectories. The requirement placed on the phases is $\varphi_n = \varphi_0$ in order for multipacting to occur.

COMMON MULTIPACTING SCENARIOS   185

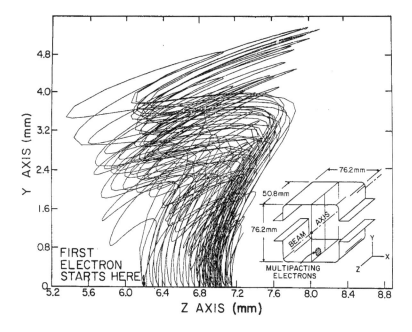

Figure 10.6: One-point multipacting trajectories in a muffin-tin cavity [160].

10.5.1 One-Point Multipacting

The most frequent type of multipacting in $\beta = 1$ cavities used to be one-point multipacting, where the charges impact the cavity wall at, or very near, the emission site itself. Emitted electrons are accelerated primarily by the electric field perpendicular to the rf surface while, simultaneously, the surface magnetic field forces the electrons along quasi-cyclotron orbits, so that they return to their point of origin. To satisfy $\varphi_1 = \varphi_0$ in this simplified model, the rf period must be an integer multiple of the cyclotron period. Furthermore, the requirement that $\delta(K) > 1$ must be fulfilled for the impact energy. One-point multipacting trajectories in muffin-tin cavities are depicted in Figure 10.6.

The number of rf periods required by an electron to return to the point of origin denotes the multipacting *order*. The trajectories for first-order multipacting therefore roughly follow a simple closed orbit, second-order trajectories are a figure eight, and so on. Such trajectories are depicted in Figure 10.7 for the first three orders. Multipacting orbits are usually well-localized compared to the overall cavity size, because the impact energy must be less than a few thousand electron-volts so that $\delta(K) > 1$. This simplifies the analysis of multipacting in some cases.

If we assume that the charges follow simple cyclotron orbits, then the orbit frequency scales as

$$\omega_c \propto \frac{\mu_0 H e}{m}, \tag{10.2}$$

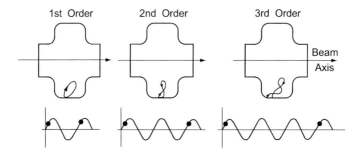

Figure 10.7: Typical one-point multipacting trajectories for orders one, two, and three.

where e and m are the positron charge and mass, respectively, and H is the magnitude of the local magnetic field. If multipacting of order n is to occur, we need

$$\omega_g = n\omega_c, \qquad n = \text{integer}.$$

The fields at which multipacting *may* occur thus scale as

$$H_n \propto \frac{m\omega_g}{n\mu_0 e}. \tag{10.3}$$

Equation 10.3 predicts an infinite number of multipacting barriers up to a maximum field proportional to $m\omega_g/\mu_0 e$. However, the number of barriers is limited by the requirement that $\delta(K)$ has to exceed unity at the impact energy. We also see from (10.3) that low-frequency structures and $\beta < 1$ cavities, which also tend to be at low frequency, are more likely to be plagued by multipacting at low fields than high-frequency structures.

The magnetic field, of course, varies throughout a cavity, so the question arises: To which field does (10.3) apply? Generally, multipacting will only occur in a region of the cavity where H does not vary much along the cavity wall and stable trajectories are possible. This usually occurs in the high-magnetic-field regions near the equator, where H approaches H_{pk}, so that the fields given by (10.3) apply to the peak cavity fields.

Given this simple model where the electrons are accelerated by the electric field while the magnetic field turns them around, the impact energy scales as

$$K \propto \frac{e^2 E_\perp^2}{m\omega_g^2}, \tag{10.4}$$

where $E_\perp$ is magnitude of the component of **E** perpendicular to the surface, which we assume to be constant along the cyclotron orbit. This is a reasonable approximation, since one-point multipacting orbits are not spatially very extended. For example, in typical $\beta = 1$ cavities, when the magnetic field is on the order of 300 Oe, the cyclotron frequency is close to 840 MHz. At an energy of 500 eV, an electron moves at 1.3×10^7 m/s. Thus, at most the total length of the electron orbit is about $1.3 \times 10^7 / 8.4 \times 10^8$ m = 1.5 cm, which is much smaller than the cavity dimensions. Charges with energies significantly greater than 1000 eV do not need to be considered since at these energies the SEC is less than one.

One should bear in mind that (10.3) and (10.4) are only simple approximations which yield generic information as to whether a cavity is susceptible to one-point multipacting. For more detailed information, trajectory calculations must be done, using the actual electromagnetic field distribution in the rf structure. The field distribution can be obtained from codes such as SUPERFISH and SUPERLANS as discussed in Chapter 2.

In quasi-pillbox cavities, for example, numerical trajectory calculations show that the multipacting orbits are governed not only by $E_\perp$ but also by the component of **E** parallel to the rf surface, which increases linearly with distance from the wall. Calculations show that multipacting of order n occurs at fields [161]

$$H_n = c_1 \frac{m\omega_\text{g}}{n\mu_0 e}, \tag{10.5}$$

where $c_1 = 0.64 \pm 0.08$. In this case, the mean impact energy is

$$\langle K \rangle = c_2 \frac{e^2 E_\perp^2}{m\omega_\text{g}^2} \tag{10.6}$$

with $c_2 = 2 \pm 0.4$. Nevertheless, we see that the scaling of (10.3) and (10.4) is preserved.

In these simulations, a distribution of emission energies was included which we have neglected so far. Typically, the energy at emission ranges from 0 to 10 eV, the peak being around 2 to 3 eV. This distribution results in a spread of impact energies about $\langle K \rangle$. In certain geometries, however, the trajectories may critically depend on the initial energy, which will lead to deviations from the behavior of (10.3) and (10.4).

If the outer wall of the quasi pill-box cavity is surrounded with thermometers, the equator region of the cavity shows strong heating correlated with the onset of multipacting, confirming the above model of one-point multipacting in a region of high magnetic field.

Several solutions to the one-point multipacting problem described above were explored. One approach was to reduce $E_\perp$ by making subtle alterations to the shape of the cavity so as to push K below K_1. This was successfully done with the HEPL quasi-pill-box cylindrical cavity [162] where the corners were sharpened to reduce the electric field. The method was successful in reducing multipacting for a muffin-tin cavity [163]. Another approach was to place grooves in the surface where multipacting normally occurs. This was used with some success in muffin-tin and cylindrical cavities [164]. At 1.5 GHz, magnetic field levels up to 430 Oe were reached, exceeding H_1 in equation (10.5). Trajectory calculations for rectangular grooves show that $E_\perp$ is strongly attenuated

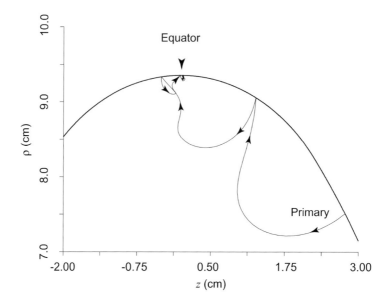

Figure 10.8: Electron trajectories in an elliptical cavity. The charges drift to the equator where multipacting is not possible.

at the bottom of the grooves, and secondaries generated there remain trapped, being unable to gain sufficient energy to create further generations. For other trajectories the grooves produce changes in the path lengths, altering the return phase of the electrons in a random way and thus destroying the resonance conditions.

By far the most successful solution to multipacting was to round the cavity wall to make a "spherical" cavity [165]. In this shape, the magnetic field varies along the entire cavity wall so that there are no stable electron trajectories, as electrons drift to the equator within a few generations. At the equator, $E_\perp$ vanishes, so that the secondaries do not gain any energy and the avalanche is arrested.

An elliptical shape is preferred over the spherical one due to the improved mechanical stability and easy drainage of liquids during cavity preparation [166]. But the basic antimultipacting principle is the same as for the spherical cavity. Figure 10.8 depicts trajectory calculations for an elliptical cavity.

Spherical and elliptical cavity shapes are now universally adopted, and one-point multipacting is no longer a serious problem for $\beta = 1$ superconducting cavities. In particular, 350-MHz and 500-MHz cavities have been successfully shown to be free of one-point multipacting to field levels beyond the first order.

COMMON MULTIPACTING SCENARIOS

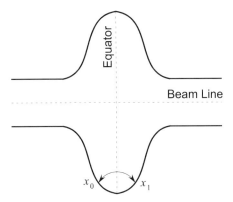

Figure 10.9: Two-point multipacting trajectories in an elliptical superconducting cavity.

10.5.2 Two-Point Multipacting

Another type of multipacting occurs when the electron trajectories include two impact sites, $\mathbf{x}_0$ and $\mathbf{x}_1$, which generally are two points opposite a symmetry plane. The resonance condition typically requires that the time between impacts is a half-integer ($[2n-1]/2$) multiple of the rf period. The integer n denotes the order of multipacting. Consider, for example, electrons emitted at $\mathbf{x}_0$ when the electric field changes sign, which then impact at $\mathbf{x}_1$ just as the electric field changes sign again. If $\mathbf{x}_0$ and $\mathbf{x}_1$ are on opposite points of a symmetry plane, the newly emitted electrons at $\mathbf{x}_1$ will return to $\mathbf{x}_0$ along an orbit which mirrors the trajectory of the primaries.

The only place in a $\beta = 1$ cavity where two-point multipacting has been observed is between opposite points of the equator, as shown in Figure 10.9. This form of multipacting has been found in 350-MHz LEP cavities, as shown in the temperature map of Figure 10.10 [149]. At 350 MHz, the first order occurs at an accelerating field of 5 MV/m. Trajectory calculations show that the impact energy is only about 30 to 50 eV for a starting energy of 2 to 4 eV. This type of multipacting therefore yields to processing and can be avoided altogether with special care in cleanliness and drying.

In general, two-point multipacting in (large-gap) $\beta = 1$ cavities is rare, because for most trajectories the impact energy is greater than the second crossover of $\delta(K)$. However, in other rf components such as coaxial lines, couplers, low-β cavities, and parallel plate geometries, two-point multipacting is far more common due to the small gaps between rf surfaces.

As an example of two-point multipacting amenable to analytic treatment, consider the circular parallel plate arrangement in Figure 10.11. If we ignore the fringe fields, then $\mathbf{E} = E_0 \sin(\omega_g t)\hat{x}$, where $E_0 = V/d$. V is the applied voltage, and d is the plate separation. The electric field thus is independent of position. We can ignore the azimuthal magnetic field provided that the frequency is low

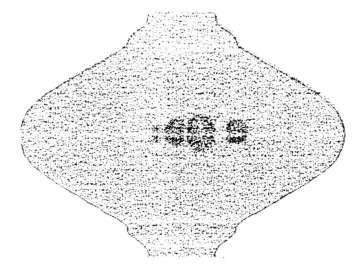

Figure 10.10: Temperature map of a 350-MHz LEP cavity. Dark regions denote high temperatures (full scale = 1.4 K). The hotspots on opposite sides of the equator reveal two-point multipacting in progress [149]. (Courtesy of CERN.)

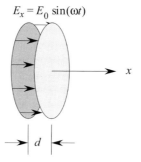

Figure 10.11: Parallel plate arrangement used to illustrate two-point multipacting. The right plate is at $x = 0$.

enough. In this case, the force acting on the electrons is entirely electric, and a one-dimensional analysis suffices. The acceleration of an electron emitted at time $t = 0$ into the gap from the right plate with a phase φ_0 is thus

$$\mathbf{a}(t) = -\frac{eE_0}{m}\sin(\omega_{\mathrm{g}}t + \varphi_0)\hat{x}. \tag{10.7}$$

Integrating twice and imposing the condition that $x_0 = 0$ yields the electron's position at time t:

$$x(t) = \frac{eE_0}{\omega_{\mathrm{g}}^2 m}\left[\sin(\omega_{\mathrm{g}}t + \varphi_0) - \sin\varphi_0\right] - \frac{eE_0}{\omega_{\mathrm{g}}m}t\cos\varphi_0. \tag{10.8}$$

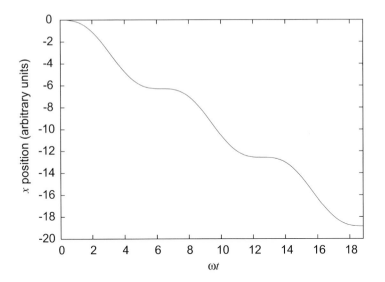

Figure 10.12: Motion of an electron emitted at time $t = 0$ with phase $\varphi_0 = 0$ from the right plate in Figure 10.11.

If we consider, for example, an electron emitted just when **E** begins to point in the $+\hat{x}$ direction, then $\varphi_0 = 0$ and

$$x(t) = \frac{eE_0}{\omega_g^2 m}(\sin \omega_g t - \omega_g t). \tag{10.9}$$

We see that the electron follows a drifting oscillatory motion as shown in Figure 10.12.

The resonant condition for multipacting requires that the charge traverses the gap in a half-integer rf period. Hence, we require

$$x\left(t = \frac{[2n-1]\pi}{\omega_g}\right) = -\frac{(2n-1)\pi eE_0}{\omega_g^2 m} = -d, \qquad n = 1, 2, 3 \ldots \tag{10.10}$$

The gap voltage at which two-point multipacting of order n can occur is thus

$$V_n = E_0 d = \frac{d^2 \omega_g^2 m}{(2n-1)\pi e}. \tag{10.11}$$

The voltage scales quadratically with frequency.

Integration of (10.7) yields the electron's velocity as a function of time. For $\varphi_0 = 0$, we find

$$v(t) = \frac{eE_0}{m\omega_g}(\cos \omega_g t - 1), \tag{10.12}$$

so that the nonrelativistic impact energy at $t = (2n-1)\pi/\omega_g$ is

$$K_n = \frac{2e^2 V_n^2}{m\omega_g^2 d^2}. \tag{10.13}$$

(Note that again we ignored any velocity the electrons may possess at the time of emission.) Equation 10.13 can now be used in (10.1) to determine whether multipacting occurs at V_n for the electrode materials used.

10.6 NUMERICAL MULTIPACTING SIMULATIONS

Although we derived a few simple expressions for multipacting thresholds, we already alluded to the fact that most structures preclude any analytical analysis of the multipacting orbits, and numerical codes are usually needed.

Due to the complexity of the process, different approaches may be used. We will limit ourselves to outlining one of these. It is important, however, to bear in mind that this is only meant to be an example of techniques that have been used in the past.

10.6.1 Multipacting Thresholds Determined with Electron Tracking

The most success in analyzing multipacting in rf structures has been achieved with electron trajectory calculations, using the field distribution for the structure being investigated. The following is an example of this approach [167].

The electric and magnetic fields are determined with codes such as SUPERFISH (see Chapter 2). The fields must be known very accurately, especially near the surfaces, and this requirement can prove to be demanding on numerical codes. Since codes only calculate the fields at discrete points, attention must be paid to the selection of the mesh.

Once the field distribution is known, electron orbits are calculated by integrating the equations of motion at various field levels for a set of representative emission sites, energies, and phases. Typical emission energies range from 0 to 5 eV. For each emission site $\mathbf{x}_0^{(j)}$ and emission phase $\varphi_0^{(j)}$ a series of n impact sites $\mathbf{x}_1^{(j)}, \mathbf{x}_2^{(j)}, \ldots, \mathbf{x}_n^{(j)}$ and impact phases $\varphi_1^{(j)}, \varphi_2^{(j)}, \ldots, \varphi_n^{(j)}$ are calculated. Those sites $\mathbf{x}_0^{(j)}$ for which any $\mathbf{E}(\mathbf{x}_m^{(j)}, \varphi_m^{(j)})$ ($m \leq n$) prohibits secondaries from escaping the wall at $x_m^{(j)}$ are removed from consideration. (When determining whether a secondary electron does escape the wall, the emission velocity is ignored). The remaining $c_n(|\mathbf{E}|)$ sites are retained for further analysis.

The process is repeated at sufficiently small intervals of $|\mathbf{E}|$ to ensure that no resonances are missed. The resulting plot of $c_n(|\mathbf{E}|)$ will generally peak at certain values of $|\mathbf{E}|$. At these fields, a distance

$$d_n^{(j)} = \sqrt{\left|\mathbf{x}_0^{(j)} - \mathbf{x}_n^{(j)}\right|^2 + \kappa \left|\exp(i\varphi_0^{(j)}) - \exp(i\varphi_n^{(j)})\right|^2} \qquad (10.14)$$

is calculated for each pair $(x_0^{(j)}, \varphi_0^{(j)})$, with κ being some suitably chosen scaling function. The emission pairs for which $d_n^{(j)}$ is minimized lead to near stationary trajectories — that is, trajectories which repeat after p impacts, where n is an

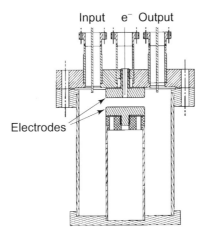

Figure 10.13: Resonator used to study multipacting for various electrode materials. The electric field between the electrodes is nearly uniform, whereas the magnetic field in this region is negligible [159]. (Courtesy of DESY.)

integer multiple of p. These are potential multipacting sites, and their trajectories can be analyzed to reveal the nature of the multipacting. One also must determine whether N_e given by (10.1) exceeds one. Sites for which this is not the case are also eliminated from consideration.

The approach outlined above was tested by studying the resonator in Figure 10.13 operating at 500 MHz [159, 167]. A coaxial line through the central port permits the direct measurement of multipacting currents. The geometry of the resonator is such that the magnetic field between the electrodes (spaced 10 mm apart) is negligible compared to the near-uniform electric field, and therefore the analytical result (10.11) can be used for comparison with simulation results.

Figure 10.14 illustrates the results obtained for $c_{30}(|\mathbf{E}|)$ up to a gap voltage of 2000 V. The threshold voltages for two-point multipacting of order 1–4, as calculated by (10.11), are also indicated by asterisks. The counter function peaks at voltages slightly lower than the first- and second-order thresholds, because an emission velocity was included in the trajectory calculations but was not included in (10.11). We also see that third- and fourth-order multipacting is not predicted by this simulation.

A comparison of the predictions with measured data is given in Figure 10.15 for first-order multipacting during processing. The electrodes, in this case, were made of copper coated with titanium. The agreement is quite reasonable. The broadness of the measured peak is primarily due to the range in starting velocities and emission angles of the secondaries which was not included in the simulations.

Figure 10.16 depicts the multipacting current collected as a function of time as the resonator was being processed. As is to be expected, aluminum, with its

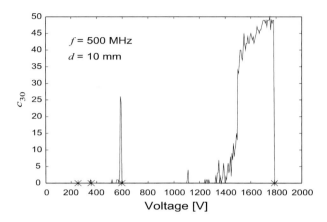

Figure 10.14: Calculated counter function c_{30} versus gap voltage for the resonator of Figure 10.13. The asterisks mark two-point multipacting thresholds predicted by Equation 10.11 [167]. (Courtesy of University of Helsinki.)

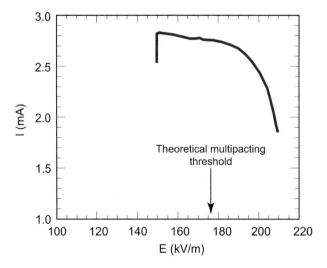

Figure 10.15: Measured multipacting current versus electric field during processing. The electrode material was titanium on copper [159]. (Courtesy of DESY.)

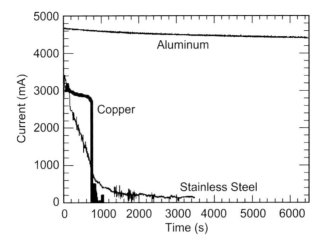

Figure 10.16: Measured multipacting current versus time for various electrode materials. When copper or stainless steel was used, the multipacting processed rapidly; processing was not possible with aluminum electrodes [159]. (Courtesy of DESY.)

natural oxide, yields the largest currents. Other materials produce less current, and multipacting barriers easily process within tens of minutes. Such materials are responsible for soft multipacting barriers.

The same numerical approach was applied to coaxial lines in standing wave operation [167]. This is of interest because many couplers can be approximated by a coaxial geometry. In this case, it was found that both one- and two-point multipacting occurs primarily at the electric field maxima. Away from the maxima the trajectories tend to drift into "shadow" regions where the phase of **E** prohibits secondary emission. Thus, multipacting is determined by the electric field alone. The average incident power flow at which multipacting occurs is found to scale as

$$P^{(1)} \propto (fD_2)^4 Z \qquad (10.15)$$

for one-point multipacting and

$$P^{(2)} \propto (fD_2)^4 Z^2 \qquad (10.16)$$

for two-point multipacting. Here Z is the impedance of the line, D_2 is the diameter of the outer conductor, and f is the generator frequency. The average incident power flow is related to the voltage drop between the inner and outer conductor by

$$P = \frac{\pi V^2}{4\eta \ln(D_2/D_1)}, \qquad (10.17)$$

where η is the wave impedance of the vacuum, and V is the voltage between the inner conductor (radius D_1) and the outer conductor.

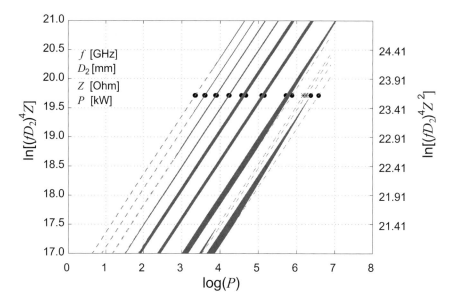

Figure 10.17: Incident power levels at which multipacting in a coaxial line is predicted. The left ordinate applies to one-point multipacting and the right ordinate applies to two-point multipacting. Bands enclosed by solid circles mark power levels for one-point multipacting, and the band enclosed by asterisks marks power levels for two-point multipacting. The portion of each band for which the electron impact energy satisfies 100 eV $\leq K \leq$ 1500 eV is shaded black [167]. (Courtesy of University of Helsinki.)

The multipacting powers are plotted in Figure 10.17, where the left ordinate applies to one-point multipacting (marked by the bands enclosed by the solid circles) and the right ordinate applies to two-point multipacting (marked by the band enclosed by the asterisks). Altogether, eight one-point multipacting bands and one two-point multipacting band are shown. The rightmost band is the first-order one-point multipacting band. Next, as one moves to the left, is the first-order two-point multipacting band, followed by the second- through eighth-order one-point multipacting bands. For both one-point and two-point multipacting, the bands are shaded black at the power levels for which the impact energies satisfy 100 eV $< K <$ 1500 eV.

10.7 AVOIDING MULTIPACTING

As we will discuss in Chapter 18, the geometry of an rf structure, such as an input coupler, output coupler, or window, generally evolves to satisfy many different criteria, and as such it is impractical to aim a priori for a multipacting free design. One can minimize the risk of multipacting in new structures by

AVOIDING MULTIPACTING

applying some of the lessons learned from the simple geometries. If parts of the structure can be approximated by a parallel plate arrangement, for example, the spacing may be arranged to push the threshold for low-order multipacting above the operating voltage. Recent efforts to improve the understanding of multipacting in coaxial lines have provided some guidelines to minimize the risk of multipacting in coaxial couplers. Further studies along these lines for other types of structures — windows in particular — would be very desirable.

If multipacting is still encountered despite such efforts, experience shows that progress can be made if the multipacting orbits can be localized by appropriate diagnostic tools (e.g., thermometers or light detectors). With knowledge of the multipacting location and of the field distribution, electron tracking calculations can be carried out. The results may suggest geometric modifications to shift the barriers or to change the resonant conditions altogether.

A completely different approach is to use materials with a low SEC. While this method is attractive for coating normal conducting structures, such as couplers and windows, it may be impractical for extended structures or hard-to-reach regions. In rf windows, for example, alumina, which has an intolerably high SEC, is often used. In this case, one may coat the surface with low-SEC materials, such as titanium and titanium nitride. These coatings need not be thicker than a few nanometers to reduce the SEC of the window, which is fortunate, since thicker coatings result in high ohmic losses due to the high electrical conductivity of titanium and titanium nitride (see Chapter 18).

Even when low-SEC materials are used, cleanliness is imperative. Contaminants on the surface can cancel out any benefits gained from low-SEC materials. Hydrocarbons are especially problematic. A vacuum bakeout is useful in cleaning up the surface; and once this is completed, the structure should be kept under continuous vacuum. Other possibilities include argon discharge cleaning and electron bombardment during multipacting. The effect of the former on the SEC can be very dramatic, as can be seen in Figure 10.5.

Yet another approach is to destabilize the multipacting trajectories at the operating fields by applying a dc electric or magnetic field. Its magnitude and direction are chosen to alter the multipacting trajectories so that the resonance conditions are no longer met. This approach has been successfully modeled in coaxial lines [168] and is used in the input coupler for LEP (see Chapter 18).

CHAPTER 11

Thermal Breakdown

11.1 INTRODUCTION

In this chapter we turn to cavity performance limitations arising from regions of high magnetic fields. We examine phenomena that limit the achievable field below the theoretically expected critical magnetic field discussed in Chapter 5. Over the years, several measures have been developed to surpass these limitations. We will discuss the evolution of these techniques as well as their efficacy.

11.2 THERMAL BREAKDOWN OF SUPERCONDUCTIVITY

In Chapter 1 (Figure 1.4) we showed that structures reach average accelerating fields of 9 MV/m in laboratory tests, but span a large range, between 5 and 20 MV/m. The corresponding average surface magnetic field is 400 Oe, with a maximum of 900 Oe. One phenomenon that limits the achievable field is thermal breakdown of superconductivity, also known as a "quench." The other phenomenon is field emission, which we will cover in the next chapter. Thermal breakdown originates at sub-millimeter-size regions that have rf losses substantially higher than the surface resistance of an ideal superconductor. These regions are called "defects."

In the dc case, supercurrents flow around defects. But at rf frequencies, the reactive part of the impedance causes the rf current to flow through the defect, producing Joule heating. When the temperature at the outside edge of the defect exceeds T_c, the superconducting region surrounding the defect becomes normal conducting, resulting in greatly increased power dissipation. As the normal conducting region grows, the power dissipation increases, resulting in a thermal instability as sketched in Figure 11.1.

An earlier interpretation [169] of quench was that the breakdown of superconductivity originates with a phase transition. If the rf magnetic field exceeds the local rf critical field, which was presumed to be depressed due to the presence of the defect, a quench results. This interpretation was ruled out as a common cause of quench by experiments [164] in which the local rf magnetic field at the breakdown spot was increased by superposing the field of a second cavity mode. A 2-cell cavity was excited in the two modes (called π and

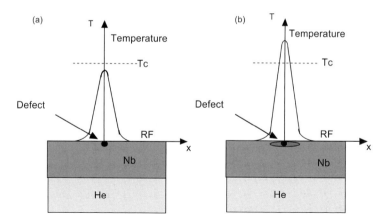

Figure 11.1: Thermal breakdown of a niobium cavity. (a) At low field, the temperature in the vicinity of the defect is higher than that in surrounding areas, but it remains below T_c. (b) As the field is raised, the temperature exceeds T_c, so that the niobium near the defect becomes normal conducting, and the power dissipation increases unstably.

$\pi/2$) of the fundamental passband, and the breakdown field for each mode was measured separately. The same breakdown spot for both modes was observed by thermometry. Subsequently, both modes were simultaneously excited, and different ratios of field amplitudes were adjusted to obtain breakdown at the same spot. If the magnetic instability model were applicable, one would expect $H_\pi + H_{\pi/2} = constant$. If a temperature instability was responsible for breakdown, one would get $H_\pi^2 + H_{\pi/2}^2 = constant$. In several test cavities, for which the quench field ranged from 150 to 500 Oe, the second result was unambiguously obtained.

These experiments definitively showed that the breakdown level depends not on the local H, but on the local H^2. Furthermore, when equal power was applied in both modes, the surface magnetic field at the breakdown location was found to be $\sqrt{2}$ times higher than the quench field in either mode alone, ruling out the magnetic instability mechanism. If a phase transition were to occur at a $\approx$ millimeter-size defect, we would also expect a small drop in Q at a lower rf field, to be followed at a higher field by a thermal breakdown at the normal conducting site. This Q switch is not normally seen. On the other hand, thermal maps can detect ohmic heating at a defect and can track the temperature rise all the way from low fields to the quench.

A quench produces characteristic traces for the transmitted and reflected power, as shown in Figure 11.2. When an input rf pulse is turned on, the cavity fills and the reflected power decreases toward zero if the input probe is unity coupled. The stored energy in the cavity and the transmitted power increase until the quench field is reached. At this point, the power dissipation at the defect drives a large portion of the cavity surface normal. Unity coupling to

EXAMPLES OF DEFECTS

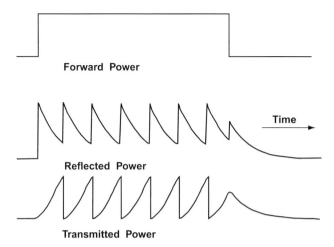

Figure 11.2: RF traces for a cavity in thermal breakdown. Note how the stored energy in the cavity (proportional to the transmitted power) refills after the quench if the rf power stays on, because the region around the defect cools down to return to the superconducting state.

the input coupler is lost, and all power is reflected. Furthermore, the field in the cavity falls abruptly as the stored energy dissipates in the large normal conducting region around the defect. Once the field level is sufficiently low, and the cavity cools, the Q returns to its original value and the cavity fills again. The entire process then repeats. The cavity jumps back and forth spontaneously between the superconducting and the quenched states.

It is important to note that because the stored energy involved is only a few joules to a few tens of joules, a quench in a superconducting cavity does not have as severe an effect as a quench in a superconducting magnet, which involves the loss of kilojoules. As a result, a superconducting cavity can recover from the quench within milliseconds, whereas it takes a magnet about an hour. Also, a quench does not result in any permanent damage to a superconducting cavity as it may to a magnet.

11.3 EXAMPLES OF DEFECTS

It is possible to locate defects that lead to thermal breakdown by using the temperature-mapping technique discussed in Chapter 8. In Figure 11.3 we see a possible defect near the equator, labeled site #1 [170]. When plotted as a function of H^2, the temperature rise at this location is found to be linear, corresponding to ohmic heating. After locating the hot site, the cavity is cut cleanly at the equator into two halves, which are able to fit into a scanning electron microscope (SEM) as shown in Figure 11.4. When we examine the

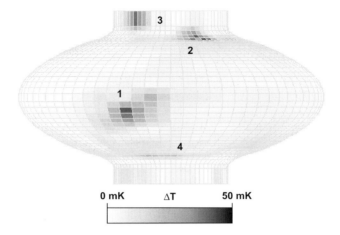

Figure 11.3: Temperature map at about 400 Oe of a 1.5-GHz, single-cell cavity showing heating at a defect site, labeled #1.

Figure 11.4: SEM chamber with a single-cell cavity half about to be inserted for analysis.

vicinity of the predicted site, we find the defect and its associated contaminants. A chemical analysis can also be carried out, using an energy dispersive x-ray (EDX) system.

In case of the defect of Figure 11.3, a search of the area under suspicion revealed a 50-μm copper particle, which appeared to be attached, perhaps sintered, to the niobium surface via a root-like structure, as shown in Figure 11.5. The sintering at the base gives us a hint about the particle's origin. It is likely that a foreign copper particle fell onto the cavity equator and then heated up in the rf magnetic field. Initially the particle had poor thermal contact to the nio-

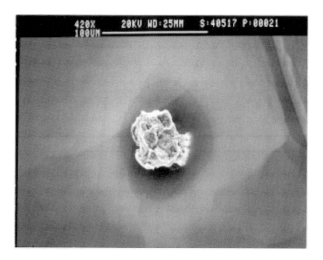

Figure 11.5: SEM micrograph of a defect found in a high magnetic field region of a 1.5-GHz single cell cavity.

bium wall, so that in the rf magnetic field the temperature rose rapidly to near the melting temperature of copper. At this stage the base of the particle sintered to the underlying niobium. With the improved contact, the temperature of the particle stayed in the liquid helium temperature range for the remainder of the rf test. Subsequently, the maximum temperature rise was only a few hundred millikelvins.

The effective area presented by the copper to the rf field is about 4×10^{-9} m^2. The rf magnetic field at the equator is about 427 Oe (34,000 A/m) Assuming a typical surface resistance of 5 mΩ for copper, we can estimate the power dissipation in this defect

$$P_{\text{diss}} = \frac{1}{2} A R_s H^2 = 11.6 \text{ mW}. \tag{11.1}$$

Thermal calculations (discussed below) predict that such a power input on the inner cavity surface should result in a temperature rise of 154 mK on the outer surface. The measured temperature rise was 39 mK. Considering that the efficiency of the thermometers is only about 25–30%, as discussed in Chapter 8, the agreement is reasonable. The consistency also confirms our estimate for the surface resistance of the defect.

Two other examples of defects are shown in Figure 11.6 [171]. There are many opportunities for such defects to enter a superconducting cavity during the various stages of cavity production and preparation: manufacture of sheet metal, deep drawing of cups, electron-beam welding, chemical etching to remove the surface-damage layer, rinsing, drying, insertion of coupling devices, and in the final attachment of the cavity to the vacuum system of the test stand or

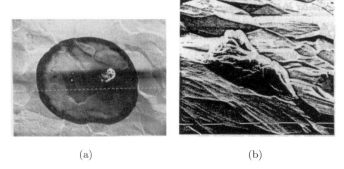

Figure 11.6: SEM micrographs of defects that caused thermal breakdown. (a) A chemical or drying stain 440 μm in diameter. The small crystal on the right side contains K, Cl, and P. This defect quenched at $E_{\text{acc}} = 3.4$ MV/m. (b) A 50-μm crystal containing S, Ca, Cl, and K. This defect quenched at $E_{\text{acc}} = 10.7$ MV/m. Note the general grain boundary structure of the niobium surface in both (a) and (b). (Courtesy of CERN.)

the accelerator. The many types of defects associated with TIG welds shown in Chapter 6 were also located by thermometry and examined with an SEM and EDX. On a statistical basis, we expect the number of defects present to increase with the cavity area, so that larger area cavities will break down at lower fields. The copper defect of Figure 11.5 indicates that large foreign particles which fall into a cavity can become attached by rf heating.

Experience shows that the quench field level does not change during a test or after cycling to room temperature. Only the step of re-rinsing or re-etching a cavity has an effect on the quench field. This suggests that most defects are permanently lodged on, and strongly attached to the superconducting surface.

On rare occasions, when raising the rf power, there can be a small Q drop, called a "Q switch," as shown in Figure 11.7. In most cases, a Q switch is caused by a region of niobium that is poorly attached to the surface, such as a blister in a niobium film (see Chapter 14), or a poorly adhering, overlapped layer of niobium embedded during forming operations. Niobium balls from weld spatter will also cause a Q switch. At low fields, the niobium blisters or beads are superconducting and do not adversely affect the Q. At the switching field, the loosely attached region, which is thermally isolated, becomes normal conducting, dropping the Q. If the field is lowered, the Q does not recover since the detached region does not cool down until the rf field is lowered well below the switching field. More recently, a tantalum-rich area of niobium was found at a Q switch site [172]. It is possible that this is the first example of an rf magnetic switch due to the reduced critical field of Nb-Ta.

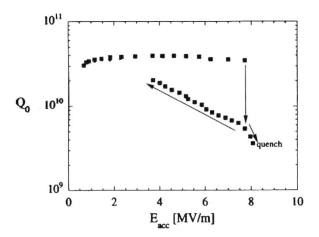

Figure 11.7: An example of a Q switch. Above a certain field level, the Q drops sharply. Lowering the power does not recover the high Q until the field level is lowered substantially. Raising the power gives a further sharp drop in Q until there is a quench. (Courtesy of DESY.)

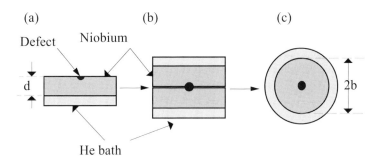

Figure 11.8: A hemispherical defect on the rf surface may be modeled by a heat source inside a sphere. The defect radius is a and the radius of the niobium sphere is b. The thickness of the niobium sheet is d.

11.4 A SIMPLE MODEL FOR THERMAL BREAKDOWN

We now present a simple model and calculation that illustrates the essential features of thermal breakdown at a normal-conducting defect. Suppose, as shown in Figure 11.8(a), there is a hemispherical defect of radius a and surface resistance R_n embedded in a sheet of thickness d, cooled by a bath at temperature

T_b. The power dissipation (in watts) at the defect is

$$\dot{Q}_T = \frac{1}{2} R_n H^2 \pi a^2. \tag{11.2}$$

The geometry of the problem can be changed to a spherical heat source with a rate of heat generation $2\dot{Q}_T$, embedded inside a block, thickness $2d$, surrounded by the bath at temperature T_b, as in Figure 11.8 [173]. If $a \ll b$, the block is not much different from a sphere, with the spherical defect at the center. At any radius r, the rate of heat flow (in watts) out through a spherical surface is

$$-4\pi r^2 \kappa \frac{\partial T}{\partial r}, \tag{11.3}$$

where κ is the thermal conductivity in W/m-K. We want to make this equal to $2\dot{Q}_T$, which leads to the simple expression (when κ is constant)

$$-4\pi r^2 \kappa \frac{dT}{dr} = 2\dot{Q}_T, \frac{1}{r^2} dr = -\frac{2\pi\kappa}{\dot{Q}_T} dT, \int_a^b \frac{dr}{r^2} = -\frac{2\pi\kappa}{\dot{Q}_T} \int_{T_a}^{T_b} dT. \tag{11.4}$$

Since $b \gg a$, this reduces to

$$\frac{1}{a} = \frac{2\pi\kappa(T_a - T_b)}{\dot{Q}_T}. \tag{11.5}$$

Using Equation 11.2 and solving for H, we get

$$H = \sqrt{\frac{4\kappa(T_a - T_b)}{aR_n}}. \tag{11.6}$$

When the defect reaches T_c, the field reaches its maximum value

$$H_{\max} = \sqrt{\frac{4\kappa(T_c - T_b)}{aR_n}}. \tag{11.7}$$

For example, at a bath temperature of 2 K, a 50-μm radius defect with $R_n = 10$ mΩ will break down at $H_{\max} = 820$ Oe, if the RRR of the niobium is 300, i.e., the average thermal conductivity is 75 W/mK. This model is certainly oversimplified, because it ignores many physical aspects: the temperature- and frequency-dependent BCS surface resistance of the surrounding superconductor, the residual resistance, the temperature dependence of the thermal conductivity, and the details of the heat flow between the niobium cavity wall and the helium bath. In the next section, we will discuss computer codes based on an iterative solution of the heat flow equation to include these factors and to calculate the equilibrium temperature distribution at the rf surface and elsewhere. Remarkably, the salient conclusions reached by the simple model remain essentially true. For a given defect (a, R_n), $H_{\max}$ increases as $\sqrt{\kappa}$. Apart from the defect parameters and the thermal conductivity, the other factors do not play

SOLUTIONS TO THERMAL BREAKDOWN

Figure 11.9: A setup for defect removal. The cavity sits on Teflon rollers so that it can rotate freely. The Teflon arm on the right carries a dc motor holding a small grinding abrasive wheel. The arm is articulated so that the grinding tool can reach any part of the cavity surface via adjustment of the penetration and the angle.

as significant a role, because the heating at the defect dominates the power dissipation at the rf surface, and because the relatively low thermal conductivity of the niobium essentially isolates the defect from the bath.

Generally one does not have access to both (a) an SEM micrograph of a defect (to estimate its size) and (b) the thermometry data (to estimate its surface resistance). Typically, we know only the quench field level. For convenience, if we take a defect surface resistance of 10 mΩ, typical for normal conducting niobium, we can characterize the "strength" of a defect purely in terms of its size. In the future, we will refer to the strength of a defect only in terms of the size of a normal conducting defect.

11.5 SOLUTIONS TO THERMAL BREAKDOWN

11.5.1 Guided Repair

If there are one or two gross defects (e.g., diameter $\approx$ one millimeter) in a cavity due to manufacturing errors (usually rare if good practice is followed), these can be located by thermometry and removed by mechanical grinding by an arrangement similar to that shown in Figure 11.9. By repeated application (e.g., four times) of the guided repair method, the accelerating gradient of a 350-MHz single-cell niobium cavity (RRR = 40) was increased from 5 to 10 MV/m [174]. But it is not so easy to eliminate smaller, more frequently occurring defects em-

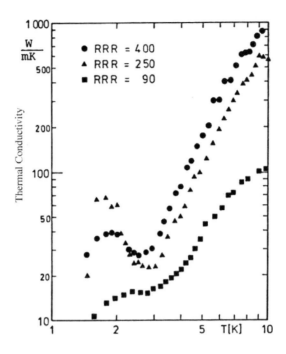

Figure 11.10: Thermal conductivity of niobium with RRR = 90 (as received), RRR = 400 after postpurification with yttrium, and RRR = 250 after annealing the postpurified sample for 6 hours at 1400°C, which degraded the purity. (Courtesy of Wuppertal.)

bedded in the cavity material or introduced into the surface during preparation and handling.

11.5.2 Raising the Thermal Conductivity of Niobium

The most effective cure for small defects is to raise the thermal conductivity of the niobium to increase $H_{\max}$ as suggested by Equation 11.7. Then defects will be able to tolerate more power before driving the neighboring superconductor into the normal state.

Figure 11.10 shows the thermal conductivity of three samples of niobium between 1.5 and 10 K [104]. The samples have different RRR and different histories of heat treatment. The common feature of all three curves is the sharp drop below $T_c = 9.2$ K. However, the higher the RRR, the higher the thermal conductivity. As discussed in Chapter 3, the reason for these features is that electrons are the dominant carriers of heat. The phonons (lattice vibrations) also play a role in heat conduction, but this component is significant only at $T < 4$ K. Below T_c, the thermal conductivity drops precipitously, as more and more electrons condense into Cooper pairs. Because the depairing energy is not available from random thermal motion, the pairs are not scattered by the lattice vibrations and therefore cannot conduct heat from one part of the niobium to another. At high temperatures (4 K $< T <$ T_c), a significant, though small, fraction of electrons is not frozen into Cooper pairs and can carry heat effectively, provided that the electron-impurity scattering is low. Since the temperature in the neighborhood of the defect is between the bath temperature and T_c, it is the high-temperature (4 to 9.2 K) thermal conductivity which is the most important and which will have the strongest effect on thermal breakdown.

As electrons condense into Cooper pairs, electron–phonon scattering also decreases. Below about 4 K, the thermal conductivity from phonons dominates and begins to increase, leading to the phonon peak near 2 K. With decreasing temperature, the number of phonons decreases $\propto T^3$. Ultimately, the value of the phonon conductivity maximum is limited by phonon scattering from lattice imperfections, of which the grain boundary density is the most important. If the crystal grains of niobium are very large, because of annealing at high temperature, one observes a large phonon peak, as shown in the thermal conductivity behavior of the sample with RRR = 250, which was annealed at 1400°C. The most significant electron scattering impurities are the interstitial ones, such as O, N, C, and H. The ideal RRR of niobium is 35,000 i.e., when there are no impurities, only electron-phonon scattering. The interstitial impurities have an equivalent effect on the low temperature electrical and thermal conductivity. Therefore, one can gauge the thermal conductivity and the purity of niobium by measuring the low-temperature resistivity in the normal state. At 4.2 K, the thermal conductivity of niobium is given approximately by

$$\kappa = 0.25 \text{ (W/m-K)} \cdot \text{RRR}. \tag{11.8}$$

11.5.3 Thin Films of Niobium on Copper

Yet another weapon in the arsenal against thermal breakdown is to use a micron-thick film of niobium on a thermally stabilizing copper substrate. The thermal conductivity of copper [175] is shown in Figure 11.11 for a variety of purities and annealing states. For comparison, we also give the low-temperature thermal conductivity of niobium with low and high RRR.

The technique of sputter coating niobium has been developed by CERN for 350-MHz structures and applied successfully to hundreds of structures [98]. Nb/Cu cavities have rarely been seen to quench. Although Q_0 values greater than 10^{10} are obtained at low fields, the rf losses of Nb/Cu cavities increase steadily with field. This effect is attributed to intergrain losses in the niobium films, which become more severe at higher frequency. We will return to niobium coating techniques and results in Chapter 14. Structures for heavy-ion accelerators which are made from thin electroplated Pb films on Cu are also not observed to quench.

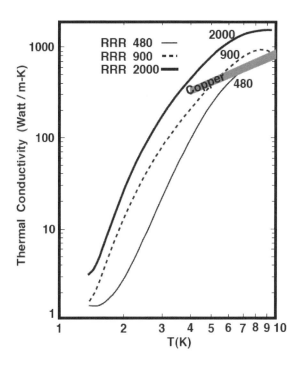

Figure 11.11: Thermal conductivity of high-purity copper samples compared to the low-temperature thermal conductivity of niobium samples of various RRR. The large range in thermal conductivity of copper samples is due to differences in purities and annealing states. Note that at RRR ≈ 1000, the thermal conductivity of niobium begins to approach that of copper.

11.6 HEAT TRANSPORT AT THE HELIUM INTERFACE

The thermal model calculations discussed in the next section show that the heat transfer coefficient at the niobium–helium interface does not play as strong a role as the thermal conductivity in determining the quench field. The temperature rise outside the defect is strongly dependent on the heat transport in the niobium near the defect, especially since the thermal conductivity of niobium is only several tens of mW/mm-K. For a cavity wall thickness of a few mm, the defect is, to first order, isolated from the bath. However, one can expect the importance of heat transfer to the bath to increase as the thermal conductivity improves with purity, as the wall thickness is decreased, or if there are no defects. Therefore it is worthwhile to discuss the physics of heat transport at the outer wall of the cavity. The latter also determines the temperature rise at the outer wall of the

HEAT TRANSPORT AT THE HELIUM INTERFACE

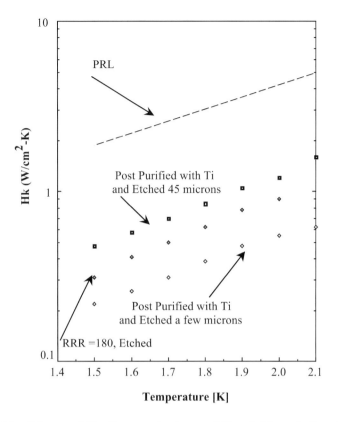

Figure 11.12: Measured Kapitza conductance of the niobium–helium interface, compared to the theoretical phonon radiation limit (PRL) (Courtesy of Saclay.).

cavity and hence determines the sensitivity of thermometers used to study the behavior of cavities.

Unlike thermal conductivity, the heat transport across a solid–helium interface is not as well understood. For a brief overview, see [176]. When heat flows from a solid body into liquid helium, there is a small temperature difference ΔT across the interface. The temperature discontinuity was first observed for the solid-He-II interface by Kapitza [177]. The Kapitza conductance, H_k is defined as

$$\lim_{\Delta T \to 0} \frac{\dot{Q}_T}{A \Delta T}, \qquad (11.9)$$

where $\dot{Q}_T/A$ is the power density (W/m^2) of heat flow at the interface. H_k depends on temperature, pressure, the elastic properties of the solid, and the structure of the surface, as well as the properties of liquid helium. To a good approximation, heat transport across a solid-helium interface takes place predominantly via phonons. Since He-II is an insulator, the electrons in the metal must give up their energy to the phonons to transmit energy across the interface.

By analogy with blackbody photon radiation, phonons are treated as bosons; it is therefore possible to derive an upper bound for phonon transmission across the interface. This is called the "phonon radiation limit," or PRL. Figure 11.12 shows measured values [178] of the Kapitza conductance of niobium along with the conductance in the phonon radiation limit, which is $\propto T^3$ for small ΔT. The data follow a T^3 to T^4 power law but they are typically an order of magnitude below the PRL.

The hypothesis to explain the reduced phonon transmission is that there is an "impedance mismatch" across the interface [179]. As a result, a large fraction of phonons impinging on the surface are not transmitted. However, the acoustic mismatch (AM) theory predicts H_k values that are more than an order of magnitude lower than the data. The agreement with the AM theory is improved by additional hypotheses [180], such as one that postulates a dense layer (15 Å) of helium atoms that improves the impedance match.

There is a large variation in H_k data due to differences in surface preparation. Surface strains, damage layers, and chemical impurities usually lower H_k. For example, in Figure 11.12, note how H_k increases upon removal of the Ti layer deposited on the niobium surface during the postpurification titanium treatment. High-power tests on cavities show that the removal of 40–50 μm of niobium is necessary after Ti treatment to avoid premature quenches in the presence of heating due to field emission.

When the heat flux exceeds a critical value, $\dot{Q}_T^*$, film boiling occurs and the surface is covered with a film of vapor. This results in a very high thermal boundary resistance. The reported values for $\dot{Q}_T^*$ range from 1 to 10 W/cm^2, depending on sample shape and configuration in the helium bath. These effects are reviewd in [91].

With their lower BCS surface resistance at 4.2 K, low frequency cavities ($f < 500$ MHz) are operated in He-I. The temperature drop across a metal to He-I interface is primarily determined by the properties of the liquid helium and not by the properties of the metal. Above 5 mW/cm^2, bubbles start to form to help carry away the heat quite effectively [181]. This condition is known as nucleate pool boiling. In the steady-state nucleate boiling regime we have

$$\dot{Q}_T = C\Delta T^n, \tag{11.10}$$

where the exponent n ranges from 1.4 to 2 and the constant, C, is typically 1 W/cm^2 [91]. Nucleate boiling persists until a peak flux of about 1 W/cm^2, above which the bubbles are so numerous that the surface is covered with a thin film of vapor [182]. This is the film boiling limit which can bring about thermal breakdown at areas of intense heat deposition. In the nucleate boiling regime, He-I cools about as effectively as He-II, but the peak heat flux that may be sustained before film boiling begins is lower for He-I. Thus He-II is more effective for high-field operation of superconducting cavities.

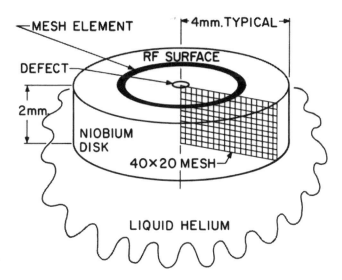

Figure 11.13: Simple model used to simulate the onset of thermal breakdown at the rf surface.

11.7 THERMAL MODEL SIMULATIONS

Computer codes based on the iterative solution of the heat flow equations have been developed [130] to simulate the onset of thermal breakdown. They calculate the equilibrium temperature distribution along the surface of a niobium disk in the presence of an rf magnetic field. Typically, the codes assume a finite cylindrical disk of thickness, e.g., 2 mm, and radius, e.g., 4 mm, with liquid helium at the lower surface and a uniform rf magnetic field H on the top surface, as shown in Figure 11.13. Liquid helium also surrounds the outer rim of the disk, a reasonable boundary condition to use for defects with radius < 0.2 mm. For a defect-free calculation, the boundary condition at the outer rim is inadequate, as it permits lateral heat flow. To handle this case, the problem is reduced to a one-dimensional calculation by setting the lateral thermal conductivity to zero. Only heat flow normal to the rf surface then takes place, as expected for 1-D.

The disk is divided into a series of ring-like mesh elements. The power dissipated in a mesh element of area A at the rf surface is $0.5 R_\mathrm{s}(T) H^2 A$ where $R_\mathrm{s}(T)$ is the surface resistance (BCS + residual). For the defect, located on the center element, we use a higher surface resistance, typical for a normal-conducting metal (e.g., 10^{-2} Ω). In the presence of the rf magnetic field H, the transition temperature is given by

$$T_\mathrm{c}(\mathrm{K}) = 9.2\sqrt{1 - H(\mathrm{Oe})/2000}. \qquad (11.11)$$

When an rf surface element reaches the field-dependent critical temperature, the resistance of that element is set to that of the normal conducting state. At the lower surface of the disk and the outer rim, both of which are in contact

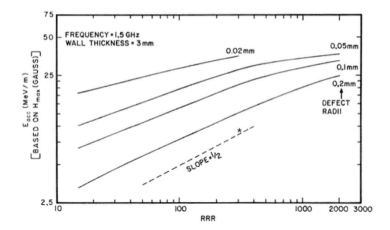

Figure 11.14: Thermal model predictions for the breakdown field of various normal conducting defects. For comparision, the quench field calculated from Equation 11.7 is shown as (*). Note that the full simulations show a higher field value because they include the temperature dependent thermal conductivity, the average of which is higher than the thermal conductivity at 4.2 K. For these cavities $H_{\rm pk}/E_{\rm acc} = 47$ Oe/MV/m.

with helium, we chose the boundary condition

$$\kappa(T)\frac{\partial T(r,z)}{\partial z} = \frac{\dot{Q}_{\rm T}(\Delta T)}{A}, \qquad (11.12)$$

where $\dot{Q}_{\rm T}(\Delta T)/A$ is the heat flux density from the outer niobium surface to the helium bath for a temperature difference ΔT. For example, in superfluid helium ($T < 2.18$ K), we use

$$\frac{\dot{Q}_{\rm T}}{A} = H_{\rm K}\Delta T, \qquad (11.13)$$

where $H_{\rm K}$ is the measured Kapitza conductivity of Nb [178].

During the course of the heat flow calculations, one keeps track of the heat flux density at the niobium–helium interface and one reduces the heat transfer coefficient by an order of magnitude when the film boiling limit (e.g., 1 W/cm^2) is reached. This procedure is intended to simulate the case of thermal breakdown triggered by film boiling.

In a representative result of the thermal model calculation, shown in Figure 11.14, we see that the breakdown field for a 200-μm radius defect increases roughly as $\sqrt{\rm RRR}$, as expected from the simple model and Equation 11.7. We compare the results of the simulation with experiments in order to estimate the size of defects. Here we assume that thermal breakdown takes place near the equator, i.e., at the highest magnetic field region of the cavity. Experience

THERMAL MODEL SIMULATIONS

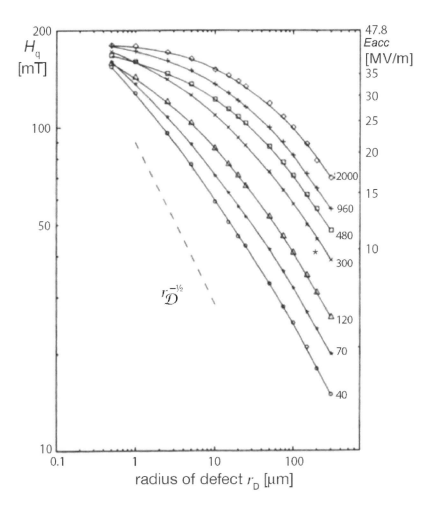

Figure 11.15: Calculated thermal breakdown field vs. defect size for niobium of various RRR values, shown next to each curve. The simple model estimate is indicated by (*). (Courtesy of Wuppertal.)

shows that the typical breakdown field is 200 Oe for niobium of RRR = 40, which translates to 5 MV/m for a velocity-of-light structure. Simulations reveal that if the defect is a normal conducting region of 200 μm radius, it will break down at 5 MV/m. To quote the defect size alone is, of course, a very rough, but useful, characterization of a defect, as mentioned. One would normally expect the resistivity and size of defects to span broad ranges. Note that as the defect size gets smaller, the exponent of RRR is lower than 0.5. The calculation predicts that larger increases in RRR are needed to improve upon the breakdown fields for cleaner cavities.

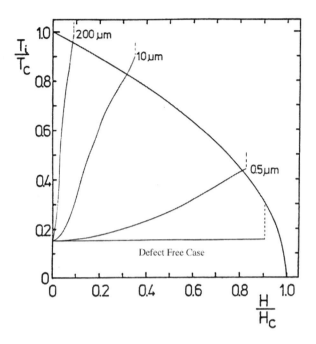

Figure 11.16: Temperature vs. rf field strength for the region just outside the defect. The calculations are for various defect sizes, including a case with no defect. The niobium selected has RRR ≈ 30. (Courtesy of Wuppertal.)

For reliable calculations with the smallest defects, a variable mesh density must be used [183]. Figure 11.15 shows the results for a large range of defect sizes and RRR values. The thermal conductivity of niobium used for these calculations are given in Ref. [183].

It is interesting to track the temperature rise just outside the defect for large defects, small defects, and the defect free case, as shown in Figure 11.16. Here the calculation is carried out for low-RRR Nb. Note how the temperature outside the defect rises steadily to near T_c at low field for a large defect. Even for a 0.5-μm defect the temperature rises steadily with increasing field, until it crosses T_c. But in the defect-free case, the surface temperature remains low and becomes unstable at a high field due to the exponential temperature rise of the BCS surface resistance. We will return to this type of instability in Section 11.10. Finally, we compare in Figure 11.17 the temperature rise at the rf surface with the temperature rise at the outer wall for a low-RRR niobium cavity and a large defect. A thermometer can easily detect the large temperature rise at the outer wall. The temperature near the defect depends only weakly on the bath temperature, as expected for low thermal conductivity niobium.

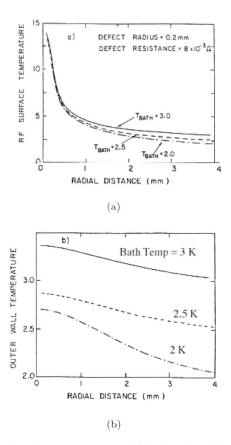

Figure 11.17: Calculated temperature of (a) the rf surface and (b) the outer wall of a 2.86-GHz RRR $\approx$ 30 niobium cavity.

11.8 METHODS TO IMPROVE NIOBIUM PURITY

We now turn to measures developed to improve the thermal conductivity of niobium, and thus to overcome thermal breakdown. These topics have been touched upon before in Chapter 3, in connection with the fundamental properties of niobium, as well as in Chapter 6, in the discussion of the material requirements for the fabrication of cavities.

Several methods have been explored to improve niobium purity. The first involves heating niobium in an excellent vacuum (10^{-9} torr) at $T > 1900°C$ and for long periods ($\approx$ 5 to 10 hours) [77]. Since niobium has a very high affinity for oxygen, very high temperatures are necessary to degas oxygen into the vacuum. Even at 1900°C the removal of O is slow as it takes place only indirectly by evaporation of NbO and NbO_2 from the metal surface.

A series of 8.6-GHz niobium cavities were purified by degassing at 2000°C us-

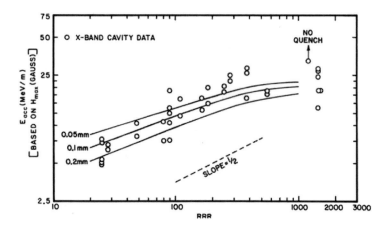

Figure 11.18: Measured quench fields of 8.6-GHz cavities after varying the niobium RRR by outgassing at high temperature. The solid lines are calculations for various defect sizes. $H_{\mathrm{pk}}/E_{\mathrm{acc}} = 47$ Oe/MV/m.

ing resistive and induction heating [184]. RRR values up to 1200 were obtained. The rf test results from these cavities provided the first proof-of-principle that improving niobium purity leads to higher quench fields, as shown in Figure 11.18. Here we compare the measured quench fields with thermal model calculations for various defect sizes. Although valuable as a research technique, the high-temperature outgassing method is not practical for accelerator structures due to the severe deformations, creep, and loss of yield strength that accompanies the high-temperature treatment.

The most effective approach to increasing the thermal conductivity of niobium is to remove the interstitial impurities by improving electron beam melting practices used for refining the ingot. This requires a moderately good vacuum ($< 10^{-5}$ to 10^{-6} torr) in the electron-beam furnace at the melting temperature of niobium (2470°C). It is also important to melt the ingot slowly in order to achieve equilibrium [185]. Niobium is now available with RRR = 250 to 300 from U.S. and European suppliers who use the techniques of multiple and slow melting. New Russian niobium is available with RRR = 500 to 700 [122].

Yet another technique for improving niobium purity is "solid state gettering." It was first applied [131] to raise the RRR of niobium cavities by using yttrium to coat the niobium surface. Yttrium has a higher affinity for oxygen than does Nb [186, 187, 188]. The coated niobium is heated to a temperature $> 1200°$C so that oxygen diffuses rapidly. The mobile interstitial impurity atoms sink into the foreign metal layer when they arrive at the surface of the niobium. The coating and purification operations can be combined into one step because the vapor pressure of yttrium is large enough to form an evaporated layer at the diffusion temperature. With solid state gettering for 4 hours, a factor of 2 to 3 improvement in RRR is possible for commercially available niobium, corresponding to nearly complete removal of the oxygen, which is the dominant impurity. After the purification stage, the getter material and the underlying compound layer are chemically etched away. Commercial niobium of RRR = 30 can thus be improved to RRR = 90, and commercially prepared niobium of RRR = 250 to 300 has been improved to RRR = 500. Russian niobium now available with starting RRR = 500 to 700 has been improved to RRR = 1000 to 1400. As mentioned in Chapter 6, the postpurification technique must take place after the half-cell forming stage or later, because the grain growth at high temperature will destroy the mechanical workability of the niobium.

Soon after the application of solid state gettering by yttrium to niobium cavities, it was found that titanium is also an effective solid state getter [132]. However, the vapor pressure of titanium is lower than for yttrium, so that higher temperature or longer times are needed. For example, to remove oxygen in a few hours, titanium must be used at 1350°C to 1400°C. Since titanium diffuses into niobium to a substantial depth ($\approx 100~\mu$m) along the grain boundaries [189], heavy chemical etching becomes necessary after the postpurification step. The outside surface of a cavity must also be etched about 50 μm to reestablish a good Kapitza conductance [134]. Titanium does have the intrinsic capability to remove nitrogen and carbon by solid state gettering due to its appreciable affinity to these impurities. A comparison of affinities can be found in [185]. But the diffusion rates of nitrogen and carbon are much lower than for oxygen. Therefore very long gettering times are necessary. Using titanium for more than 50 hours, RRR values > 1000 have been achieved in samples of starting RRR = 200 [133].

One must keep in mind a few negative side effects from high RRR niobium. The BCS surface resistance increases with the electron mean free path, by about a factor of 2 (Figure 4.6). Therefore high RRR cavities have a lower BCS Q_0. Of course, this effect is not important at operating temperatures below 2 K when the BCS component is negligible, and residual losses dominate the Q. As discussed earlier, high RRR niobium cavities are more sensitive to Q degradation from hydrides, due to the lack of interstitial impurities which serve to trap diffusing hydrogen and prevent the formation of lossy hydride clusters on the rf surface. Precautions are therefore necessary during the chemical etching stage of high-RRR niobium to avoid hydrogen contamination. Niobium suppliers have found that crystal grains grow rapidly when annealing high-RRR niobium sheets for final recrystallization. Therefore the annealing temperature and time must be kept low (near 800°C). This requirement competes with the demands for complete recrystallization, which is necessary to obtain good drawing properties for cavity fabrication. As a result, very careful control of time and temperature are necessary to achieve complete recrystallization as well as small grain size. If the high RRR is obtained by the high temperature, solid state gettering treatment, it lowers the yield strength of the material. In addition, as already pointed out, titanium diffuses into niobium to a substantial depth along the grain boundaries [134]. Therefore heavy chemical etching becomes necessary. The outside surface of a cavity must also be etched. The overall effect is to thin down the wall of the cavity. Appropriate measures must be taken to avoid collapsing a thin-wall cavity whose yield strength has been lowered by purification.

11.9 QUENCH SUPPRESSION WITH HIGH-PURITY NIOBIUM

Over the years 1980 to 1990, as the RRR of niobium has improved, so has the performance of cavities built by various labs. Figure 11.19 shows the improvement for single-cells. (The benefits to multicells will be discussed in Chapter 13.) Before thermal breakdown was understood and the high thermal conductivity cure was implemented, the typical RRR of niobium was 30–40, and the typical thermal breakdown field for niobium cavities was about 200–300 Oe. The corresponding $E_{\rm acc}$ ($\beta = 1$ cavities) was 6 to 7 MV/m for 1500-MHz and 3000-MHz single cells, and 5 MV/m for larger area, 500-MHz single-cell cavities. Large-area cavities usually quench at lower field due to the higher probability of encountering defects. The *best* gradient for the low-RRR cavities was 10 MV/m, and it was only 8 MV/m for 500-MHz single cells. A comparison with the simulations of Figures 11.14 and 11.15 suggest that the typical normal conducting defect radius is 100–200 μm.

Between 1980 and 1990, the RRR of commercially available niobium increased steadily from 40 to 300. Even at RRR = 100, there was a factor-of-2 improvement in average field, to 10 MV/m, with best values of 15 MV/m. When the RRR reached 300 the best gradients reached 20 MV/m, but the average gradients were limited by field emission. The best measure for the expected quench field limit for RRR = 300 niobium is obtained from the distribution of quench fields [190] in a large number of TJNAF 5-cell 1500 -MHz cavities, as shown in Figure 11.20.

These results are only a subset of the 338 TJNAF cavities that have been tested. Only those cavities which were limited by a quench, and not by field emission, are presented. The large spread is indicative of the wide range in size or resistance of defects that one can encounter. The average magnetic field is ≈ 600 Oe. A comparison with the simulations of Figure 11.15 suggests that for a quench field of 600 Oe the typical normal conducting defect has a radius of 100 μm, not much different from the single-cell results discussed in connection with Figure 11.19.

More recently [191], there has been some progress in reducing the severity of defects. Data from 1-cell 1300-MHz cavities tested at KEK show $E_{\rm acc} = 18$–20 MV/m for the modest RRR = 100, rising to $E_{\rm acc} = 25$–35 MV/m for RRR = 200 (Figure 11.21). These results suggest that KEK is using better preparation techniques or better material, which leads to reduced defect strength. A comparison with the thermal model calculations of Figures 11.14 and 11.15 indicate that to reach the corresponding 1200-Oe fields with RRR = 200 niobium a normal conducting defect would have a diameter of 10 μm, instead of the more usual size of ≈ 100 μm. It is possible that high-pressure water rinsing, a technique adopted for removing field emitters (see next chapter), also helps to remove

QUENCH SUPPRESSION WITH HIGH-PURITY NIOBIUM

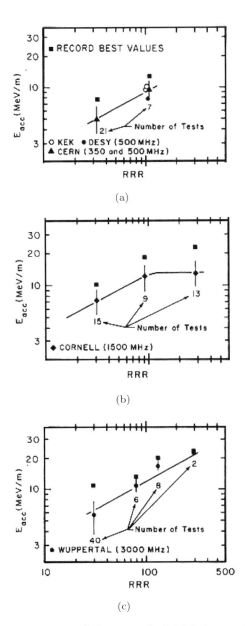

Figure 11.19: Improvement of the quench field between 1980 to 1990 with increasing RRR for single-cell cavities at (a) 350–500 MHz (b) 1500 MHz, and (c) 3000 MHz.

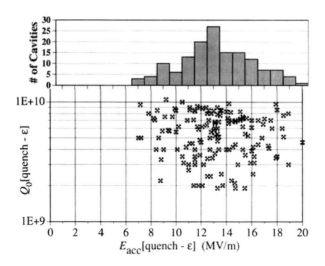

Figure 11.20: Distribution of quench fields for about 100 5-cell, 1.5-GHz TJNAF cavities. The Q value shown is just below the quench field. $H_{\rm pk}/E_{\rm acc} = 47$ Oe/MV/m. (Courtesy of TJNAF.)

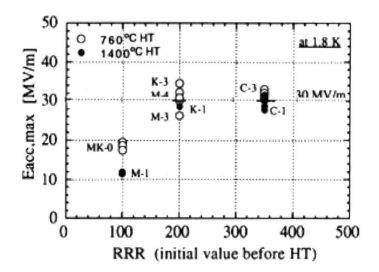

Figure 11.21: Recent data on 1-cell 1.3-GHz cavities prepared by KEK. Note the high accelerating fields reached with RRR = 200 niobium, suggesting reduced defect strength. Note also that the achieved fields do not improve much with RRR = 300, as expected for small defects according to the simulations of Figures 11.14 and 11.15. (Courtesy of KEK.)

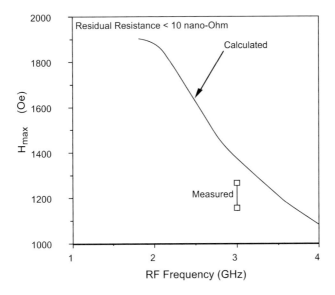

Figure 11.22: Results of thermal model calculations for the global thermal instability for 3 GHz niobium cavities. This type of instability was seen in two single-cell cavity tests at the measured field values shown.

large defects. Because of the better rinsing, there may be fewer foreign particles left in the equator region. Another difference is that KEK cavities are usually electropolished (Chapter 6.)

The simulations of Figure 11.14 also show that for minor defects the expected improvement with higher RRR is not be as substantial as it is for larger defects. This behavior is indeed borne out by the KEK data of Figure 11.21, which show $E_{acc} = 25$–35 MV/m for RRR = 350. Although the results give an excellent prognosis for the future, it remains to be seen whether this record in defect-free preparation can be extended to large-area, multicell cavities.

The effort to further increase the RRR of niobium continues. We already mentioned that solid state gettering niobium with RRR = 250–300 provides an improvement of a factor of 2. However, above $E_{acc} = 10$ MV/m, field emission takes over as the dominant performance limitation. Therefore, we postpone the discussion of the impact of RRR = 500–600 until Chapter 13 so that we may first cover the topic of field emission.

11.10 DEFECT-FREE CAVITIES

What happens if there are no defects at all? Thermal model calculations predict that under certain circumstances, quench will occur just due to the exponential temperature dependence of the BCS surface resistance. For example, if the rf frequency is above 2 GHz or if the bath temperature is high, so that the BCS

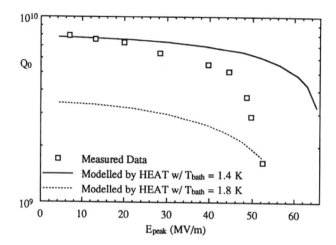

Figure 11.23: Calculated and measured values of the Q decrease due to the temperature rise of the rf surface arising from global heating in a 3-GHz single-cell cavity. There was no field emission in this test. During the test, the bath temperature drifted from 1.4 to 1.8 K.

contribution dominates, there will be global thermal heating at a sufficiently high field, and the end result is a "global thermal instability" (GTI). The residual resistance component also plays a role. The calculated [192] field for the onset of thermal instability as a function of the rf frequency is shown in Figure 11.22. For typical residual resistance values and frequencies > 2 GHz, the field at which a global thermal instability can occur is significantly less than the rf critical magnetic field, $H_{sh}(0) = 2300$ Oe. Figure 11.22 also shows the quench field data [193] on two 3-GHz single-cell cavities tested that reached the global thermal instability limit.

One symptom of GTI is a drop in Q due to the increased rf surface temperature, as shown in Figure 11.23. Curves are shown for two different bath temperatures. The maximum field reached was 1200–1300 Oe, close to the GTI prediction. In the tests, there was no field emission (no x rays), but the Q nevertheless dropped at high field, as shown by the data points of Figure 11.23. We attribute the Q drop to global heating of the rf surface. Because of the large dissipated power and limited helium bath pumping capability, the bath temperature changed from 1.4 to 1.8 K. The drift in bath temperature explains why the data fall between the predicted high-field curves.

Temperature mapping (Figure 11.24) shows global heating of the equator region. When we compare this map to one taken in a different experiment, where ohmic heating at a defect was dominant, we see a big difference in the character of global heating and local defect heating. In both cases, the temperature maps are taken at a field level below breakdown.

The calculations and experiments were conducted for cavities with a wall

DEFECT-FREE CAVITIES   225

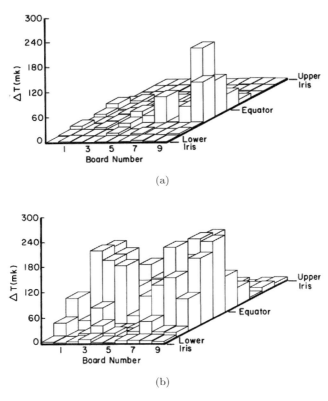

Figure 11.24: A comparison of (a) defect heating and (b) global heating, detected by thermometers on single-cell 3-GHz cavities.

thickness of 1.4 mm. Cavities with thinner walls should reach higher fields before GTI because the temperature difference between the rf surface and the helium bath plays the determining role when the heating of the rf surface is uniform (no defects). The thinner the cavity wall, the closer the helium bath is to the rf surface, and the less the rf surface temperature rise at a given power flux. However, thin-wall cavities have other problems associated with mechanical instability.

The best solution to GTI is to lower the rf frequency. As a practical matter, for applications demanding the highest possible gradients, such as linear colliders, it is necessary to keep the rf frequency below 2 GHz as indicated by Figure 11.22.

CHAPTER 12

Field Emission

12.1 INTRODUCTION

In the previous chapter, we discussed the most important field-limiting mechanism related to the surface magnetic field. We now turn to the chief limitation associated with the surface electric field, namely, the emission of electrons from the regions of high electric field on the cavity surface. Earlier reviews of the subject can be found in References [161, 194, 149, 195, 196, 197]. There are also recent books that cover a wide range of topics related to field emission [198, 199].

We start in Section 12.2 with a presentation of the general symptoms of field emission. The temperature mapping diagnostic technique for superconducting cavities shows that emission always arises from particular spots, called "emitters," located in high electric field regions. The pattern of temperature rise as a function of position along a given meridian contains implicit information about the location and characteristics of the source. We relate the symptoms to the trajectories of the electrons that emerge from the emitters, travel in the rf fields of the cavity, and impact the rf surface. (More details of the electron trajectory calculation and results are given in the last section of this chapter.) The power deposited by the impacting electrons depends not only on the trajectory but also on the intrinsic properties of the emitter — i.e., on the field emission current. Accordingly, we turn to the basic theory of field emission (Section 12.3). Fowler and Nordheim (FN) showed that, in the presence of an electric field, electrons tunnel out of the metal into the vacuum because of their quantum wave-like nature. However, a comparison with the observed currents reveals that at a given field, emission is substantially higher than the FN predictions. Traditionally, the excess has been attributed to a "field enhancement factor," which is believed to be related to the physical properties of the emitter.

Much has been learned about the nature of emitters from superconducting rf cavity studies in which (a) the emitters are located by temperature maps and (b) their field enhancement factors are characterized from a measurement of effects related to emission current, such as temperature increments or x ray emission. The rf tests are followed by a dissection of the cavity to examine the emitter with surface analytic instruments. These studies and emitter characteristics are discussed in Section 12.4. There have also been substantial advances in understanding the nature of field emission through dc high-voltage studies that locate emission sites with a needle-shaped electrode, followed by electron microscopy studies of the sites, as covered in Section 12.5. By and large, both rf and dc studies reveal that emitters are micron- to sub-micron size contaminant particles. Sensitized by these results, new approaches have been adopted to strive for a higher level of cleanliness in cavity surface preparation, leading to fewer emission sites and better cavity performance. We briefly touch upon the improved cleanliness methods in Section 12.6, but postpone the discussion of the accompanying improvements in cavity performance to the next chapter.

A remarkable finding from the dc studies on emitters is that not all microparticles turn out to be field emitters. Section 12.7 discusses additional physical aspects that appear to play a role in determining whether a particle is, or is not, a field emitter: the detailed geometry, the nature of the interface with the underlying substrate, the interaction between the conducting particle and the insulating layer on the surface, and the condensed gas adsorbates.

Having covered the topics of field emission and the nature of emitters, we turn to the rich subject of "processing." When raising the rf electric field in a superconducting cavity for the first time (i.e., with a freshly prepared surface), the field emission often decreases abruptly; the cavity is said to "process" or "condition." There has been much progress in characterizing processed emtters at a microscopic level using techniques such as SEM, EDX, Auger, and AFM. Both rf processing (Section 12.8) and dc processing (Section 12.9) studies are covered. Based on the results of these studies, our understanding of conditioning has considerably improved. Emitter processing is an explosive event that accompanies what we usually refer to as a "spark" or a "discharge," or the "electrical breakdown" of the insulating vacuum. A key aspect in this process is the role of the gases that form a plasma, as discussed in Section 12.10. We conclude the main discussion of this chapter with a summary Section 12.11 that attempts to form a comprehensive picture of field emission and processing. The very last section details the methods for analysis of field emission heating in cavities.

Based on the improved knowledge about processing, new approaches have been developed to destroy emission sites that still occur despite all efforts to produce and maintain a clean surface. Having discussed the mechanism of processing, we follow in the next chapter with a presentation of the benefits of enhanced processing methods to the performance of superconducting cavities.

12.2 DIAGNOSING FIELD EMISSION

When E_{pk} exceeds 10 to 20 MV/m, the Q_0 of a niobium cavity typically starts to fall steeply due to exponentially increasing electron currents emerging from the surface. Figure 12.1 shows the typical field emission dominated Q vs. E behavior for TJNAF cavities [190]. Unlike multipacting, which occurs at discrete field levels due to resonant production of electrons, when field emission dominates, the field can be increased gradually with more rf power. A probe

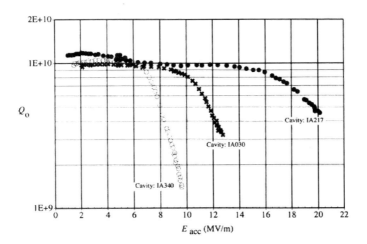

Figure 12.1: Sample of vertical test results of TJNAF 5-cell, 1.5-GHz cavities that show field emission. Many cavities show the onset of field emission at $E_{\rm pk} = 10\text{–}20$ MV/m, but a few best cavities remain field emission free up to $E_{\rm pk} = 30$ MV/m. The $E_{\rm pk}/E_{\rm acc}$ ratio of these cavities is 2.6. (Courtesy of TJNAF.)

placed near the beam axis of the cavity will pick up an electron current. Detectors placed outside the cavity will show x rays due to bremstrahlung from the field-emitted electrons when they hit the cavity wall. Thermometers placed on the exterior wall of the cavity will detect heating from the bombardment of the cavity wall by field-emitted electrons. All these observations are generally correlated [200]. In particular, the temperature maps indicate that electrons originate from particular spots in the high electric field regions of the cavity surface [147].

When the rf power is increased for the first time, there may be small, abrupt reductions in emission, called *processing* (or *conditioning*). Occasionally, electron emission will suddenly "turn on," resulting in a decrease of Q and of the field level. Eventually, the processing and turn-on events subside, and the emission becomes stable. If the emission becomes sufficiently intense, the heat deposited via electron bombardment can initiate thermal breakdown. If field emission processes, the thermal breakdown will cease. This is quite distinct from the stable defect-induced thermal breakdown described in the previous chapter. Field emission induced thermal breakdown may recur at a higher field level, when the current increases once again with field.

To understand the symptoms of field emission, we track the electrons that start from an emitter located in a high electric field region. Figure 12.2 shows calculated trajectories followed by electrons from a hypothetical emitter on the surface of a 5-cell, 1.5-GHz cavity. Each line represents the trajectory of a field-emitted electron originating at a particular phase of the rf period. Due to the

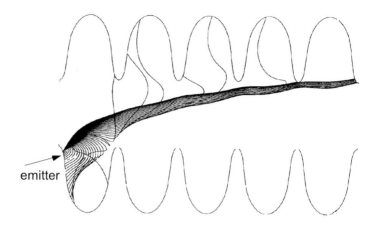

Figure 12.2: Calculated electron trajectories in a 5-cell 1.5 GHz cavity operating at $E_{\text{acc}} = 20$ MV/m. The emitter is located in the end cell, where the surface electric field is 45 MV/m. Note that a significant number of field-emitted electrons bend back in the magnetic field and strike the wall near the emitter.

axial symmetry of the accelerating mode, electrons are confined to travel in the ρ–z plane of the emission site. Some electrons are captured by the accelerating field and can traverse the entire structure. Such electrons may also strike peripheral devices, such as the window or antenna of a power coupler, and may cause problems in these devices. The electrons that travel along the beam tube can be collected by a pickup probe placed on the beam tube for diagnostics. A large fraction of the electron trajectories are bent by the rf magnetic field and strike the wall of the cell in which the emitter resides, producing heat and bremsstrahlung x rays. The heat deposited by the impacting electrons can be detected by an array of thermometers placed on the exterior of the cavity wall. We have already shown the line heating patterns associated with field emission in our discussion of thermometry in Chapter 8. As discussed in the last section of this chapter, the heating profile can be unfolded, with the help of the calculated electron trajectories, to yield the location of the emitting site to within a few millimeter. A comparison between temperature maps and simulations also gives the emissive properties of the site.

12.3 THEORY OF FIELD EMISSION

Electrons are confined inside a metal by a potential well, so that under normal circumstances the energy of an electron is insufficient to enable it to escape from the metal. The electron must be given extra energy in the form of thermal energy (thermionic emission) or radiation energy (photoemission) to

THEORY OF FIELD EMISSION

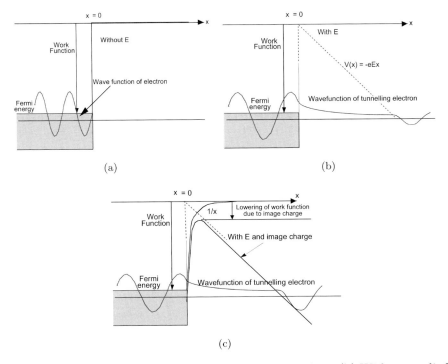

Figure 12.3: (a) Electrostatic potential at a metal surface. (b) With an applied field, electrons can tunnel from the metal into the vacuum because the barrier is lower. (c) Effect of image charge on the potential.

escape from the metal. In the quantum mechanical picture, the wave function of the electron is attenuated rapidly outside the surface potential barrier. But if the barrier is thin enough, the attenuation is not complete and there is a finite probability that some electrons will tunnel through the barrier and escape into the vacuum.

In one of the first applications of quantum wave mechanics [201], Fowler and Nordheim (FN) showed that when the work function barrier at the metal surface is lowered by an applied surface electric field (Figure 12.3), electrons can tunnel through the resulting triangular barrier. Due to the electric field, the electron will see the potential

$$V(x) = -eEx. \tag{12.1}$$

An electron experiences an attractive force due to the presence of the conducting surface. This force can be represented by a positive "image" charge, producing a potential

$$V(x) = -\frac{e^2}{16\pi\varepsilon_0 x}. \tag{12.2}$$

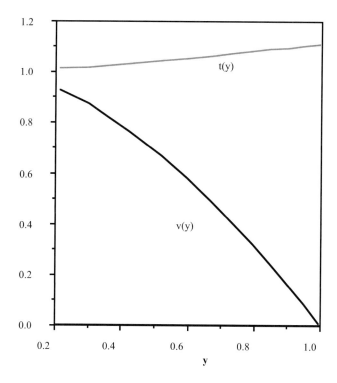

Figure 12.4: Fowler–Nordheim functions $v(y)$ and $t(y)$.

The net potential is thus

$$V(x) = -\frac{e^2}{16\pi\varepsilon_0 x} - eEx. \tag{12.3}$$

The maximum potential occurs where the image force equals the force of the electric field.

Fowler and Nordheim obtained the following expression for the tuneling current density

$$j = \frac{e^2}{8\pi h}\frac{E^2}{\phi t^2(y)}\exp\left(-\frac{8\pi\sqrt{2m(e\phi)^3}v(y)}{3heE}\right), \tag{12.4}$$

where e is the electron charge, m is the electron mass, h is Planck's constant, ϕ is the work function of the metal (in eV), and E is the instantaneous electric field in V/m,

$$y = \sqrt{\frac{eE}{4\pi\varepsilon_0\phi^2}}. \tag{12.5}$$

Here ε_0 is the permittivity of free space. The image force introduces the two slowly varying functions $v(y)$ and $t(y)$ which are calculated in terms of elliptic integrals (Figure 12.4).

THEORY OF FIELD EMISSION

In the case of the triangular potential barrier, $v(y) = 1$ and $t(y) = 1$, so that the FN expression reduces to

$$j(E) = \frac{A_{\text{FN}} E^2}{\phi} \exp\left(-\frac{B_{\text{FN}} \phi^{3/2}}{E}\right), \qquad (12.6)$$

where $A_{\text{FN}} = 1.54 \times 10^6$, $B_{\text{FN}} = 6.83 \times 10^3$, E is in MV/m, and j is the current density in A/m^2.

The FN theory has been experimentally confirmed for the case of a dc electric field established at a sharp point electrode, as for example a single-crystal tungsten needle, opposing a planar surface [202].

In a superconducting cavity, the current collected by the pickup probe and the x ray intensity outside the cavity are also observed to follow the functional dependence of the FN law. But the function has to be modified by an important factor, β_{FN}, customarily referred to as the field enhancement factor

$$I(E) = \frac{A_{\text{FN}} A_e (\beta_{\text{FN}} E)^2}{\phi} \exp\left(-\frac{B_{\text{FN}} \phi^{3/2}}{\beta_{\text{FN}} E}\right), \qquad (12.7)$$

where $I = j A_e$ is the current from the emitter, and A_e is the effective emitting area. Without the help of this enhancement factor, the predicted magnitudes of the current and x ray intensity are many orders of magnitude lower than observed. To a large extent, the science of field emission is devoted to understanding the physical origin of β_{FN} and A_e.

If $\beta_{\text{FN}} E$ replaces E everywhere in the FN equations, including the expression for y, Figure 12.5 gives the field emission current density as a function of $\beta_{\text{FN}} E$ [197]. There is a large increase in current if the image charge effect is also enhanced by β_{FN}.

The average current has a slightly modified dependence on $\beta_{\text{FN}} E$ in the rf case. When Equation 12.7 is averaged over one rf period, one obtains

$$I \propto (\beta_{\text{FN}} E)^{2.5} \exp\left(\frac{-B_{\text{FN}} \phi^{3/2}}{\beta_{\text{FN}} E}\right). \qquad (12.8)$$

For "broad" area dc electrodes, where the lateral dimensions are comparable to the spacings between electrodes, excessive currents are seen for anomalously low applied dc fields (10 MV/m), but the I–V characteristics are still found to fit a modified FN law. Plotting $\ln(I)/E^2$ vs. $1/E$, β_{FN} is obtained from the slope of a fitted straight line, and A_e from the intercept, provided that one is plotting the total current from the emitter, and not just a fraction of the current intercepted by the probe.

The classical interpretation of β_{FN} is that there are certain sites on the surface, such as whiskers or sharp projections, which enhance the electric field. Field enhancement factors for different shapes of microprotrusions have been calculated. For a semiellipse type of projection, shown in Figure 12.6(a) β_{FN} is determined in terms of the aspect ratio $r = h/b$. Figure 12.6(b) gives β as a function of r for large values [203]. For the case $r = 1$, which corresponds to the geometry of a hemishpere on a plane, $\beta \approx 3$.

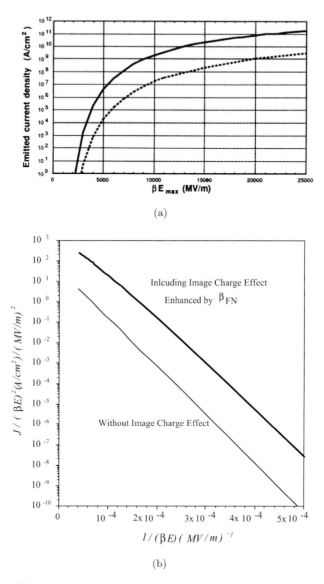

Figure 12.5: (a) A comparison of the emission current density with and without image charge effects. (b) Same functions as (a) except plotted to show the exponential behavior of the FN current.

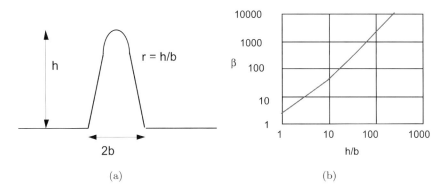

Figure 12.6: (a) A semiellipsoid shape microprotrusion, with semimajor axis h and semiminor axis b. (b) The calculated geometric field enhancement factor.

When we discuss the results of microscopic examination of emission sites, we will see that the expected whiskers are usually *not* found. As yet, the physical mechanisms of enhanced field emission are not completely understood, but there has been much progress, which we will describe shortly.

In light of this situation, it should be recognized that the FN parameters, $\beta_{\rm FN}$ and $A_{\rm e}$, should be used only as parameters to express the dependence of the emitted current on the field. The physical significance of $\beta_{\rm FN}$ and $A_{\rm e}$ is still a matter for debate. For this reason, throughout this chapter (except where noted) we will use the simplified FN expression which ignores the lowering and rounding of the potential barrier due to image charge effects, as well as the slightly different exponent of E between the dc and rf cases.

12.4 FIELD EMITTERS IN SUPERCONDUCTING CAVITIES

As mentioned earlier, emission is observed to occur from localized sites. Apart from the global indicators of field emission, such as x rays, temperature maps are used to localize particular sites in superconducing cavities, and to deduce the $\beta_{\rm FN}$ and $A_{\rm e}$ values. Information from companion x ray mapping systems is often used to corroborate the location of emitters. Figure 12.7 shows a pair of correlated temperature and x ray maps [152]. Where there is a peak in the heating, one also detects a peak in the x ray flux. However, x ray backscattering can yield additional spurious peaks, so that temperature mapping is the cleanest method for studying field emission sites.

We now turn to the detailed characterization and the microscopic examination of field emission sites found in superconducting cavities. Temperature mapping is, once again, a powerful diagnostic tool. In Figure 12.8 (a) temperature map taken [170] at $E_{\rm pk} = 17.2$ MV/m shows several emission sites near the iris of the cavity (locations 4, 5, 6 and 7). The same map revealed a hot

FIELD EMISSION

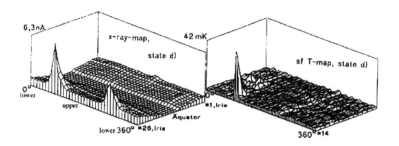

Figure 12.7: Temperature and x ray maps show that the heating from impacting electrons and bremstrahlung from x rays are well correlated. There is a broader second x ray peak from backscattering, 180° from the emitter. (Courtesy of Wuppertal.)

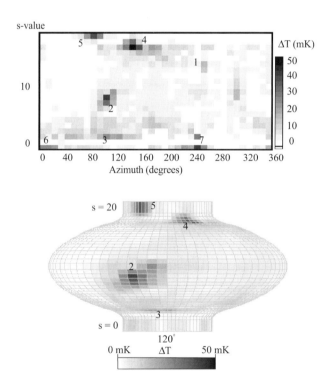

Figure 12.8: Temperature map of a 1-cell, 1.5-GHz cavity showing several hot spots, some of which are field emission sites. The lower figure only shows part of the cavity. The top map is a flattened view of the entire cavity.

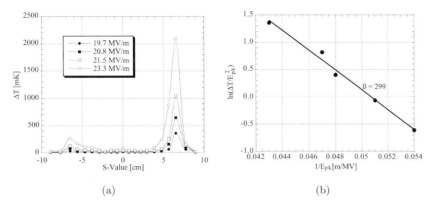

Figure 12.9: (a) Measured heating profiles at site #4. The abscissa is the distance along the cavity wall, starting at $S = 0$, the equator. (b) A FN plot of the peak temperature vs. peak electric field. Field values quoted are E_{pk}.

spot at the equator, where a 50 micron copper particle was found, as discussed in the previous chapter. Site #4 was still emitting at the end of the test. Many of the other sites processed. We will discuss processed sites later.

To determine the FN properties of site #4 we measured several temperature profiles, each at progressively higher fields, as shown in Figure 12.9(a). The peak temperature is observed to follow a FN behavior, as shown in Figure 12.9(b). From the β_{FN} value deduced, we find that the equivalent maximum current density at this site was 10^{12} A/m^2.

As we will detail in Section 12.12, the heating due to impacting field emitted electrons is modeled via a numerical solution of the relativistic equations of motion of the electrons in the cavity fields, combined with a heat-transfer model for the niobium–helium system. A comparison of the measured and simulated signals yields not only the location of the emission site but also its FN parameters. By applying the procedures discussed in Section 12.12 to the data for site #4 it was determined that the site had a $\beta_{\text{FN}} = 299$ and $A_e = 2.7 \times 10^{-17}$ m^2.

After the temperature map identifies the interesting sites, the cavity is dissected in a class 1000 clean room, and the surface is examined in an SEM. During dissection, which is carried out with a large pipe-cutter, the cavity is pressurized with filtered nitrogen to prevent dust contamination. Vacuum suction is maintained near the cutting tool. Machining is not used so as to minimize dust.

Figure 12.10 shows particles containing Fe and Cr (probably stainless steel) at the predicted emission site. Note that there are a couple of small balls present in the outlined region, suggesting that the particles suffered local melting. We will discuss partial melting of emission sites later. The rest of the site is a collection of jagged particles. Following the initial SEM examination, the area was cleaned with a high pressure carbon dioxide jet to ensure that the particles

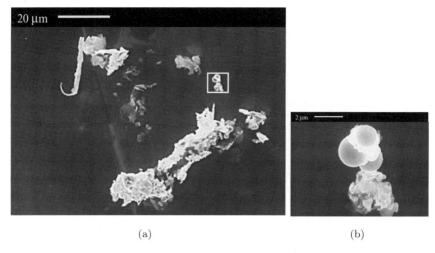

Figure 12.10: (a) SEM micrograph of particles at site #4. Note the cluster of small spherical balls in the framed portion which indicate that a part of the site melted. (b) The melted cluster is expanded. EDX analysis show that the particles are stainless steel.

were not loose debris that landed on the cavity during dissection, despite the precautions taken to avoid dust. Reexamination in the SEM showed that the site was unaltered, except that the molten balls were missing. It is clear that all the stainless particles had strongly adhered to the cavity surface. We surmise that the balls became weakly attached to the surface due to necking upon solidification, so that the balls were dislodged by the gas jet.

The geometrical aspect ratio (height/diameter) of the stainless steel particles is typically < 10 and is therefore not consistent with the observed β_{FN} value of 380. However, the particles do have a jagged structure, which may produce additional field enhancement, a topic to which we will return when we discuss dc field emission studies.

Figure 12.11 shows another emitter [204], this time an indium particle which was found in a one-cell, 3-GHz cavity at the location of a field emission site. In this case, the best fit to the temperature maps give $\beta_{\mathrm{FN}} = 350$ and $A_{\mathrm{e}} = 1.9 \times 10^{-15}$ m^2. This particle is not shaped like a needle either, but does have jagged features. Like the stainless particle, the indium particle also shows a small molten spot. A possible reason for these melted regions in the stainless and indium particles is rf heating, but this would melt the entire flake. The more likely explanation is that the stainless steel and indium particles field emit only from a small region, which heats up due to Joule heating from the emission current. At a high surface field, the exponentially increasing emission current can melt the superficial particle. It is possible that when one region melts, emission from another region takes over. This is one possible explanation for the observed instability in the emission current, when the fields are raised for

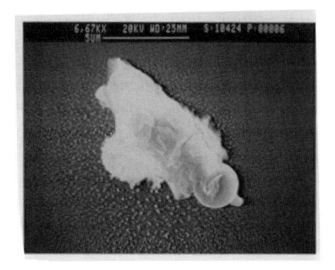

Figure 12.11: SEM micrograph of an indium metal flake field emitter. A small melted region can be recognized by its spherical shape. The particle was subjected to a maximum electric field of 26 MV/m in the rf test of the 3-GHz cavity in which it was found.

the first time.

In the dc study, previously mentioned [202], the tip of the sharp tungsten needle was found to melt above a certain field value. Estimates for such a sharp-tip emitter, with a known geometric field enhancement factor, have predicted that the tip of the emitter melts when the current density reaches 10^{11} to 10^{12} A/m^2. The superconducting cavity temperature maps also indicate a current density of 10^{11} to 10^{12} A/m^2 for the partially melted stainless steel and indium emitters at the maximum field. A rudimentary estimate [197] shows that a FN current density $> 10^{11}$ A/m^2 is sufficient to melt a μm-radius, μm-length cylindrical metallic particle (e.g., normal conducting niobium) within 10 ns, provided that the particle is thermally isolated from the substrate.

Recent studies at Saclay [205] suggest, however, that the thermal contact between a typical particle and the substrate may be high enough so that the FN current alone is insufficient to cause melting. One must consider other possible heat sources, in addition to the ohmic heating from the FN current. One possible source for extra heat is the ionized residual gas, or the gas desorbed from the heated tip and its surroundings, resulting in an ion current which bombards the emitter. We will return to other evidence that points to the presence of gases, and the important role played by gases in heating up a field emission region.

For reasons we will discuss later, processed emitters are much easier to locate in the SEM than unprocessed emitters. Upon examining more than a hundred processed emitters in niobium cavities, the Cornell studies found a number of foreign metal residues (e.g., copper, iron, and titanium). Metallic particles are therefore common examples of emitters in superconducting cavities. Besides metallic field emitters, residues of semiconducting materials have also been found (e.g., carbon and silicon). In a study of artificial emitters conducted at Saclay [205], particles of insulating aluminum oxide, deliberately introduced into a copper rf cavity, were also found to emit.

One of the characteristic features of field emission is the occasional abrupt turn-on and turn-off of current. Sensitive thermometry studies [206] have revealed cases when emitter turn-on is associated with the arrival of a particle. Studies at Saclay [205] on surfaces intentionally contaminated with iron particles showed that particles can become detached from the surface in the presence of the rf electric field, presumably when the repulsive force due to the surface electric field exceeds the adhesive force. This could be a simple mechanism for emitter turn-off, although we will discuss another one later.

Many of the contaminants found in field emitters can be traced to various stages of cavity preparation. Carbon and silicon are likely due to airborne dust particles. Silicon may also come from the borosilicate glass found in high efficiency particulate air (HEPA) filters, which are used in clean rooms. Copper may come from rf probe tips, copper gaskets for vacuum seals, and copper scrapers for cleaning the indium from flanges. Indium flakes may be generated from In-wire vacuum joints. Experience with 3-GHz cavities [204, 207] shows that an hour-long presoak of a cavity in nitric acid greatly reduces the occurrence of indium emitter sites. Titanium may originate from sputter-ion vacuum pumps. The cavity assembly tools, vacuum pipes, and TIG weld joints of stainless steel vacuum pipes are all possible sources of stainless steel and iron particles.

Depending on the level of cleanliness in surface preparation, there may be a large number of emission sites present in a superconducting cavity, with a variety of β_{FN} and A_e values. As determined from temperature maps, the β_{FN} and A_e values of emitters found in a number of tests at Cornell and CERN are shown in Figure 12.12 for 500-MHz [149], 1500-MHz [196] and 3000-MHz [204] single-cell cavities. The β_{FN} values were between 100 and 700, and the emissive area values were between 10^{-9} and 10^{-18} m^2. Judging from the field values at which these emitters were found, it is clear from Figure 12.12(a) that to reach high fields, surfaces must be prepared in such a way as to avoid the high β_{FN} emitters. On the other hand, Figure 12.12(b) suggests that there is no strong correlation between the emitter area, A_e and maximum field reached.

We now turn to the density of emitters found in superconducting cavities [196]. Figure 12.13 shows a sequence of temperature maps taken as the field level was raised from low fields to $E_{pk} = 31$ MV/m. The number of sites increases with field level. Note the emitters observed at 18 MV/m and 26 MV/m that processed. At the maximum field in this test, there were eight emitters identified that did not process. On the basis of many temperature mapping experiments on several cavities, the density of emission sites is shown in Figure 12.14. In the next section we will discuss the other data shown in Figure 12.14, — i.e., the density of sites found in dc studies. An important feature is that the density of observed sites increases exponentially with increasing field. Thus it is not only the field emission current which grows exponentially with field, it is also

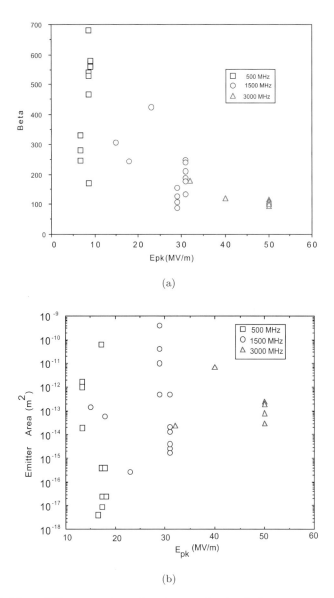

Figure 12.12: FN properties of emitters obtained from temperature maps of single-cell 500-MHz cavities studied at CERN, and single-cell 1500 and 3000-MHz cavities studied at Cornell. (a) β_{FN} (b) A_e.

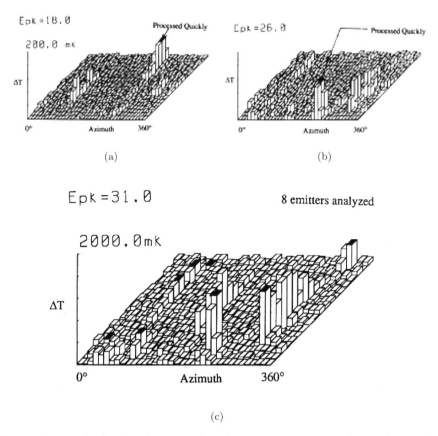

Figure 12.13: A selection from a series of temperature maps taken at increasing field levels in a cavity prepared by standard chemical treatment. Note that the emitters which appear at 18 and 26 MV/m process away. At the highest field of 31 MV/m, eight emitters are stable.

the number of active emitters. Therefore, in the presence of field emission, it is very difficult to raise the field with a fixed amount of rf power, unless the emitters process. Consequently, it is important to invent ways of reducing emitter density (i.e., cleanliness) and also to understand the factors that govern emitter processability, so that we may eliminate emitters when they do appear.

12.5 DC STUDIES OF FIELD EMISSION

Motivated by the desire to reduce field emission in superconducting niobium cavities, dc field emission studies were conducted on niobium cathodes at the University of Geneva [208], CEA Saclay [209, 210], and at the University of Wuppertal [211, 212]. In all, the emission properties of many hundreds of emit-

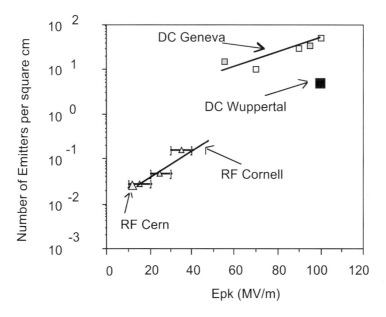

Figure 12.14: Density of field emission sites identified in dc (Geneva and Wuppertal) and rf studies (CERN and Cornell).

ters were characterized. A large number of emitters were subsequently studied with microscopy. Emitters on copper and gold cathodes were also studied. Artificial emission sites were introduced to further improve understanding of the nature of field emitters. These were most often carbon, iron, nickel, molybdenum disulfide, alumina, and silica particles. The lessons learned from the rf and dc studies as well as the models that have emerged from both the rf and dc studies are general enough to apply to most field-emitting surfaces. The studies also suggest important future directions to follow for reducing emission in cavities.

Figure 12.15 shows the pioneering Geneva field emission scanning apparatus using dc fields up to 200 MV/m in a UHV environment. A similar apparatus was used at Wuppertal. At Saclay, the field-emission scanning device was installed inside a commercially made SEM chamber. The vacuum for the Saclay studies was not as good as the Geneva and Wuppertal studies, so that their results may have been more strongly influenced by the presence of gases.

Generally, the sample (cathode) is moved in a raster pattern while a high voltage is applied to the needle-shaped anode. When the current exceeds 40 nA, the voltage is regulated to keep the current constant. In the plots of voltage vs. position, the emission sites are represented by two dimensional peaks. The sample can be scanned with a variety of microtips with different diameters. For low-resolution scans, the tip is 1 mm in diameter and the gap is 0.2 mm. A high-resolution scan is made with a micron-radius tip and a gap of 20–50 μm. The gap spacing is always much greater than the tip radius for the high-resolution

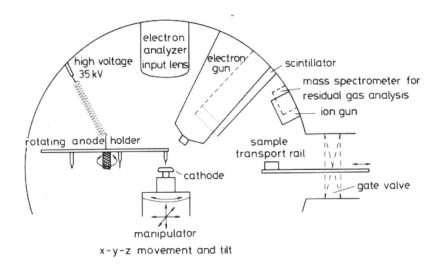

Figure 12.15: Apparatus for scanning a surface with a high-voltage needle, and analyzing emission sites. (Courtesy of Geneva.)

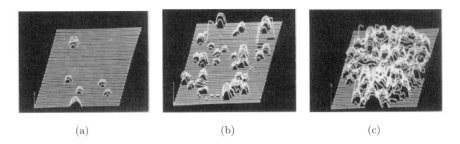

Figure 12.16: Emission sites found on a cm^2 sample of niobium using dc fields (left) 50 MV/m, (middle) 90 MV/m, (right) 100 MV/m. (Courtesy of Geneva.)

studies.

Figure 12.16 illustrates the increase in the number of emitting sites observed as the field increases. The number of sites grows rapidly with field, as was found in the rf cavity studies. Actually the dc studies chronologically preceded the rf emitter microscopy studies.

Comparing the density of sites found in dc at 50 to 100 MV/m with the density of sites observed in rf studies at less than 40 MV/m (Figure 12.14), we can infer that field emission will continue to be a major obstacle for reaching fields of 100 MV/m. Note that the Wuppertal dc samples, which were prepared in a class 100 clean room, in a manner similar to superconducting cavities, show a smaller density of sites compared to the Geneva dc samples, which were exposed to ordinary room air. Nevertheless, the Wuppertal samples still

DC STUDIES OF FIELD EMISSION

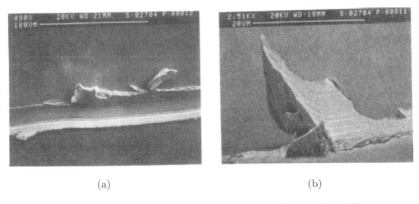

(a)   (b)

Figure 12.17: SEM micrographs of a scratch on a Nb surface. The associated projections were found to be strong field emitters. The melting, visible in (b), occurred at the tip of the scratch (Courtesy of Saclay.)

show a significant density of sites, even though they were prepared in a clean room. Considering how emitter densities grow from ≈ 0.1 cm^{-2} at 50 MV/m to ≈ 10 cm^{-2} at 100 MV/m, we must be prepared to expect a very large number of sites in a multicell accelerating structure, with a typically enormous surface area of $\approx$ m^2.

Emitters found in dc studies had β_{FN} values ranging from 50 to 500, with 100 as the most probable value. Emissive areas spanned the range from 10^{-9} to 10^{-18} m^2, with 10^{-12} as the most probable value. These emission characteristics are in rough agreement with emitter properties found in rf studies. A more careful study at Wuppertal revealed that A_e values span 18 orders of magnitude, from 10^{-4} m^2 to 10^{-22} m^2. The largest emitter areas are totally inconsistent with the physical size of emitters, which are 20 μm at the most. The smallest emitter areas are equally unphysical, i.e., smaller than atomic dimensions. Thus the effective emitting area must involve other physical parameters. A relatively new result from Wuppertal is that large β_{FN} values are correlated with small A_e values, and vice versa. Their studies showed that sites with $\beta_{\text{FN}} > 100$ almost always have $A_e < 10^{-13}$ m^2.

Once a site has been identified in the dc field emission scanning apparatus, it can be studied by the surface analytical tools incorporated in the field emission scanning device or by off-line instruments. One of the most important results is that the anticipated sharp whiskers are not found at emission sites. There is also no correlation between the location of emitters and grain boundaries, ruling out the frequently quoted possibility that a step at a grain boundary is a cause for field emission via geometric enhancement.

Certainly, as the classical, geometric interpretation suggests, sharp, metallic projections, when found, are strong emitters. For example, if the surface is scratched, the projections at the edge of the scratch (Figure 12.17) are field emitters [210]. However, scratches are rare on carefully prepared surfaces.

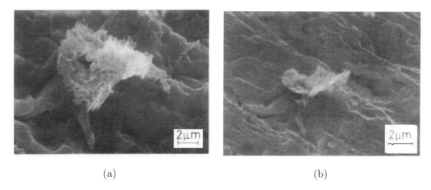

Figure 12.18: (a) An emission site analyzed to be a carbon particle (b) same as (a) but viewed at a different angle. (Courtesy of U. of Geneva.)

Observed emitters almost always turn out to be "metallic" microparticle contaminants. By metallic we mean that they are usually found to be electrically conducting. Figure 12.18 shows an electron microscope picture of a carbon flake emitter found in the Geneva dc studies. A particle was almost always found at an emission site. *When a region free of particles was probed, there was no intrinsic field emission up to 200 MV/m.*

Figure 12.19 shows field emission sites found in dc studies at Saclay. In all cases, the particles are conducting. The foreign elements found in the emitters are listed in the captions. Note that none of the particles constitute the sharp asperity that would be needed to explain a β_{FN} value of several 100. However, the particles do have jagged subfeatures, the importance of which we will discuss later.

A compilation of foreign elements found in particulate sites from the studies at four laboratories is shown in Table 12.1. The Cornell data is primarily from processed emission sites found in superconducting niobium cavities [197]. The physical size of the particles ranges from 0.5 μm to 20 μm, with 1 μm as the most probable value. There was no correlation between A_e, the FN area, and the size of the particles. A remarkably important finding is that only 5–10% of the total number of foreign particles present on the surface actually emit, a topic we will discuss in depth in the following section.

So far, the range of FN properties, the typical morphology and the elemental composition of emitters in rf and dc studies are very similar. To test the exact same emitter under rf and dc conditions, iron particles were placed on a cathode and studied in a dc field emission microscope at Saclay. Subsequently, the cathode (and the particles) were transferred into a copper rf cavity. In both dc and rf fields, the β_{FN} and A_e values found were very nearly the same.

Insulating particles on the other hand were found to behave differently in rf and dc fields. In artificial contaminant studies at Saclay, Al_2O_3 and SiO_2 particles on niobium or gold surfaces did not emit in a dc field, but were found

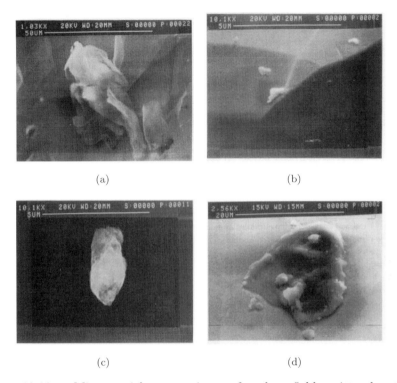

Figure 12.19: Microparticle contaminants found at field emitter locations. Foreign elements found were (a) C, N, O, (b) Ag (c) Si, O, (d) Ni (Courtesy of Saclay.)

to emit strongly when placed in a copper rf cavity. In the dc studies at Geneva, insulating particles, identified by their tendency to charge up during SEM examination, also did not emit. One possible explanation for this difference is that the insulating particles heat up to very high temperatures in rf fields and become more conducting due to excitation of electrons across the insulating band gap. Another possibility is that they become thermionic emitters. Within this interpretation, the "metallic" nature of particles remains the common denominator for all emitters.

12.6 A BRIEF LOOK AT THE IMPACT OF FIELD EMISSION STUDIES ON CAVITY PERFORMANCE

The observations on dc and rf emitters support the widely emerging view that micron-size "metallic" contaminant particles are predominantly responsible for field emission. Given the absence of naturally growing whiskers on the rf surface, it does not help to invest effort to obtain a smooth surface by mechanical

Table 12.1: Foreign elements found at emission sites

Element	Geneva DC	Saclay DC	Wuppertal DC	Cornell RF[a]
Ag	yes	yes		
Al	yes	yes	yes	
C	yes	yes		yes
Ca	yes	yes	yes	yes
Cl		yes		
Cr	yes	yes		yes
Cu	yes		yes	yes
Cs			yes	
F		yes		yes
Fe		yes		yes
In				yes
K		yes		
Mg		yes		
Mn	yes			yes
N		yes		
Na		yes		
Ni		yes		yes
O	yes	yes	yes	yes
S	yes		yes	
Si	yes	yes	yes	yes
Ti		yes	yes	yes
W	yes		yes	
Zn		yes		

[a] Elements found in processed sites in rf cavities

polishing. The polishing grit may even be harmful because of the additional contaminants that it may introduce. Instead, cavities should be assembled in dust-free clean rooms to minimize contaminants. The rinsing of the cavity after etching should be done with high purity liquids, filtered for micron and submicron particles.

A controlled exposure study [213] showed that Class 100 quality air and high-purity filtered methanol are not a serious source of emitters. Heat-treated Nb cavities (1400 °C), previously shown to be field emission free up to $E_{pk} = $ 30–40 MV/m, were deliberately exposed to ordinary room air passed through a 0.3-μm filter. Emission free cavities were also exposed to clean methanol. In these exposure tests, the cavities remained emission-free up to the same field levels. However, if the niobium surface was freshly etched with acid, the observed density of sites increased, and heavy field emission set in at $E_{pk} > $ 20 MV/m. Clearly, there are challenges in cleaning a freshly etched niobium surface. Perhaps foreign particles stick better to a newly etched surface. Perhaps

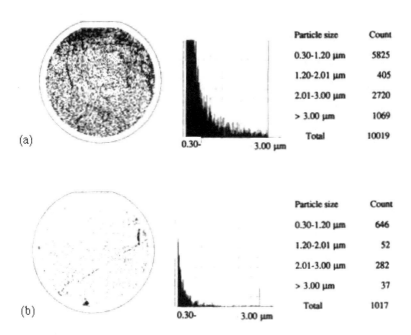

Figure 12.20: (a) A 100-cm^2 silicon wafer disk prepared by exposure to standard chemicals and cleaning techniques used for superconducting cavities shows a large number of contaminant particles detected by a laser scanner. (b) The same disk, after high-pressure rinsing, shows a substantial reduction in particle count. (Courtesy of KEK.)

there are minute impurity particles within the bulk that are exposed by chemical etching. Perhaps the water rinsing procedures are inadequate. We are still in need of additional cleaning measures, or emitter destruction measures.

High-pressure ($\approx$ 100 bar) water rinsing has recently emerged as a promising candidate for particle removal [140]. A jet of ultrapure water is used to dislodge surface contaminants resistant to conventional rinsing procedures. The effect of high-pressure rinsing was studied at KEK [214] using a silicon wafer as the test surface and laser scanning to detect the superficial particles. Figure 12.20(a) shows a large number of foreign particles found when a silicon disk was exposed to cavity treatment chemicals outside the clean room, rinsed with water used to clean niobium cavities, and dried in a Class 100 clean room. The idea was to expose the disk to the same environment as the cavities.

There are more than 10,000 particles accumulated over an area of 100 cm^2 as in Figure 12.20(a). This number is consistent with the observed emitter density of 10 cm^{-2} at 100 MV/m provided we consider that only a few percent of the particles present are likely to be emitters. As the distribution of Figure 12.20 shows, most of the particles are between 0.3 to 1 μm in size. When a similarly prepared disk was subjected to high-pressure rinsing (HPR) as the

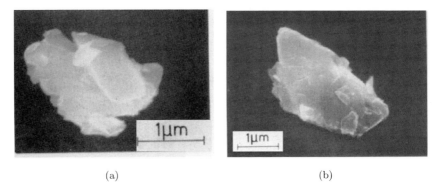

Figure 12.21: SEM micrographs of two particles of MoS$_2$ from an artificial field emitter study: (a) emitting, (b) nonemitting. (Courtesy of Geneva.)

final treatment, the number of particles was drastically reduced, as shown in Figure 12.20(b). Corroborating evidence for the benefits of HPR comes from the Wuppertal: dc field emission scans also show fewer emission sites on cm^2 Nb samples after HPR. We will discuss the benefits of HPR for niobium cavities in the next chapter [141].

12.7 NATURE OF FIELD EMITTERS

It is remarkable that in all three dc studies, only 5–10% of the total number of foreign particles actually present on the surface were found to emit. When particles of MoS$_2$ were deliberately introduced, some were found to emit in a dc field, as for example the particle of Figure 12.21(a); but other particles, which appeared to have very similar geometrical characteristics did not emit, as for example the particle of Figure 12.21(b).

12.7.1 The Tip-on-Tip Model

What makes a micron-size conducting particle into a field emitter?

The jagged structure of the stainless steel and indium particles (Figures 12.10 and 12.11) suggests a simple geometrical field-enhancement interpretation. In a definitive dc field emission experiment conducted at Saclay [215], with intentionally introduced particles, it was found that smooth spherical iron or nickel particles do not emit, but jagged iron particles emit strongly. One example of each particle is shown in Figure 12.22. Iron particle emitters are frequently found in cavities. Therefore, iron is a suitable candidate for artificial emitter studies. This experiment shows that microgeometry plays an important role in field emission.

NATURE OF FIELD EMITTERS

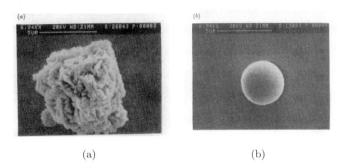

(a)   (b)

Figure 12.22: SEM micrographs of two nickel particles from an artificial field emitter study: (a) jagged shape and emitting (b) smooth and non-emitting up to 100 MV/m. (Courtesy of Saclay.)

A simple interpretation for β_{FN} would be that the particle as a whole enhances the field by $\beta_1 \approx 10$ and smaller protrusions on the particle further enhance the field by $\beta_2 \approx 10$. The product $(\beta_1 \beta_2)$ is sufficient to explain observed values of $\beta_{FN} \approx 100$. Static electric field calculations for a tip-on-tip geometry support the idea that field enhancement factors can be cascaded [216]. All the emitting particle pictures we have presented show jagged features. As previously mentioned, the microtip features also offer a simple explanation of current instabilities; namely, when one tip melts and becomes smooth, the local β_{FN} value decreases, and emission from another tip takes over.

Another geometric effect can arise when an uncharged metallic particle polarizes in the presence of an electric field. In a study conducted at Saclay [209], iron particles were found to align in the rf field so that the particle's longest axis became perpendicular to the cavity wall (Figure 12.23). The erect particles produce a geometrical field enhancement and also weld themselves to the surface at the base, presumably during field emission. The stand-up effect may also explain emission turn-on at a certain onset field. In extreme cases, several particles originally lying flat on the surface, adjacent to each other, were found to be arranged one above the other after application of rf. It is yet not known whether these effects are related to the magnetism of the iron, or whether the same behavior can be expected for all conducting particles.

12.7.2 The Role of the Interface

Results from field emission studies that involve heat treatment [208, 211, 212] suggest that the interface between the particle and the substrate may also play an important role. For example, as shown in Figure 12.24, heating a Nb surface with emitting particles to 1400 °C renders the surface emission-free. In some cases, the responsible particles also disappear, presumably by evaporation or

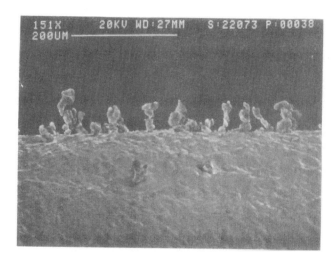

Figure 12.23: SEM micrograph of iron particles sprinkled on a niobium surface and placed in the high field region of a copper cavity. Strong emission was observed at low fields. Subsequently, many particles were missing. The remaining particles had aligned themselves perpendicular to the surface. Some piled up on each other. In all cases, the base of the particle appeared welded to the Nb surface. Note that these particles are rather large compared to emission sites naturally found in cavities. (Courtesy of Saclay.)

dissolution. But, in other cases the original particles are still present on the surface. One may argue that the jagged edges on the surviving particles are made smooth by heating to 1400 °C. However, a surprising result is that heating an *emission free* Nb surface to temperatures between 200 °C and 600 °C converts some nonemitting particles into emitters. It is unlikely that heating to 200 °C will make a smooth particle into a jagged one.

One possible explanation for the new activation is that the interface between the particle and the underlying surface is affected by the heat treatment. Another possibility is that reheating produces new adsorbates; for example, excess sulfur was found on reactivated sites [208]. But we still need a deeper understanding of the influence of the interface and adsorbates on β_{FN}.

12.7.3 The Metal–Insulator–Metal Model

Helpful information about the emission process comes to light from studies at the University of Aston on the electron spectra from field emitters using an anode probe-hole technique [217]. A high dc voltage is established between a plane cathode and an anode which has a 0.5-mm probe hole. The cathode is moved in front of the anode until a current from an emission site is observed through the hole. The electrons transmitted through the hole are either energy analyzed, or an emission image is obtained on a phosphor screen. A magnified emission image of a typical site is shown in Figure 12.26(a). They find that the electron spectra from emitters are very different from spectra expected, and obtained, for emission from a metallic sharp point. The natural emitter spectra are broader and shifted significantly below the Fermi energy (Figure 12.26(b)). The emission images show sharply defined crescent moon segments which increase in size with increasing current [217, 218]. The shifted and broad spectra suggest the involvement of an insulating medium in the emission process. The Aston group has proposed several models which involve an insulating interface between the emitter and the base metal; we will discuss one that is based on a metallic particle on a metal surface, with an insulating layer in between.

The metal–insulator–metal (MIM) model is based on a small metal flake insulated from the surface by a thin layer [218]. The metal flake acts as an antenna, leading to a considerable field enhancement across the insulating layer. The insulator modifies the electric potential as shown in Figure 12.25. Here χ is the electron affinity of the insulator. In the presence of the locally high field, electrons tunnel into the insulator from the metal substrate. Here they acquire kinetic energy from the penetrating electric field and are "heated." The electron population at high energy is similar to that of a metal surface at a high temperature.

Those electrons with enough energy to pass over the surface potential barrier are emitted into the vacuum as in the standard thermionic emission process which follows the Richardson–Dushman law

$$j = \alpha T_e^2 \exp\left(-\frac{e\chi}{k_B T_e}\right). \tag{12.9}$$

Here α is a constant $= 1.2 \times 10^6$ Am^{-2}K^{-2}. The electron temperature T_e is given by the voltage drop dE/ε_r in the insulating layer.

$$T_e = \frac{2ed}{3k_B \varepsilon_r} E, \tag{12.10}$$

where d is the insulator thickness and ε_r is the dielectric constant. Substituting T_e in Equation 12.9, the emission becomes

$$I = \alpha A_e \left(\frac{2ed}{3k_B \varepsilon_r}\right)^2 E^2 \exp\left(-\frac{3\chi \varepsilon_r}{2dE}\right). \tag{12.11}$$

A_e is the effective emitter area. The dependence of I on E follows the FN relation, but β_{FN} takes on a new significance in terms of insulator properties (dielectric constant, thickness and electron affinity). A strong feature of this model is that the half-width of the emitted electron spectrum agrees with the observed field dependence. At the top metal layer the electrons are not thermalized, but coherently scattered, so that they are emitted into the vacuum without losing the kinetic energy which they gained from the field in passing through the layer. The crescent moon structure of the emission image (Figure 12.26(a)) is explained by Bragg scattering of emitted electrons from the metal layer above.

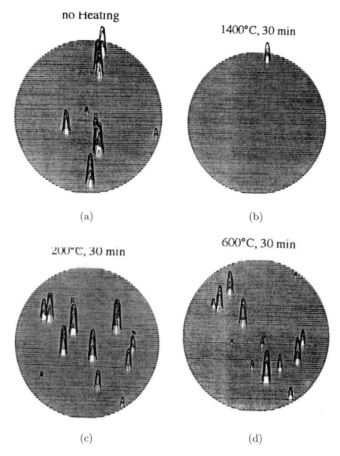

Figure 12.24: DC field emission scans to study the effect of heat treatment on emitters. In these plots of voltage vs. position (similar to Figure 12.16), the emission sites are represented by two-dimensional peaks. (a) Without heat treatment. (b) An emission-free niobium surface obtained by heating to 1400 °C. (c) Reheating a 1400 °C treated field emission-free surface to 200 °C. (d) Reheating a 1400 °C treated field emission-free surface to 600 °C. (Courtesy of Wuppertal.)

A useful feature of the model is that it provides a mechanism for emitter turn-on and for the instability of emission current. The total charge on the conducting particle is altered if some emitted electrons are captured or if secondaries are emitted from the metal particle. A sudden change of total charge could switch an emitter on or off. The MIM model has some attractive features and successes, but it is rather complex, involving several mechanisms.

For the case of niobium, we know that there is a 60-Å-thick natural oxide layer of Nb_2O_5 at the surface. The fact that all types of emitting sites can be

NATURE OF FIELD EMITTERS

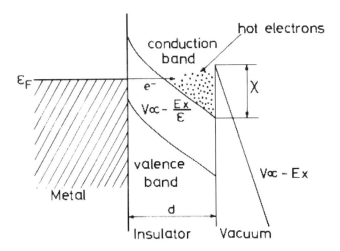

Figure 12.25: The presence of an insulating layer modifies the electric potential of Figure 12.3. (Courtesy of Aston.)

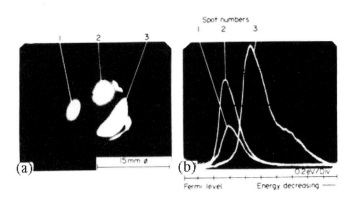

Figure 12.26: (a) Image of a field emission site obtained by the anode probe-hole technique. The site is composed of several subsites, reminiscent of multiple contaminant particles. (b) Individual emission spectra of subsites. (Courtesy of Aston.)

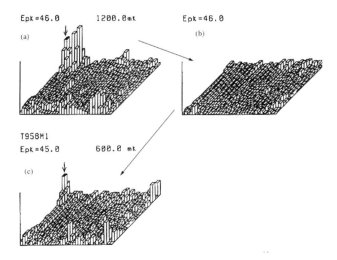

Figure 12.27: Temperature maps from a 1.5-GHz single-cell cavity. (a) Several field emitters are active at 46 MV/m. (b) On cycling to room temperature, emission disappears, presumably due to removal of condensed gas. (c) On admitting He gas into the cold cavity, field emission reappears at some of the same sites as in (a), presumably due to condensation of gases at the dormant sites.

deactivated by heating at 1400 °C for 30 minutes may be due to the dissolution of this oxide layer. It is well known that this temperature is high enough to yield an oxide free surface, as we discussed in Chapter 9. We have already mentioned that heat treatment between 200 °C and 800 °C results in strong activation of emission from formerly inactive particles, as for example on surfaces heat-treated at 1400 °C. Perhaps the insulating layer re-forms during the re-heat stage. The Wuppertal study showed that more sites are re-activated on low-RRR samples [212]. Since these samples are richer in interstitial oxygen impurities, it may be easier to form a more uniform insulation layer. However, the additional observation that thicker oxide layers have no influence on the density and properties of emitters presents a difficulty for the MIM model.

12.7.4 Condensed Gas and Adsorbates

There are several experiments [219] which show that condensed gases can activate field emission, presumably by adsorbing on the surface of dormant particulate sites. The emission landscape observed by temperature maps was occasionally found to change on warming a cavity to room temperature and re-cooling. Two emission sites apparent in Figure 12.27(a) are no longer active, or become dormant, after cycling to room temperature and cooling down again to 1.6 K, as shown in Figure 12.27(b). However, when helium gas was admitted into the cold cavity, one of the two former sites reactivated as shown in Figure 12.27(c).

In another experiment, an emission-free superconducting cavity (Fig-

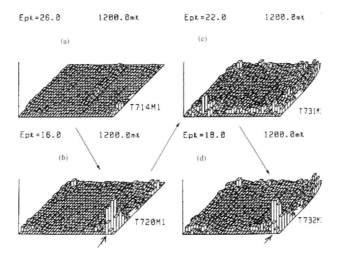

Figure 12.28: (a) Temperature maps of a 1500-MHz cavity. (a) The surface is emission-free at 26 MV/m. (b) Emitters appear when the cavity is exposed to oxygen. (c) Emitters are no longer active after cycling to room temperature. (d) The same emitters reappear when exposed to oxygen a second time.

ure 12.28(a)) was exposed to a steady stream of oxygen gas while the cavity was cold. Most of the oxygen probably condensed on the vacuum pipes, but some reached the cavity surface, as evidenced by a sudden increase in field emission, accompanied by a drop in the Q and the field. Figure 12.28(b) shows a strong emitter activated by the condensed gas. To rule out the possibility that the new emitter was a new particle, introduced accidentally with the oxygen stream, the cavity was cycled to room temperature. On returning to 2 K, emission at the previously activated site was absent. However, readmission of oxygen reactivates the *same* site. It is highly unlikely that a particle would land on the same spot as two separate doses of gas are introduced.

At CERN, cavities were exposed to a mixture of gases (H, H_2O, CO, and CO_2) typically found in an accelerator vacuum system. The onset of field emission changed from 10 MV/m to 7 MV/m if more than one monolayer of gas was adsorbed [220].

There are theoretical models [221] to show how emission properties are amplified by a superficial layer of absorbed gas. The presence of an adsorbed atom (adatom) can introduce an attractive potential within the triangular barrier as shown in Figure 12.29. Certain adsorbates can provide broad energy levels that lie above the conduction band minimum. Quantum tunneling calculations show that if an electron arrives from the metal with an energy approximately equal to a bound state of the adatom, its tunneling probability is greatly enhanced. The current can increase by as much as four orders of magnitude but the β_{FN} value does not change, i.e., only A_e changes. This process is called "resonant tunneling." The large range of A_e values observed for emitters may be due to a

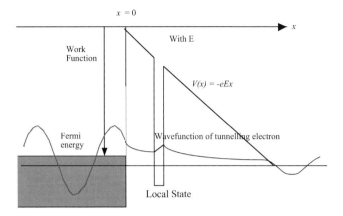

Figure 12.29: Attractive adsorbed atom state inside a potential barrier. Such states can lead to resonant tunelling.

variety of condensed gas species and condensed layer thicknesses.

12.8 INVESTIGATIONS ON PROCESSED EMITTERS IN RF CAVITIES

The term "processing" refers to the event when field emission is irreversibly reduced at a given field level. Microscopic studies conducted on processed emitters in superconducting cavities coupled with observations from other rf and dc field emission studies allow us to construct a picture of how an emission site processes. We will first present the evidence that has accumulated to form our picture, which we then bring together in the summary of Section 12.11.

12.8.1 Dissecting Single-Cell Test Cavities

A processing event [204] is shown in Figure 12.30 in terms of the rf behavior. As usual, when field emission starts, the Q falls sharply due to the exponentially increasing field emission current. Temperature maps record the original field emission, as shown in Figure 12.31(a). At about $E_{\text{pk}} = 29$ MV/m, when the incident rf power is increased, the peak field in the cavity jumps to 39 MV/m. There is a processing event. Figure 12.31(b) shows a temperature map taken after the processing event. Note that the "before" and "after" maps, which were both recorded at the same field, show that the field emission heating is substantially reduced at 29 MV/m. Upon dissecting the cavity and examining the predicted location in the SEM, the site shown in Figure 12.32 is found. The 200-μm site has a "starburst" shape with a 10-μm molten crater-like core region accompanied by micron-size molten particles within and near the crater. EDX analysis shows that the starburst region and the molten crater are all pure

INVESTIGATIONS ON PROCESSED EMITTERS IN RF CAVITIES   259

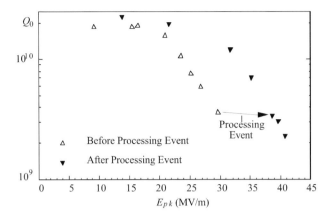

Figure 12.30: A processing event in a 1-cell 3-GHz cavity, realized by increasing the cw rf power. At 29 MV/m, the Q suddenly increased from 3×10^9 to $> 10^{10}$. After the event, it was possible to raise the surface field to 40 MV/m.

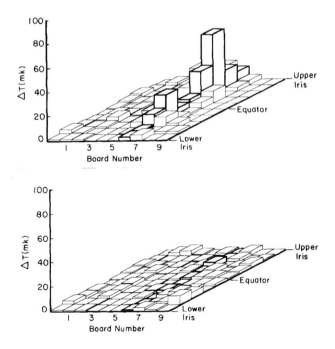

Figure 12.31: (Above) Temperature maps at 29 MV/m before the processing event in Figure 12.24. (Below) After the processing event and again at 29 MV/m.

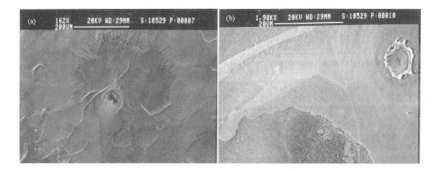

Figure 12.32: SEM pictures of the processed site found at the location predicted via temperature maps. (a) Low magnification and (b) high magnification of the crater region within the starburst of (a). A particle at the center of the molten crater as well one outside the crater were found to contain copper.

niobium, within detection limits. The particulate matter in the crater region, visible more clearly in the expanded Figure 12.32(b), reveals copper as the only contaminant. Presumably, a μm-size copper particle was originally responsible for the field emission.

The molten crater and splash-type features at the edges of the crater make it clear that emitter processing is an explosive event. The fact that the event can melt niobium on the cold surface suggests that the explosion takes place on a time scale much shorter than the thermal relaxation time. Later we will discuss studies that help to clarify the nature of the starburst feature.

It is interesting to note that the field emission decreased and higher field levels were reached, despite the appearance of molten droplets surrounding the craters. This is not surprising in view of the fact that studies at Saclay show that smooth particles do not emit, as we previously discussed. In some cases it has been found that processing events can cause a small increase in residual rf losses. In most cases, Q values $\approx 10^{10}$ can be maintained. We will return to this topic in the next chapter.

More than 20 one-cell, 1.5-GHz and 3-GHz cavities have been dissected after field emission sites were located with temperature mapping, and processed away. The processing event of Figures 12.31 and 12.32 was realized using about 10 W of cw rf power, which turns out to be barely sufficient to process emitters. In most cases, the processing of 3-GHz, one-cell cavities was carried out using 50 kW of pulsed ($\approx$ms) rf power. We will discuss other aspects of high-power processing (HPP) of rf cavities further in the next chapter.

In most cases, starbursts and molten craters were present in processed sites found in the dissected cavities. About a hundred starburst/molten crater sites have been examined. These sites are usually found near the iris of the cavity–i.e., in the high electric field region. In general, the higher the rf electric field, the larger the number of starbursts/craters found. The starburst has turned out to be a convenient means of locating processed sites because of its large

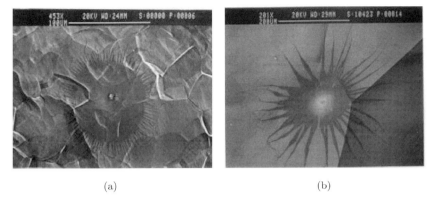

Figure 12.33: SEM micrographs of two distinctly different starburst shapes. No foreign elements could be identified with EDX in the melted central crater of site (a). The melted region of the processed site in (b) has indium droplets.

size, typically 100–500 μm. However, it is discernible only in the SEM; it is not visible in an optical microscope. Another interesting aspect about the starburst is that it tends to fade away after a few hours of air exposure, but it is stable in a vacuum or a dry nitrogen atmosphere. Of course, the molten craters remain unaltered on exposure to air. There are intriguing variations in starburst shapes as shown by the selections of Figure 12.33: One starburst is a compact dark disk with a fringe of short rays at the outer periphery; the other has a small central core and long discrete rays. Roughly speaking, the size of the starbursts scales inversely with the rf frequency, from 100 μm at 6 GHz to 500 μm at 1.5 GHz.

12.8.2 Demountable Mushroom Cavity Studies

In the early stages of the study of processed emitters, a special 6-GHz superconducting cavity, the "mushroom" cavity (Figure 12.34), was developed in order to expose a small region, the dimple, to a very high rf electric field [197, 110]. If field emission occurs in the cavity, the sites responsible are more likely to be on the dimple than elsewhere, because the electric field over 80 mm^2 of the dimple is greater than $0.5E_{\text{pk}}$. A TM$_{020}$ operating mode was chosen for the mushroom cavity because it provides a higher electric field at the center of the end plate than the other modes.

Another very useful feature of the mushroom cavity is that the dimple is part of a demountable end plate so that the cavity does not have to be destroyed for SEM examination. The entire plate fits into the SEM, so that the central dimple can be easily examined. Emitters that occur naturally in the dimple region were examined after the region was exposed to a high rf electric field. In more than 50 tests, surface fields at the dimple between 30 and 90 MV/m were reached. SEM examination revealed starbursts with molten craters and, in many cases,

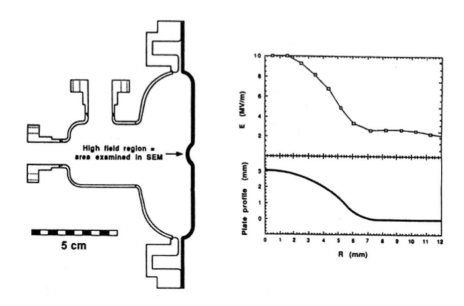

Figure 12.34: (a) Mushroom cavity operated at 6-GHz. The cavity is fabricated from a beam tube and a half-cell of the 3-GHz accelerating cavity. Coupler ports are introduced into the beam tube. The demountable end plate has a central dimple where the electric field is highest. A choke joint at the "equator" helps reduce the current at the indium joint. (b) Electric field distribution near the dimple for the TM_{020} operating mode.

foreign molten particles near the craters. In fact, the mushroom cavity study preceded the one-cell, 3-GHz and 1.5-GHz studies. Thus the first starbursts and molten craters were found in the mushroom cavity. In all, several hundred starburst/molten crater sites have been studied via the mushroom cavity.

Although Figure 12.32 shows a single crater from the processing event, it is more usual to find multiple craters. Sometimes a central crater within a starburst is surrounded by a number of "satellite" craters. Figure 12.35 shows a variety of crater patterns discovered.

Atomic force microscopy (AFM) images were obtained [222] for the cratered region and surrounding ripples in Figure 12.35(d). With the topological imaging capability of AFM, the large craters are found to be depressions 1 μm deep with raised edges of the same height. The ripple patterns were found to be topographical features which emanate from a point near the central crater and propagate on the order of 100 μm. The ripple wavelength is several hundred nm and depends on the rf frequency: it is 400 nm at 3 GHz and 200 nm at 6 GHz. The crest-to-trough amplitude is 10–35 nm. One possibility is that the ripples are created by a disturbance, modulated by rf forces, that produces oscillations in a molten niobium surface. The ripples freeze as the surface cools below the

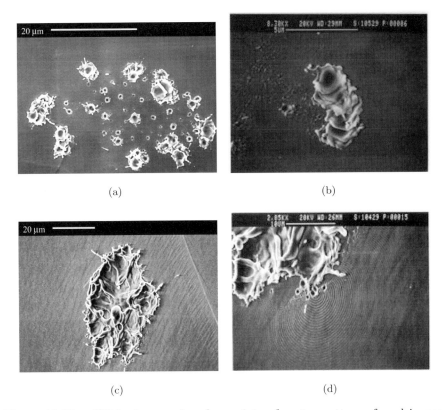

Figure 12.35: SEM micrographs of a variety of crater patterns found in processed emission sites (a) A cluster of craters seen as a tight ensemble about a central point. (b) Crater tracks suggesting a path defined by a sequence of craters. (c) Chaotically overlapped craters. (d) Ripples emerging from molten craters.

melting point. As with the craters, the ripples also survive exposure to air.

When molten particulate debris is found near craters, it can usually be analyzed by EDX. But often no debris is found. In such cases, thin layers of foreign metals have been detected [222] by the more surface-sensitive Auger method. Most of the impurities are seen within 2 or 3 crater radii of the crater center.

The comprehensive list of foreign elements found in the starbursts (Table 12.1) is based on the analysis of more than 100 processed sites. As mentioned before, there is a substantial overlap with the foreign elements found in dc field emission studies. Many of the contaminants elements can be traced to various stages of Nb cavity preparation.

It is worth pointing out that Auger studies show that a majority of the satellite craters have no impurities, especially when the secondary craters are

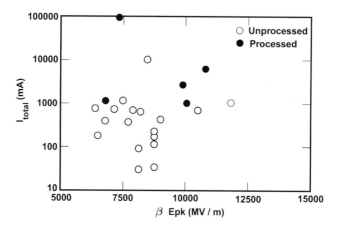

Figure 12.36: Total emission current for sites that do and do not process. There is no strong correlation between $\beta_{FN} E$ (current density) and the processability of an emitter. However, emitters that reach a large *total* current appear more amenable to processing. Values of $\beta_{FN} E_{pk}$ and total current are based on β_{FN} and A_e values obtained for emitters in 3-GHz one-cell cavities.

more than a few crater radii away. This observation suggests that an explosion at a primary contaminant-loaded emission site can trigger explosions on the Nb surface nearby without the need for particulate contaminants at the secondary sites. We will suggest a mechanism that is responsible for producing secondary explosions in Section 12.11.

12.8.3 Copper Cavity Studies

Craters have been found in copper rf cavities after high-power conditioning to reach very high surface fields. Extensive tests have been carried out at SLAC [223] on 3-GHz and 11.4-GHz copper cavities using high power (50–100 MW) pulsed (μsec) rf. Record surface electric field levels between 200 and 600 MV/m were reached after many tens of hours of conditioning. The processing is characterized by a sequence of vacuum breakdown events. During these events the emitted current is observed to increase by a factor of 20–30 and the vacuum degrades from a pressure of 10^{-8} torr to 10^{-7} torr. There is a significant increase in CO and CO_2 among the residual gases. The pressure bursts suggest that gas evolution plays a role in breakdown. Subsequent visual inspection of high field regions shows numerous crater areas, with many overlapping craters. After processing to very high fields (200 MV/m) the entire iris is pock-marked by overlapping craters.

12.8.4 Emitter Processability and Fowler–Nordheim Properties

We have seen that partial emitter melting can occur when the field emission current density exceeds 10^{11} A/m^2. A key question is whether melting alone is sufficient to trigger the processing event. The answer appears to be no. Many emitters in one-cell, 3-GHz cavities were characterized [204] by obtaining their β_{FN} and A_e values via temperature maps, as discussed before. The emitters were then processed with pulsed rf power of 2–50 kW, and 0.2–1 ms pulse length, as will be discussed in the next chapter. Even when large $\beta_{FN}E$ values were reached, certain emitters did not process, as shown in Figure 12.36. Conversely, some emitters did process even though the $\beta_{FN}E$ values were relatively small. Therefore it appears that a high current density and local melting alone are *not* sufficient for an emitter to process. If, however, the emitters are classified according to the *total extrapolated current* emitted at the processing field, then (in almost all cases) emitters which exceeded a total extrapolated current of $\approx$ 1 mA fell into the processed category. The extrapolated current was obtained in the following manner. The FN properties β_{FN} and A_e were first measured at the cw field for which the emitter was active and the emission current subsequently calculated from these properties at the *processing field*. These results have strong implications for the model of processing and the role played by gases, as we will discuss in the next section.

12.9 DC VOLTAGE BREAKDOWN STUDIES

Craters and molten areas have also been observed in dc emission studies, when vacuum breakdown with a high dc voltage results in arcing. Extensive studies have been performed in dc, and comprehensive reviews are available [199]. As with cavities, the craters found in the dc studies have characteristic sizes on the order of microns. There are often multiple overlapping craters. DC field emission studies using very short pulses show that spark formation times are between 10^{-9} and 10^{-6} s. In such a short time scale, the emission site is thermally isolated from the surface. As a result, it is possible to find molten metal in superconducting cavities which operate at liquid He temperatures. It is reasonable to surmise that the processing of emitters in superconducting cavities also results from vacuum breakdown and is accompanied by the formation of an arc, or spark, produced by the rf electric field.

Examination of cathode spots during dc arcs [224, 225, 226, 227] reveals luminous spots, attributed to the presence of a plasma. However, the starburst feature has never been reported in dc studies. Part of the reason for this is that niobium is not a usual cathode material for dc studies. In addition, as we shall explain below, the formation of the starburst depends on the special chemical composition of the niobium surface. Also, starbursts tend to fade away in a few hours on exposure to air, so that the niobium surface must be preserved in dry nitrogen or vacuum.

Pursuing the close similarities between rf processing and dc high-voltage

Figure 12.37: SEM micrograph of a starburst and central molten crater found on a niobium surface at the location of a dc spark at 100 MV/m.

breakdown, experiments were carried out in UHV to initiate a spark across a small gap between two niobium electrodes with a high dc voltage [197]. Using a field of about 100 MV/m, a millimeter-radius needle was scanned across a niobium cathode until there was a spark at a particular location. The niobium surface was prepared in the same fashion as cavity surfaces. SEM examination of the niobium cathode in the sparked area showed starbursts with molten cores (e.g., Figure 12.37), similar to those found in niobium superconducting cavities. This result supports the conclusion that starbursts in rf cavities are also produced during a spark or microdischarge. The size of the dc starburst is smaller (≈ 80 μm) than that found in superconducting cavities, which may be related to the ≈ 100-μm anode-to-cathode gap used to set up the high field. In rf cavities there is no anode to limit the spreading of the plasma, so that the starburst can extend over hundreds of microns. The physical factors that determine the total extent of a starburst are not well understood. However, the starburst size is correlated with the rf frequency [170].

Auger studies shed light on the nature of the starburst [222]. Comparing Auger spectra inside and outside a starburst, fluorine is found outside, but not inside. A 2-D fluorine map (Figure 12.38) of the surface near the starburst clearly showed a good overlap between the geometric shape of the starburst and the region of fluorine depletion. Following argon sputtering of the surface, we determined that the fluorine layer present everywhere on the niobium surface was at least 50 Å deep, but not more than 1500 Å. Fluorine presumably comes from the HF in the acid etching solutions used to prepare the niobium surface. Later, dc spark studies on niobium cathodes at Saclay confirmed the presence of starbursts and craters. They found that starbursts are easily produced on

THE ROLE OF GAS IN PROCESSING

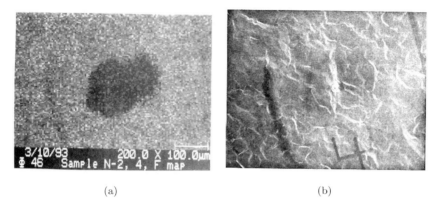

(a) (b)

Figure 12.38: (a) A two-dimensional Auger scan of a starburst region restricted to look for the element fluorine. The dark region means an absence of fluorine. (b) SEM micrograph of the same region. Note that the starburst has partially faded due to exposure to air during the time it took to transport the specimen to the Auger facility. While the sample was in the SEM, the electron beam was used to burn in a set of "brackets" around the starburst, so that it would be easier to locate the same region in the Auger system.

a freshly etched cathode, but not on niobium that has been exposed to air for some time. These results suggest that the starburst is a pattern formed on the niobium by the plasma cloud that forms during the spark. The superficial residue of fluorine is preferentially removed by ion or electron bombardment in the plasma. Presumably other superficial layers, such as water or hydrocarbons, may also be removed by the plasma, but these re-grow relatively fast on exposing the cavity for examination.

There is additional evidence of intense activity in the plasma region. In Figure 12.39, we see a small crater near the center of a starburst and relatively large 5-μm iron particles distributed over the starburst region, all the way out to the periphery. It is unlikely that such large particles could have been ejected from the crater, considering the small size of the crater. Rather, it appears that the temperature over the starburst region was high enough to melt iron particles that were lying on the surface.

12.10 THE ROLE OF GAS IN PROCESSING

There is evidence to indicate that the starburst can form *before* the explosion that eventually destroys the emitter. The site of Figure 12.40(a) shows a starburst surrounding an emitter that had not yet processed: i.e., it was still emitting strongly at the end of the rf test. The core of the starburst area (Figure 12.40(b)) does not contain a major crater that would indicate the occurrence of an explosion. Instead, there is a flat region suggestive of a chemical residue.

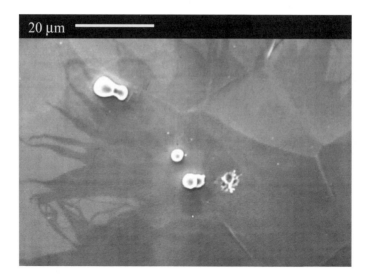

Figure 12.39: SEM micrograph of a starburst with a small molten crater found in the mushroom cavity. Large molten stainless steel particles located all the way from the crater to the periphery suggest that iron particles present on the surface were melted by the plasma. The temperature at the outer periphery of the starburst can therefore be quite high.

Another example of a starburst surrounding an active and *nonexploded* site is shown in Figure 12.41(a). Temperature maps showed that the emitter was still active at the end of the rf test. Once again, the core of the starburst does not show any major crater. At increased magnification (Figure 12.41(b)) we see a 50-μm liquefied region, with titanium and carbon as foreign elements. Unlike other processed sites, most of the contaminant particle is still intact, although it is melted. The large particle did not explode. An order of magnitude estimate shows that ohmic heating from the FN current is not sufficient to melt a region as large as 50 μm. The existence of such a large molten region suggests that some other mechanism (e.g., the high-temperature plasma) is responsible for additional heating of the emitter, even before the explosion.

Among the many starbursts found in the mushroom cavity, we found several cases where there were no molten craters within. We suspect that the responsible emitters were still active at the end of the test. It is quite likely that without an obvious exploded crater to guide the SEM exam, we missed locating the micron to submicron particle that was responsible for the emission that gave rise to the starburst. Once again, the starburst-without-crater points to the presence of a plasma surrounding an emitter before the explosion.

The gas to produce the plasma may come first from the metal atoms evaporated from the melting emitter. We have shown several cases in which a small part of the emitter melts. Additional gas could be evolved from thermal desorption of the surface layer of fluorine, water and hydrocarbons. It is well known

THE ROLE OF GAS IN PROCESSING 269

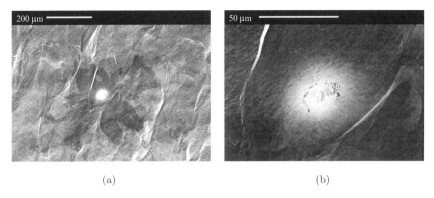

Figure 12.40: An emitter active at the end of an rf test showing a dark region resembling a starburst. (a) Entire starburst region. (b) Central region, magnified.

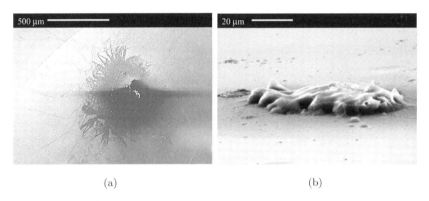

Figure 12.41: (a) SEM photograph of an active emitter surrounded by an irregularly shaped starburst. The framed portion is enlarged in (b), which shows a 70-μm large particle of Ti and C that has almost completely melted, but not exploded.

from Auger studies of a freshly prepared niobium surface that the first few monolayers are water and hydrocarbons. Once there is some gas, the field emission current will ionize the gas and the ions will bombard the surface, leading to more gas production. Indeed, the pressure may build up exponentially and eventually lead to the instability we recognize as the discharge.

A key fact that supports the importance of gas in exploding an emitter comes from helium processing experiments. In the next chapter we discuss the general experience with helium and how it has been used for many years as a tool to overcome field emission, although its efficacy is limited. In this particular example [170], the strong field emission evident in the Q vs. E curve (Figure 12.42(a)) was dominated by a single emitter as shown in the temper-

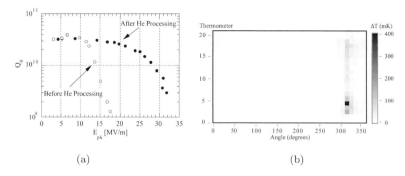

Figure 12.42: (a) Q vs. E curve before and after He processing a 1.5-GHz cavity. (b) Temperature map at 17.3 MV/m showing a strong field emission site prior to helium processing. This site was completely extinguished by helium processing. The abscissa is the azimuthal location of the thermometer. The ordinate extends from the upper iris of the cavity to the lower iris.

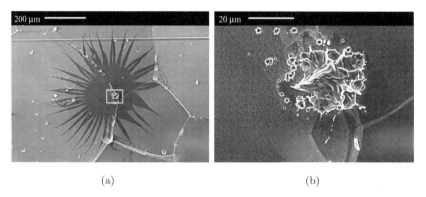

Figure 12.43: SEM picture of the helium-processed site showing (a) a starburst and (b) molten craters in an expanded view. Note that (b) is rotated by about 60 degrees relative to (a).

ature map of Figure 12.42(b). This emitter did not yield to processing at the maximum applied rf power. Once helium gas at a pressure of about 10^{-4} torr was admitted into the cavity, the emitter immediately processed, without any further increase of rf power. Upon examination of the cavity at the location indicated by the temperature map, we found the starburst and molten crater shown in Figure 12.43. We surmise that the addition of helium enhanced the plasma density to trigger the processing discharge.

12.11 SUMMARY—A PICTURE FOR FIELD EMISSION AND PROCESSING

There are a large number of particles on the rf surface, typically 100 particles/cm^2 with sizes between 0.3 μm and 20 μm. These particles are lodged on the surface during preparation, exposed from the bulk during chemical etching, or introduced at the cavity assembly or pump-out stages.

Only some ($\approx$ 10%) of these particles are field emitters between 20 and 100 MV/m. The metallic ones (i.e., the conducting particles) are most likely to emit. Particles generally have a very irregular shape, and the microprotrusions enhance the field emission characteristics. Other important factors are (a) the nature and quantity of condensed matter on the particle and (b) the interface between the particle and the substrate. Because of these many factors it is not surprising to find a large distribution in β_{FN} and A_e values. Some particles that are not field emitters may become emitters later if the interface changes, or if gas adsorbs, or if the electric field polarizes the particle and changes its orientation. Emission may stop abruptly if the repulsive force due to the electric field becomes stronger than the adhesive force so that the particle is ripped away from the surface. Emission may also activate abruptly in an emission-free cavity if a particle arrives at a high field region.

When the field increases and the emission current density exceeds 10^{11} A/m^2 the temperature at the emission region becomes high enough to melt a small region of the particle. A microprotrusion of the emitting particle may melt and cease to emit, but the overall emission from the particle will continue at some base level. When the cavity fields are raised for the first time and individual microemitters melt, the emission current is unstable until the susceptible regions are all melted.

Atoms evaporate from the melted regions. Ohmic heating from the FN current also degasses surface adsorbed atoms. The presence of gases plays a paramount role in emitter processing. A study is presently under progress [206], using the program MASK [228], to simulate the ionization of the gas by field emitted current and the attendant consequences. At 30 MV/m, emitted electrons will gain 30 eV within a micron of the rf surface, sufficient energy to ionize the gas. A chain of events then takes place on a very short time scale (nanoseconds) with the presence of gas playing a central role. As the field emission current ionizes the evaporated and/or desorbed gas, the ions are accelerated by the field toward the emission site. The ion current produces secondary ions and electrons, and heats the site further by bombardment, so that more gas is produced. A plasma is formed extending out to several hundred microns. Electron and ion bombardment from the plasma cloud cleans up the surface, leaving a physical pattern on niobium that that is devoid of surface residues, such as flourine. The plasma fingerprint has the shape of a starburst.

MASK simulations show that since the ions move slowly, a significant number can accumulate near the emitter, leading to substantial electric field enhancement. The amount of field enhancement depends on the *total current* from the emitter. Estimates based on available processed emitter data suggest that when the total current approaches the level of milliamperes the emitter will process; i.e., there are enough ions to initiate a discharge, which is the avalanche breakdown of the gas surrounding the site. At the core of the arc, the intense current can melt niobium, produce molten craters, vaporize the entire emitting particle, and leave a deposited film of the original contaminant on the crater. In many cases, the discharge event leaves behind molten debris. Plasma pressure during the discharge excavates the molten zone and ejects droplets. There may be multiple arcs between the ion cloud and the niobium, resulting in multiple craters from a single original emission site. The crater and other melted particles do not emit because they are smooth particles.

Does this model suggest ways to improve processing of emission sites? For a site with a particular value of β_{FN} and A_e, the field E must be increased to reach a $\beta_{FN}E$ value corresponding to an emission current density of $> 10^{11}$ A/m^2 to approach heating and melting at the site so that a sufficient gas density is created. But to process a site the total current must reach a threshold value near one mA. To reach the necessary field level, high rf power is required. Short pulses are sufficient, because the emitter explosion takes place very fast (nanoseconds) when the conditions are ripe. In the next chapter, we will discuss the experience of high pulsed power processing to destroy field emitters.

12.12 SIMULATING FIELD EMISSION HEATING

With thermometric diagnostics on cavities, as described in Chapter 8, one does not usually detect the power dissipated by field emission currents at the emitter location. Rather one observes the heat produced by the field emission electrons impacting the walls, after being accelerated by the cavity fields. The heat produces a temperature rise ΔT on the outside wall, which is detected by nearby thermometers. If an axisymmetric cavity is operated in the TM$_{010}$ mode, the electrons follow trajectories that lie entirely in the ρ–z plane of the emission site. Electrons emitted at different phases of the rf cycle follow different trajectories. Hence a "fan" of charge is produced (see Figure 12.2) and field emission heating is characterized by a line of heat at $\phi = \phi_{em}$ and/or at $\phi = \phi_{em}+180°$, where ϕ_{em} is the azimuth of the emission site. This situation is depicted in Figure 12.44, the line heating at 310° being very apparent. Figure 12.45 depicts the heating at 310° as a function of distance (S) from the cavity equator along the wall of the cavity. The asterisk marks the location of the emission site itself which clearly does not coincide with the peak in the heating.

To study field emitters it is important to identify the actual emitter location. We therefore outline the following three step approach as an example of how this can be done.

The field distribution of the cavity is obtained using a code such as SUPERFISH [70]. This need only be done for one field level. More details on this process are given in Chapter 2. The field distribution is then used for trajectory calculations in another program such as MULTIP [229]. The rela-

SIMULATING FIELD EMISSION HEATING 273

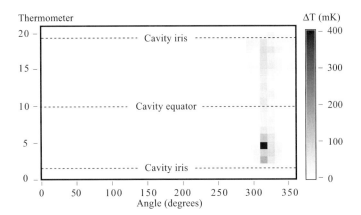

Figure 12.44: Unfolded temperature map taken at $E_{\text{pk}} = 17.36$ MV/m in a single-cell 1.5-GHz cavity. The abscissa represents the azimuthal coordinate around the cavity. Field emission heating is apparent at 310°.

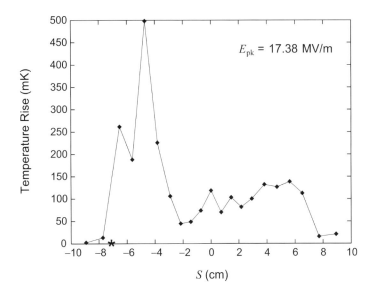

Figure 12.45: Temperature profile at 310° in Figure 12.44. The asterisk marks the location of the emitter. The low reading of the fourth thermometer from the left is probably due to an unusually poorly attached thermometer.

tivistic equations of motion for a charge q with mass m are

$$\frac{d\mathbf{x}(t)}{dt} = \frac{1}{\gamma}\mathbf{u}(t) \tag{12.12}$$

$$\frac{d\mathbf{u}(t)}{dt} = \frac{q}{m}\left[\mathbf{E}(\mathbf{x},t) + \frac{1}{\gamma}\mathbf{u}(t) \times \mathbf{B}(\mathbf{x},t)\right], \tag{12.13}$$

where $\mathbf{x}(t)$ is the position of the charge at time t, $\mathbf{E}(\mathbf{x},t)$ and $\mathbf{B}(\mathbf{x},t)$ are the electric and magnetic fields, respectively, and

$$\gamma = \frac{1}{\sqrt{1 - (|\mathbf{v}|/c)^2}}. \tag{12.14}$$

In (12.14), $\mathbf{v}$ is the velocity of the charge which is related to the proper velocity $\mathbf{u}$ by $\mathbf{v} = \mathbf{u}/\gamma$. The user supplies the peak electric field in the cavity (E_{pk}) which is used by the code to scale the cavity fields obtained with SUPERFISH. MULTIP divides the rf period into a user-determined number of time intervals of length $\Delta\varphi/\omega_0$, where $\Delta\varphi$ is the phase advance and ω_0 is the cavity angular frequency. At the beginning of each interval, an electron is emitted from location $S = S_0$ with an energy K_0. Its trajectory is then determined by integrating the equations of motion above (Equations 12.12 and 12.13) for a preset number of rf cycles, or until the trajectory crosses one of the cavity boundaries. When the latter occurs, the impact conditions are stored for later use. If desired, secondary electrons can also be included in the calculation. Figure 12.46 depicts the trajectories of electrons originating at the site marked in Figure 12.45. A two-dimensional plot is sufficient, since the trajectories lie in the ρ–z plane of the emission site. The electron trajectory "fan" is responsible for the line heating measured by thermometry.

So far, we have not included any characteristics of the field emitter itself. This is done in a third step using a program such as POWER [229], which combines several functions. It determines the total charge emitted in the various time intervals $\Delta\varphi/\omega_0$ for which trajectories were calculated by MULTIP. For niobium, the emitted current (in amperes) is given by the modified Fowler–Nordheim equation [230, 194]:

$$I_{\text{FN}} = 38.5\, A_e \frac{(\beta_{\text{FN}} E_{\text{em}})^2}{t^2(y)} \exp\left(-54640 \frac{v(y)}{\beta_{\text{FN}} E_{\text{em}}}\right), \tag{12.15}$$

where E_{em} is the electric field (in MV/m) at the emission site, A_e is the effective emission area in cm^2, β_{FN} is the field enhancement factor, $y = 0.00948\sqrt{\beta_{\text{FN}} E_{\text{em}}}$, and $v(y)$ and $t(y)$ are functions given in the literature [231]. Knowing the charge emitted in each interval, the program can compute the current density of the impacting electrons along the inside cavity wall. As an example, Figure 12.47 shows the current density along the wall, given that the emitter is located at the asterisk. Since all charges remain in the ρ–z plane, a linear density rather than an areal density is calculated.

SIMULATING FIELD EMISSION HEATING

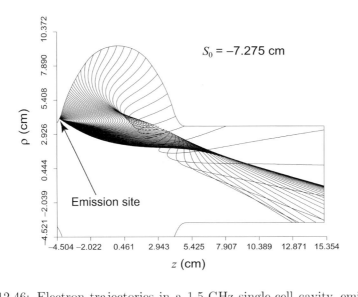

Figure 12.46: Electron trajectories in a 1.5-GHz single-cell cavity, emitted at intervals of 1/200th of the rf cycle. The trajectories lie in the ρ–z plane of the emitter. $E_{\text{pk}} = 17.38$ MV/m and the emission energy is 0 eV.

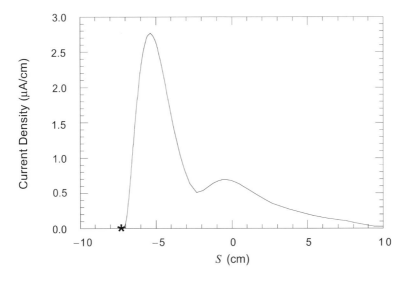

Figure 12.47: Field emission current density along the cavity wall for an emitter located at the asterisk. The cavity equator is at $S = 0$. ($E_{\text{pk}} = 17.38$ MV/m, $\beta_{\text{FN}} = 400$, $A_e = 1.58 \times 10^{-12}$ cm^2.)

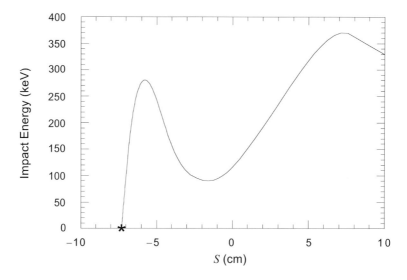

Figure 12.48: Electron kinetic impact energy along the cavity wall for an emitter located at the asterisk. ($E_{\rm pk}$ = 17.38 MV/m, $\beta_{\rm FN}$ = 400, $A_{\rm e}$ = 1.58×10^{-12} cm^2.)

Note that although some trajectories in Figure 12.46 leave the region of the cavity cell and enter the beam tubes, the current in these cases is very small. Therefore, the total energy carried by the charges into these regions is relatively small. For this reason, Figure 12.47 only depicts the current density within the cell region, where thermometers are attached at the outside surface. However, one should bear in mind that for certain emitter locations and cavity fields, the current in the beam tubes can be significant and will go undetected by a thermometry system restricted to coverage of the cell.

The electron trajectory calculations also yield the energy of the impacting electrons. A large variation along the cavity wall can be observed, as shown in Figure 12.48. POWER can determine the time-averaged linear power density deposited in the wall as a function of position (see Figure 12.49). At each point, the power density is proportional to the product of the current density and the impact energy at that location. In turn, the power dissipation is used to calculate the power flux at the helium side of the cavity wall. To achieve this, the user supplies the full width at half-maximum (FWHM) of the temperature distribution on the outside of the cavity wall due to a point heat source on the inside. This number is obtained from thermal codes which take into account the wall thickness, thermal conductivity of the wall, and the Kapitza conductance of the wall–helium interface (see Chapter 11). POWER then spreads the energy dissipated by the impacting electrons using a Gaussian distribution. By overlapping the distributions from each impact point, the approximate power flux the outside cavity surface is obtained.

Assuming the temperature measured by a thermometer for a given power

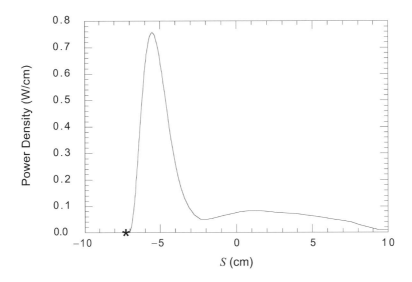

Figure 12.49: Simulated power flux along the inside of the cavity wall for an emitter located at the asterisk ($E_{\rm pk} = 17.38$ MV/m, $\beta_{\rm FN} = 400$, $A_{\rm e} = 1.58 \times 10^{-12}$ cm^2.)

flux ($\eta_{\rm T}$) across the cavity–helium boundary is known, the flux profile can easily be converted to a temperature profile. Typically, $\eta_{\rm T}$ is on the order of 2500 mK/(W/cm^2). However, $\eta_{\rm T}$ varies from one thermometer to another and from cavity to cavity, due to differing surface characteristics, variations in thermometer contact pressure, and intrinsic differences between thermometers.

We see that for a given field level, the temperature profile is primarily dependent on S_0, $A_{\rm e}$, and $\beta_{\rm FN}$. These parameters must be varied until a reasonable match between the simulated and measured temperature profiles is obtained.

It is helpful to consider the temperature profiles at several different field levels. This approach is especially useful in determining $\beta_{\rm FN}$ and $A_{\rm e}$. In many cases, $v(y)$ and $t(y)$ are approximately constant over the range of interest. Thus a plot of $\ln(\Delta T/E_{\rm pk}^2)$ versus $1/E_{\rm pk}$ for a thermometer yields a near-linear dependence, the gradient of which can be used to estimate $\beta_{\rm FN}$.[1] Analogous plots can be obtained from the simulations. A comparison of the two, as in Figure 12.50, allows us to estimate a reasonable value for $\beta_{\rm FN}$.

Once $\beta_{\rm FN}$ is known, the magnitude of the heating is used to determine the effective emission area $A_{\rm e}$. We obtain a series of simulated profiles, as in Figure 12.51(a) which match the measured data.

For cross reference it is useful to compare the total power dissipation in the cavity obtained from Q_0 versus $E_{\rm pk}$ measurements with the computed value from

---

[1] More precisely, one should plot $\ln(\Delta T/E_{\rm pk}^{3.5})$ versus $1/E_{\rm pk}$, because ΔT is a measure of the *time average* of $I_{\rm FN}E_{\rm pk}$. However, the exponential term in the Fowler–Nordheim equation dominates, so that little is lost by plotting $\ln(\Delta T/E_{\rm pk}^2)$ versus $1/E_{\rm pk}$.

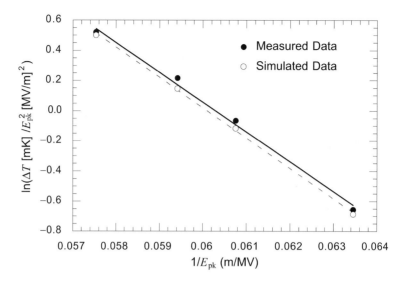

Figure 12.50: Comparison of Fowler–Nordheim plots of $\ln(\Delta T/E_{\text{pk}}^2)$ versus $1/E_{\text{pk}}$ for measured data and simulated data. In the simulation, the emitter is located at the asterisk in Figure 12.45. ($\beta_{\text{FN}} = 400$, $A_e = 1.58 \times 10^{-12}$ cm^2.)

POWER simulations. Provided that a single emitter is the dominant source of power dissipation in the cavity, the two will agree fairly well. The total emitted current can therefore be determined within about a factor of three. The current density, on the other hand, which depends strongly on β_{FN}, cannot be calculated precisely. The uncertainty in β_{FN} can be as high as ±20% for $\beta_{\text{FN}} = 250$. At typical field levels, this results in uncertainties of a factor of ten or more in the current density.

The response of different thermometers to the same power flux can vary by up to a factor of two in extreme cases, although a 30% variation is more typical. The variation in thermometer response is due to a combination of intrinsic differences between thermometers and the degree of thermal contact with the cavity surface. The low reading of the fourth thermometer from the left in Figure 12.45 is probably due to this. One might be concerned that this variation, combined with a thermometer spacing on the order of 0.75 cm, will make it exceedingly difficult to determine the location of an emitter. Fortunately, one finds that for emitters in high electric field regions (where the dominant ones usually are), the temperature profiles vary significantly for small changes in S_0. For example, a shift of the emitter in Figure 12.51(a) by 2.25 mm to the left results in a drastically different profile. This is shown in Figure 12.51(b). Generally, one is able to determine the emitter location to within a few millimeters in latitude and a few degrees in longitude.

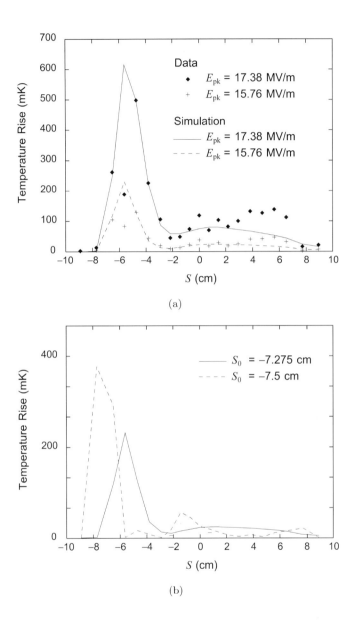

Figure 12.51: (a) Comparison of simulated and measured temperature profiles ($S_0 = -7.275$ cm). The low values measured by the fourth thermometer from the left are probably due to an abnormally low thermometer sensitivity. (b) Comparison of simulated temperature profiles at $E_{\mathrm{pk}} = 15.76$ MV/m for slightly displaced emitters. (In both figures $\beta_{\mathrm{FN}} = 400$, $A_{\mathrm{e}} = 1.58 \times 10^{-12}$ cm^2.)

CHAPTER 13

The Quest for High Gradients

13.1 INTRODUCTION

In this chapter we review the state of the art in gradients in terms of the knowledge we have gained in the previous two chapters about the field limiting mechanisms. We also examine the more recent efforts that have attempted to address the challenges before us: to overcome the limitations of thermal breakdown and field emission, to advance the gradients toward the theoretical promise of ≈ 60 MV/m, to improve the reproducibility of gradients in structures, and to preserve the gradients in an accelerator.

13.2 A REVIEW OF THE STATE OF THE ART

The state of the art in performance of sheet metal niobium cavities is best represented by the well-documented statistics of 338, 5-cell, 1.5-GHz cavities built for TJNAF [190]. The state of the art [232] for CERN niobium-on-copper cavities is reviewed in the next chapter, because this is a different technology. The TJNAF cavities were built by industry from nominal RRR = 250 niobium, and their performance was first measured in a vertical test dewar prior to installation in the accelerator. To avoid the Q disease (Chapter 9) the cool-down to 4.2 K was accomplished in 1.5 hours and the ambient magnetic field in the vicinity of the cavity was shielded to < 10 mOe.

The histogram of Figure 13.1 shows the distribution of fields at the onset of field emission. The accompanying scatter plot gives the Q_0 at the field emission onset field. On average, field emission starts at $E_{\mathrm{acc}} = 8.7$ MV/m, but there is a large spread, even though the cavities received nominally the same surface treatment and assembly procedures. In some cavities, field emission was detected as low as 3 MV/m, and in others it was found to be as high as 19 MV/m. Due to the learning curve, there was a small improvement in the average value as the production series progressed [233].

In many cases, several minutes of cw processing, with up to 100 W rf power, produced a 10–30% gain in performance for approximately 50% of the cavities. Out of the total batch, 148 cavities exceeded 9 MV/m without significant field emission loading. As mentioned earlier, the reason for the large spread

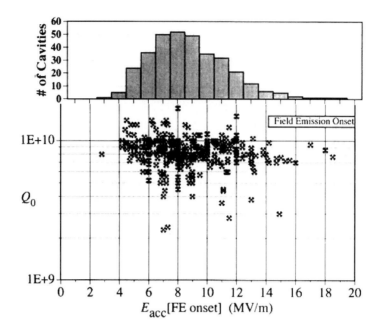

Figure 13.1: Distribution of gradients and Q_0 at the onset of field emission for more than 300 TJNAF cavities. The average gradient is 8.7 MV/m. (Courtesy of TJNAF.)

in the gradients is the large spread in emitter characteristics and the random occurrence of emitters on the rf surface.

A large number of cavities were limited by thermal breakdown. Figure 13.2 shows the distribution of quench fields, and the Q_0 values below quench for the 146 cavities. Here the average is 13.1 MV/m, and the best is 20 MV/m. The large spread in quench field reflects the spread in defect strengths and the statistical occurrence of defects.

A comparison of the field emission onset and quench histograms makes it clear that field emission is the predominant effect, but that quench limitation is not much higher. If the usable gradient is defined as the accelerating field at which there is one watt of field emission loading, or one MV/m below the quench field, then the average performance of the ensemble is 10 MV/m, and the Q_0 is about 5×10^9.

Key aspects responsible for the outstanding performance of the TJNAF cavity set are the antimultipactor, elliptical cell shape, good fabrication and welding techniques, high thermal conductivity niobium (RRR = 250), and clean surface preparation.

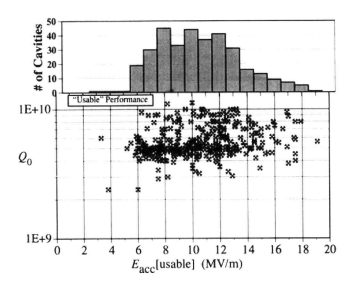

Figure 13.2: Distribution of gradient and Q_0 values for 146 TJNAF cavities that were limited in ultimate gradient by quench. The average is 13.1 MV/m. The RRR of the niobium for these cavities was nominally 250. (Courtesy of TJNAF.)

13.3 A STATISTICAL MODEL FOR THE PERFORMANCE OF FIELD EMISSION DOMINATED CAVITIES

Is the observed spread in gradients consistent with the distribution in properties of the emitters and the number of emitters found in cavities? A statistical model [234] has been developed to check for consistency. The input parameters are the number of emitters/unit area, a distribution function for $\beta_{\rm FN}$ values, a distribution function for emitter areas, $A_{\rm e}$, and a chosen emitter processing threshold. In accordance with the data of Chapter 12, the spread in $\beta_{\rm FN}$ values was chosen between 40 and 600 and the spread in $A_{\rm e}$ was chosen so that $\log(A_{\rm e} m^2)$ ranges from -8 to -18. The distribution function for $\beta_{\rm FN}$ is chosen so that the number of emitters increases exponentially with decreasing $\beta_{\rm FN}$, while the distribution function for $\log(A_{\rm e})$ is chosen to be a gaussian with a width of $\log A_{\rm e} = 2$ as follows:

$$N(\beta_{\rm FN}) \propto \exp(-0.01\beta_{\rm FN}), \tag{13.1}$$

$$N(\log A_{\rm e}) \propto \exp\left[-\left(\frac{\log A_{\rm e} + 13.262}{2.175}\right)^2\right]. \tag{13.2}$$

No correlation between $\beta_{\rm FN}$ and $A_{\rm e}$ is assumed. The surface of a superconducting cavity is divided up into a large number of segments, typically 20 per cell. Each segment is sprinkled with a random emitter density, between 0 and

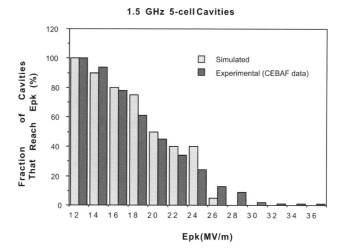

Figure 13.3: A comparison between the statistical model simulation and observed results for the fraction of 1500-MHz 5-cell cavities that reach a certain $E_{\rm pk}$.

0.3 emitters/cm^2. Here we are guided by the maximum emitter density that we find in superconducting cavities, as shown in Figure 12.14. The total number of emitters in each segment is proportional to the area of the segment. After choosing an emitter set, we calculate the power deposited by each emitter, given approximately by

$$P\ (\text{W}) = \left(\frac{A_{\rm e}\ (\text{m}^2)}{10^{-12}}\right) \left(\frac{f\ (\text{MHz})}{1500}\right)^{-1.5} 1.8 \times 10^7 \exp\left(\frac{-7.4 \times 10^{10}}{\beta_{\rm FN} E_{\rm em}}\right). \quad (13.3)$$

Here $E_{\rm em}$ is the electric field at the emitter location in V/m for the chosen value of $E_{\rm pk}$. This power function was determined by fitting the results from trajectory calculations of field emitted electrons in a single cell cavity. The computations show that, to first order, the power depends essentially on $\beta_{\rm FN} E$ and not very significantly on the detailed geometry of the electron trajectory. Emitters are considered processed if the power deposited by one emitter exceeds a specified level, which was set at 500 W per emitter for the TJNAF 5-cell unit. Strictly speaking, one should impose a total current cutoff for a single emitter (e.g., 1 mA) as discussed in Chapter 12 for the processability of emitters. But the power cutoff we used instead is roughly consistent with $\approx$ mA total current. When the simulated total power from all the emitters is *less* than 100 W, that test is declared to be a "success," since 100 W is the typical available cw rf power in a superconducting cavity test. We then calculate the fraction of tests that "successfully" reach a certain $E_{\rm pk}$.

The simulated results are compared to the data from 100 TJNAF cavities in Figure 13.3. The agreement confirms our interpretation that the spread in performance is statistical in nature and consistent with the observed distributions

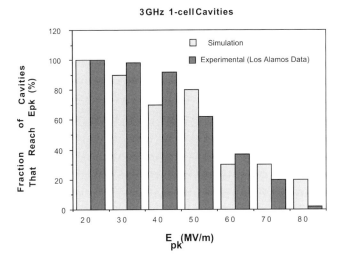

Figure 13.4: A comparison between the statistical model simulation and observed results for the fraction of 3000-MHz 1-cell cavities that reach a certain E_{pk}. For the simulation, the distribution of emitter properties and emitter density are the same as used for Figure 13.3.

of β_{FN} and A_e.

Another aspect of the performance of superconducting cavities that is confirmed by the statistical model is that the probability for reaching higher fields increases as the area of a cavity decreases [151]. For example, the same type of simulations predict that for the same number density of emission sites, there is a 20% probability for reaching $E_{\text{pk}} = 70$–80 MV/m for a single-cell 3-GHz cavity. For a single-cell 1500-MHz cavity, the probability is 20% that the cavity will reach $E_{\text{pk}} = 40$–45 MV/m. This should be compared with the 20% probability of Figure 13.3 that a 5-cell 1500-MHz cavity will reach $E_{\text{pk}} > 25$ MV/m. Therefore one should take into account the area effect in interpreting how single-cell cavity results will translate to results for multicell structures. Figure 13.4 compares the predictions of the statistical model with Los Alamos data [207] on 100 tests on single-cell 3-GHz cavities.

13.4 OVERCOMING THERMAL BREAKDOWN

As we saw in Chapter 11, the average quench field for cavities improves from $E_{\text{acc}} = 5$ MV/m to $E_{\text{acc}} = 13$ MV/m in raising the RRR of niobium from 30 to 300, roughly consistent with $E_{\text{acc}} \approx \sqrt{\text{RRR}}$. This implies that most of the benefit in quench reduction from using higher-purity niobium comes from the thermal conductivity improvement, and not from any defect strength reduction. The typical defect strength encountered in the RRR = 30 or RRR = 300 niobium cases are regions of 100-μm size with normal-conducting resistance. Note, once again, that we have no way of separating defect resistance from defect size. Therefore we refer to defect strength only in terms of defect size, presupposing a normal-conducting resistance (e.g., ≈ 10 mΩ). We also pointed out in Chapter 11 that recent 1-cell results from KEK indicate defect strength reduction to ≈ 10-μm size by better preparation methods, of which high-pressure rinsing may be an important element. KEK results are therefore encouraging for future gradient gains by defect strength reduction. It remains to be seen, however, whether the defect reductions can be achieved in multicell structures. Also, as pointed out, one must be careful about interpreting single-cell results because of their smaller area and lower probability for encountering defects.

The variable mesh thermal model calculations presented in Figure 11.15 show that to reach *quench-free* performance at $E_{\text{acc}} = 25$ MV/m for a 100-μm radius normal-conducting defect, it is necessary to raise the RRR to greater than 2000. Titanium-based solid state gettering applied to RRR ≈ 250 niobium improves the RRR to ≈ 500. As we will discuss in the next section, when coupled with the efforts to overcome field emission, niobium cavities of RRR ≈ 500 have already reached accelerating gradients between 20 and 28 MV/m for 5-cell, 1500-MHz cavities. But to reliably reach high accelerating fields, as well as in large-area accelerating structures (e.g., the 9-cell, 1.3-GHz cavities), RRR values greater than 500 values are very much desired.

What are the availability and the prospects for higher RRR niobium? There are two avenues for the future. Russian niobium producers [235] are now able to supply sheet metal niobium of RRR between 500 and 700, in small quantities. When postpurified with titanium, this material has been improved to RRR between 1000 and 1500 [122]. A 4-cell, 1300-MHz cavity was made from Russian RRR = 500 material and was improved to RRR ≈ 1000 [236]. Tests on this cavity are still limited by field emission, but also show quench at $E_{\text{acc}} = 25$ MV/m. It is not clear whether the quench was initiated by the remaining field emission which was sufficiently intense to lower the Q_0 from 10^{10} to 10^9. Therefore the final word on continued gradient improvement with higher RRR niobium is still pending. At first, single-cell 3-GHz cavities were built from higher RRR Russian niobium in order to determine whether the new material was otherwise good for cavity use [237, 193]. These cavities performed very similar to 3-GHz, RRR = 500 cavities prepared from non-Russian niobium and post-purified by solid state gettering. Note that we did not expect, and we did not get, any gradient improvement from higher RRR at 3 GHz, because of the global instability limit , as discussed in Chapter 11.

Another approach to reach high RRR is based on an extension of solid state gettering to very long times. Samples have been purified to RRR values of 2000 in 50 hours [133]. It remains to be seen how the cavities prepared by this method will perform.

For the remainder of this chapter we will see that the efforts to overcome field emission only yield gradient benefits if they are coupled with RRR improvements to commensurately raise thermal breakdown levels. Although we have not yet developed a detailed statistical model for thermal breakdown, we expect that quench-limited fields will also decrease with cavity area because, like field emitters, defects also occur randomly.

The dependence of achievable field on cavity area has been examined on several occasions over the years ([151, 131, 196, 238]) as SRF technology has continued evolve. If one defines an envelope from the maximum field reached at any stage in the technology development, the empirical rule one finds is that the maximum field decreases as (cavity area)$^{-1/4}$. Innovations such as high purity niobium, high power processing or high pressure rinsing keep pushing the envelope to higher fields.

13.5 EARLY METHODS FOR OVERCOMING FIELD EMISSION

There has been much progress in efforts to reduce field emission, starting with gains established by early techniques, such as helium processing and heat treatment (HT). More encouraging and more reproducible results are now available with the new techniques of high-pressure rinsing (HPR) and high-power pulsed rf processing (HPP).

13.5.1 Helium Processing

In this method [239], used since the early days of rf superconductivity, helium gas at low pressure (10^{-5} torr, or just below discharge threshold) is admitted into a cold superconducting cavity. The cavity is operated near the maximum possible field level for several hours. There are several indications of field emission reduction. Orders of magnitude reduction in x ray intensity along with factors of two reduction in β values have been reported. Emission heating from individual sites observed by temperature maps were reported to disappear. In high-frequency (> 1000 MHz) cavities, performance improvement occurs during the first few minutes. In low-frequency cavities, gains continue over periods between 1 and 50 hours [149].

The improvement factor from helium processing depends on the emission-limited field level, as shown by the results on a series of 1-cell, 1500-MHz cavities [134]. A gain of about 30% was seen at $E_{pk} = 20$–30 MV/m, but decreased to < 10% at 50–60 MV/m. At field values lower than those shown in Figure 13.5, improvements as high as 100% have been reported.

Helium processing was also found to be effective in suppressing field emission from an emitter source introduced intentionally: carbon flakes deliberately placed on the superconducting cavity surface [240]. As mentioned before, carbon is an excellent field emission source as found in dc field emission studies [208]. The studies have clearly established the usefulness of helium processing.

Experiments show that helium processing works in a number of different ways, or perhaps in a combination of ways. It has been established that at least part of the benefit comes from removal of gas condensates. Emitters activated by deliberately condensing gas were identified by thermometry and subsequently

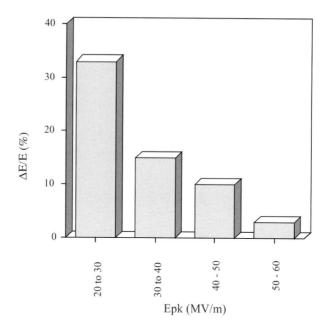

Figure 13.5: Field level enhancement due to helium processing.

removed by helium processing [219]. One expects that condensate changes would be realized by helium ion bombardment and in a short period of time (e.g., minutes). Another result [241], discussed in Chapter 12, showed that admission of helium triggers a microdischarge at a field emission site. If the field level is not high enough for rf processing, the presence of helium gas will help to initiate the discharge which subsequently destroys the emitter. Again, we expect that this effect should proceed quickly, probably immediately after admitting sufficient helium gas into the cavity.

Helium processing over time periods of many tens of hours is also known to reduce field emission. This effect has traditionally been interpreted as sputtering of the bulk emitter over longer periods of time. Long-time helium processing is more effective for low-frequency cavities because of the higher impact energy of the helium ions and the higher sputtering rate [149].

Yet another mechanism has been suggested for the effectiveness of helium processing. In a series of studies on dc field emission from copper surfaces, room-temperature high-voltage conditioning experiments have been carried out in the presence of a variety of gases, including helium. The results of these studies have been interpreted as ion implantation that alters the emitter properties to lower β_{FN} [242].

As a cautionary note, we point out that helium processing a cavity does pose a possible disadvantage. There is a danger of sputtering metal coatings on ceramic insulators present in coupling devices attached to a cavity thereby damaging the insulators.

13.5.2 Heat Treatment of Niobium Cavities

In dc studies, discussed in the previous chapter, niobium samples $\approx$ cm^2 scanned for emission sites were high-temperature annealed and subsequently rescanned at ambient temperature, but without removal from the UHV system [243, 211]. Above 1200°C, the density of emitters was drastically reduced. After heating to 1400°C, the niobium surfaces were reproducibly found to be free of emission up to 100 MV/m. Heat treatment(HT) to 1400°C is therefore one way to provide a cleaner rf surface. In many cases, particles (and emission) were observed to disappear. In some cases, the particles remained even though the emission was arrested.

In light of these encouraging results, the influence of high-temperature annealing in the *final stages* of rf cavity surface preparation has been studied [244]. (Heat treatment to eliminate field emission should not be confused with heat treatment for postpurification, where the rf surface and the exterior of a niobium cavity are surrounded by titanium or yttrium as the solid state gettering agents and where the rf surface is etched after HT to remove the evaporated layer.) Single-cell 1.5-GHz cavities were heated to 1400°C. There were some important differences between the superconducting cavity HT and the dc HT studies. The rf surface of the cavity had to be exposed to filtered air after removal from the furnace, while the samples in the dc study were always kept in UHV. In most cases, it was very difficult to keep dust from entering the cavity during removal from the furnace, so that the HT cavities were rinsed with dust-free methanol after removal from the furnace. As we mentioned in Chapter 12, exposure tests [213] have shown that high-purity, dust-free methanol is *not* an important source of emitters. At 1400°C, the cavity-treatment furnace did not have a sufficiently good vacuum (i.e., $\approx 10^{-9}$ torr) needed to maintain a high RRR. Rather, the typical pressure in the hot zone was 10^{-6} to 10^{-7} torr. Therefore, in order to preserve the RRR, the *outside* of the cavity was surrounded with titanium foil. (This approach is sometimes referred to as "single-sided" titanium treatment.) Niobium baffles were placed around the beam tube openings to prevent the titanium from reaching the rf surface.

For the superconducting cavities, the most significant reduction in field emission was observed for 4 to 8 hour heat treatments at 1400–1500°C. Figure 13.6 shows the maximum surface field reached in several tests of 1-cell, 1500-MHz cavities, and compares the results with a large number of tests on chemically treated cavities of the same type. To reach the maximum possible fields, rf processing and helium processing with about 100 W of rf power were used for both the standard chemically treated tests as well as for the HT tests. At low field, the Q_0 of the HT cavity was usually somewhat lower than before HT — e.g., 7×10^9 compared to $> 10^{10}$. The slight Q_0 degradation was attributed to a small amount of titanium reaching the rf surface, in spite of the baffles. Later, at Wuppertal [245], the single-sided titanium treatment was used together with a completely separate vacuum system for the inside of 9-cell cavities. The better vacuum technique averted the Q_0 decrease from titanium vapor. Q_0 values above 10^{10} were achieved with single-sided titanium treatment.

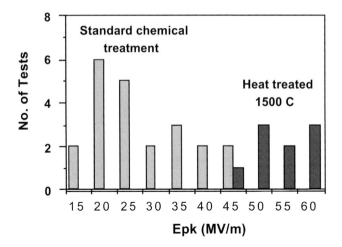

Figure 13.6: A comparison of the performance of heat-treated and chemically-treated cavities. One-cell, 1500-MHz cavities were prepared by standard chemical treatment. The same cavities were prepared by heat treatment at $T > 1400°C$. After HT, the cavities were rinsed with methanol in a Class 100 clean room. In all cases, the maximum available cw rf power as well as helium processing were also applied in an effort to reach the maximum field.

A corroboration of the benefit of HT comes from the temperature maps. Figure 13.7 compares maps between HT and non-HT cavities at 30 MV/m peak surface field, *before* application of any helium processing. Many strong emitters are seen when the standard chemical treatment is used, whereas the HT surfaces are virtually free of emitters at 30 MV/m.

From the temperature maps, statistics have been compiled on the number of emitters detected as a function of peak electric field. There is a factor-of-10 reduction in emitter density as a result of HT, as shown in Figure 13.8. As mentioned before, although these results prove that HT reduces emitters, the statistical model warns that the benefits will be less effective in larger-area, multicell cavities.

The potential benefits of a new treatment indicated by sample tests, or single-cell cavity tests, must always be checked on a large area, multicell cavity. To date, the results of HT on multicell cavities are mixed. However, the number of tests and number of cavities tested are still too low to allow a definitive negative conclusion. One 6-cell, 1.5-GHz unit was prepared and tested with and without HT [246]. Before HT, two tests were carried out. The cavity reached 18 MV/m with heavy field emission loading in both tests. In the following three tests with HT (using outside titanium protection) the results were $E_{\text{acc}} = 15, 20$, and 10 MV/m. The best results with and without HT are shown in Figure 13.9. One might say that the highest field could only be reached with HT. At high field, the emission was substantially lower, judging from the higher Q_0.

EARLY METHODS FOR OVERCOMING FIELD EMISSION 291

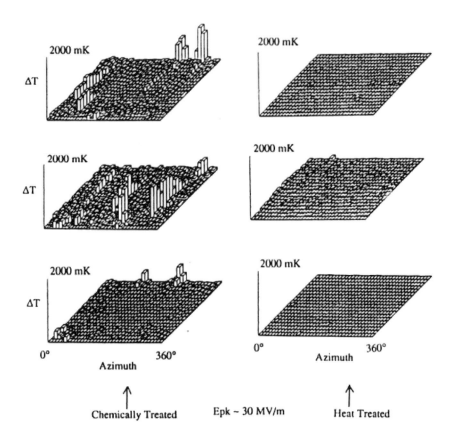

Figure 13.7: A comparison between heat-treated and non-heat-treated cavity surfaces for the density of emission sites, as recorded by temperature maps.

Two tests on 9-cell, 3-GHz cavities at Wuppertal reached $E_{acc} = 16 - 18$ MV/m [245] after HT. These should be compared to 10 tests on the same cavities without HT which showed a wide range of performance between 7 and 13 MV/m. One test without HT did reach 18 MV/m. Thermometry based scans on the 9-cell units showed a substantial reduction in the overall density of sites due to HT. Still, the presence of one or two isolated sites over the large area of the structure was found to limit the field of HT cavities, even if most of the surface appeared free of emission sites. This is an important concern about any approach that aims to reach higher fields by supercleanliness.

We stress that even though there is strong evidence to show that HT reduces the density of emitters, it remains important that additional emitters should not be introduced in subsequent preparation steps. The mixed experience with HT on multicell cavities indicates that it is indeed difficult to keep large-area multi-cell cavities clean on a reproducible basis.

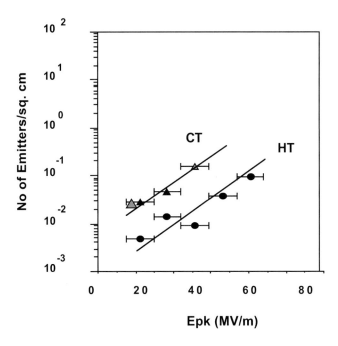

Figure 13.8: Emitter densities as a function of surface electric field in superconducting cavities prepared by heat treatment (HT), compared to emitter densities in cavities prepared by standard chemical treatment (CT). The effective area of the cavity is taken as that part of the surface over which the field is greater than 80% of the peak field.

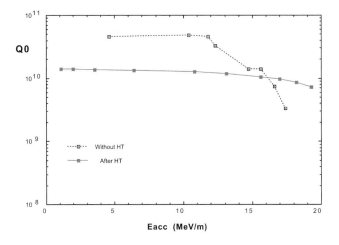

Figure 13.9: Results on a 6-cell cavity with and without HT. The lower Q_0 for the HT case is attributed to some titanium deposited on the rf surface from the titanium that surrounds the outside of the cavity to preserve the RRR.

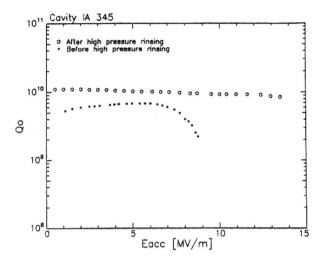

Figure 13.10: The effect of HPR (upper curve) on a 5-cell, 1500 MHz cavity. The lower curve is without HPR. (Courtesy of TJNAF.)

13.6 HIGH-PRESSURE RINSING TO AVOID FIELD EMISSION

Field emitter studies discussed in Chapter 12 show that increased vigilance in cleanliness during final surface preparation and assembly procedures is important to keep particulate contamination and associated emission under control. A technique to improve cleanliness that is now increasing in popularity is high-pressure water rinsing (HPR). The technique was originally applied at CERN [140] but the full potential of the method was best demonstrated at TJNAF [247] and KEK [214]. A jet of ultrapure water is used to dislodge surface contaminants resistant to conventional rinsing procedures. In Chapter 12, we briefly discussed the KEK studies on HPR which showed a drastic reduction in particle count on silicon wafers. DC field emission studies at Wuppertal [212] on $\approx$ cm^2 samples also show that the density of emitters is reduced by HPR.

The benefits of HPR in reducing field emission are well demonstrated in tests on single-cell and 5-cell cavities at TJNAF. The apparatus for HPR used at TJNAF was described in Chapter 6. In 10 tests on 1-cell 1.5-GHz cavities, they reached $E_{acc} = 25$ MV/m, with 32 MV/m as the best case.

The effect of HPR on multicell cavities is encouraging. Several 5-cell cavities after HPR reached the thermal breakdown limit field, without field emission, as shown in Figure 13.10. The best value was $E_{acc} = 13$ MV/m, which is typical of the quench field for RRR = 250 niobium. The same cavities, when prepared without HPR, showed strong emission (lower curve), typical of the behavior of the 5-cell cavities of Figure 13.1.

Similarly, good results for field emission reduction with HPR have been obtained [172] on large area, 9-cell 1.3-GHz cavities under preparation for the

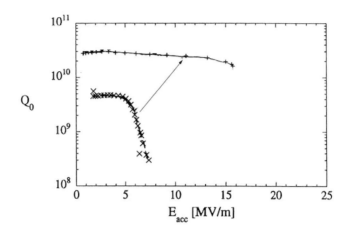

Figure 13.11: The effect of HPR (upper curve) on a 9-cell, 1.3 -GHz cavity. The lower curve shows the result on the same cavity before HPR. Notice also the substantial increase in Q at low field. (Courtesy of DESY.)

TESLA Test Facility (TTF) at DESY (see also Chapter 21). One example of the improvement with HPR is shown in Figure 13.11. This cavity which had RRR = 250 was limited by quench. Notice also the improvement in the Q_0 value at low fields, suggesting that removal of a large number of particles reduces the overall residual losses.

So far we have only presented the results of HPR on multi-cell cavities with RRR = 250. These were limited by quench, once the field emission was removed by HPR. Subsequently the RRR of several 5-cell cavities at TJNAF was improved to RRR = 500 by titanium solid state gettering. When HPR was applied to these high RRR cavities, it was possible to overcome both the field emission and quench limitation, to give the excellent results shown in Figure 13.12 [141].

Similarly good results are forthcoming from the DESY 9-cell, 1.3 GHz structures for TTF, as shown in Figure 13.13. In a spectacular best result achieved with HPR, the Q_0 remained near 4×10^{10} from low fields all the way up to $E_{acc} = 25$ MV/m. Most cavities appear field emission free up to $E_{acc} = 15$ MV/m. From the quench limitation of most of these cavities it is clear that still higher RRR is needed. At the same time, we see that HPR is not able to eliminate field emission in every cavity or every test.

The HPR technique appears more effective in field emission reduction than HT. Besides it is much simpler to implement and will therefore prove in the long run to be much less expensive than the furnace treatment.

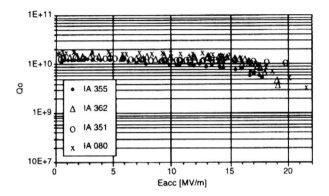

Figure 13.12: Field emission and quench-free performance of four 5-cell TJNAF cavities after improving RRR to 500 by solid state gettering with titanium and by HPR. The maximum field in these tests was limited by the available rf power. (Courtesy of TJNAF.)

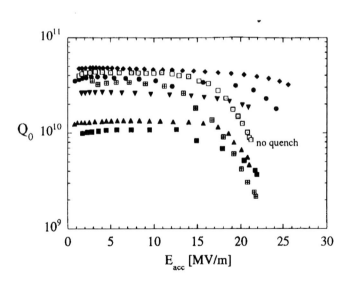

Figure 13.13: Performance of seven 9-cell TTF cavities after improving RRR to 500 by solid state gettering with titanium, and followed by HPR. The maximum field in these tests was limited by thermal breakdown, except for one cavity (labeled "no quench") which was limited by the available rf power. (Courtesy of DESY.)

13.7 HIGH-POWER PULSED RF PROCESSING

While the supercleanliness approach of HPR has unarguable potential, temperature maps have also shown that a single field emission site can degrade the Q_0 of a superconducting cavity, if the emitter will not process away at the maximum rf power available. In large area structures there is always a significant probability that a few emitters will find their way onto the cavity surface. We have already mentioned this to be the reason that the maximum achievable field decreases with cavity area. There is also the danger of dust falling into the cavity during installation of power coupling devices. The contamination threat is especially clear in the light of the experience of all laboratories that there is a 20% decrease in the performance between acceptance tests and in-accelerator tests (Figure 1.26).

Therefore a technique that eliminates emitters in situ is highly desirable for successful application of superconducting cavities to accelerators. Such a technique has been developed. Called high pulse power processing (HPP), it has been successfully applied to 3 GHz, 9-cell cavities [248], and to 1.3-GHz 5-cell and 9-cell cavities [249]. Emitters have been processed away, and the field levels raised substantially. This is *not* to say that, with HPP, the need for cleanliness is eliminated. We reemphasize that to reach the highest possible field in a reproducible manner, one must continue to be ever vigilant in the fight to avoid field emitters.

The essential idea of high power rf processing an emission site is to raise the surface electric field at the emitter as high as possible, even if for a very short time ($\ll \mu$sec). As the field rises, the emission current rises exponentially to the level at which melting, evaporation, gas evolution, plasma formation, and ultimately a microdischarge (rf spark) take place. The ensuing explosive event destroys the emitter. We have discussed in detail the evidence for this model and the probable chain of events involved.

Before presenting the results on HPP of 3-GHz and 1.3-GHz cavities, we briefly summarize several earlier studies that indicated that high power processing has strong potential for field emission reduction. We have already discussed how rf processing with 100 watts of cw power shows a limited degree of success in destroying field emitters in superconducting cavities, and raising the gradient. Unfortunately, the exponentially increasing emission current and the associated rf power dissipation quickly utilize the typical 100 watts of available cw rf power. Processing with powers of a few kilowatts has been routinely used to condition 100-MHz superconducting cavities for heavy ion accelerators [250]. However, there was no systematic study of the characteristics, parameters and the limitations of the processing. We do not know whether the success resulted from reduced multipacting or from processed field emission. In another early study [251], 3-GHz single-cell cavities were tested with MW pulses of a few μsec pulse width. Although high fields were reached in the pulsed mode, it was unclear whether there would be any accompanying benefits to the cw performance or to the long (millisecond) pulse length behavior. These are generally the desirable modes for operation of superconducting cavities.

High-power (100 MW) rf processing is also a routine procedure used in commissioning normal conducting accelerator cavities [223] to accelerating fields of 50–100 MV/m. Here a substantial amount of cratering has been observed at the iris regions of the cavities. It was not clear whether such craters would be harmful to the performance of niobium cavities. We would hope that $\approx$ 100 MW power level would not be necessary for processing superconducting cavities to ultimate gradients $\approx$ 60 MV/m (recall the magnetic field limit).

The investigations of HPP that we will now discuss have addressed many of these open issues, to wit:

1. What are the important parameters to get the best processing results: pulse length, incident power level, input coupling, surface field?

2. What are the limitations of the method?

3. Do the benefits of short pulse HPP extend to cw, or long pulse, operation of superconducting cavities, as intended for accelerator service?

4. How stable is the improvement after HPP?

5. Is the method effective against contamination introduced by vacuum accidents that may occur in an accelerator installation?

13.7.1 RF Power Supply and High-Power Test Stand

The first series of tests on HPP were done with 3-GHz cavities at Cornell [248]. The maximum rf power available was 170 kW peak, at pulse lengths of 0.2 ms. For longer pulses ($\approx$ 1 ms) the power was lowered to 10 kW due to the constraints of the high-voltage power supply. The cavities tested were 1-cell, 2-cell and 9-cell. Following the success in reducing field emission in the 3-GHz cavities, another series of tests were conducted at Cornell on 1.3-GHz, 1-cell, 2-cell and 5-cell cavities [249]. The high-power klystron and modulator system available for the 1.3-GHz study provided about 1 MW of power at a pulse length of 150 μs, later extended to 250 μs. Subsequently, at DESY, longer pulse lengths were made available to process 9-cell, 1.3-GHz cavities for the TTF [172].

Since the HPP approach to reducing field emission is to reach the highest possible E_{pk}, it is important that the test stand be capable of transmitting and coupling to the cavity as much power as possible. To achieve high fields within the short pulse length (200 μs to 1 ms) of the high power klystron, Q_e values between 10^7 and 10^5 are needed during the pulsed processing stage. As we shall see in the analysis of HPP results, due to intense field emission currents at the higher fields, the Q_0 of the cavity drops to low values (e.g., 10^6–10^7) during the pulsed processing stage. Therefore the coupler also needs to reach low Q_e to maintain unity coupling under the heavy field emission loading. Being prepared with a strong coupled input allows maximum power to be supplied to the cavity for processing emitters. Before and after HPP, we use Q_e values between 10^8 and 5×10^{10}, as usual, to obtain the customary Q_0 vs. E curve that allows us to

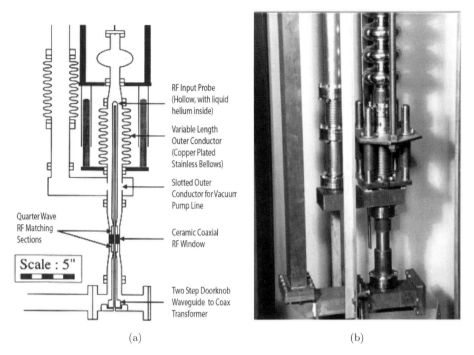

Figure 13.14: HPP test stand for 3-GHz cavities. (a) Variable coupler schematic, (b) photograph of system with 9-cell cavity attached.

assess the effect of HPP. For all of the above reasons, the input coupling needs to be variable between $Q_e \approx 10^5$ and $Q_e = $ few $\times 10^{10}$, with no break of the vacuum system.

In a typical HPP test stand (Figure 13.14), the rf input coupler is a coaxial probe aligned with the beam axis of the cavity. Coupling variation is accomplished by moving the cavity relative to the fixed input probe through a copper plated bellows. HPP is carried out at liquid helium temperature so that high fields can be reached with the available power. The variable coaxial coupler is transformed to a waveguide some distance away from the cavity. The high power from the klystron is brought into the test setup via the waveguide.

To illustrate the range of peak fields accessible, Figure 13.15 shows the theoretical maximum surface field level that a 5-cell 1.3-GHz cavity can reach using 1 MW power with a pulse length of 150 µs. We vary the input coupling, and the Q_0, which is determined by the amount of field emission loading of the cavity. For simplicity of calculation, we make the pessimistic assumption that the Q_0 falls to the field emission loaded value throughout the pulse. In actuality, the fields accessible can be somewhat higher, because the Q_0 is higher at the beginning of the rf pulse. E_{pk} in the pulsed mode is calculated from E_{acc} using the cavity properties as follows:

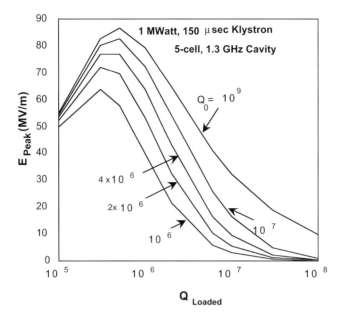

Figure 13.15: Calculated maximum surface electric field accessible with 1 MW, 150 μs pulsed rf power. Each curve is for a different Q_0 which can assume values between 10^9 and 10^6, depending on the severity of field emission loading.

$$E_{\text{acc}}(t \to \infty) = \sqrt{\frac{4\beta P_{\text{g}}(r_{\text{a}}/Q_0)Q_0}{(1+\beta)^2 L}}, \qquad (13.4)$$

$$E_{\text{pk}}(t) = E_{\text{pk}}\left[1 - \exp\left(\frac{-t}{2\tau_{\text{L}}}\right)\right], \qquad (13.5)$$

where L is the active cavity length.

The following experimental procedure has been empirically found to be effective for HPP. At first, the low-power cw performance (Q_0 vs. E) is measured, using an incident power up to 10–100 watts and $Q_{\text{e}} \approx 10^{10}$. If emission is present, Q_0 begins to fall above a certain field. Occasionally, some improvement in performance is seen through cw rf processing. We generally start HPP with a moderate incident power level ($\approx 5 - 10$ kW) and decrease the Q_{e} to establish about the same field level in the short pulse as the highest reached in the preliminary cw stage. Earlier efforts that attempted to start HPP with prematurely higher power levels resulted in damage to the Q_0 and premature thermal breakdown. Since emitter destruction is an explosive process, it is advisable to use the minimum necessary power. If the Q_0 momentarily decreases during the spark, and the Q_{e} is low, as arranged, a large amount of power may couple into the discharge, and destroy an area larger than necessary.

The peak electric field in the cavity is monitored via the fixed transmission

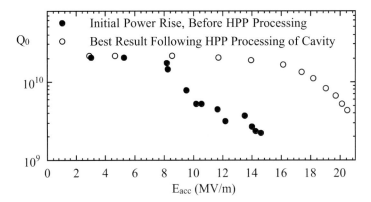

Figure 13.16: Improvement of a 9-cell cavity due to HPP processing. The solid symbols show the Q_0 vs. $E_{\rm acc}$ for the cavity before HPP. The open symbols show the results of the measurement following HPP processing with incident power up to 100 kW when the maximum pulsed surface electric field reached was 68 MV/m. The RRR of this cavity was improved to about 500 by solid state gettering with yttrium. The cavity did not reach thermal breakdown up to the maximum cw field of the test.

probe, previously calibrated during the low power cw measurements. To start HPP, the input coupling is slowly increased (Q_e lowered) so as to maximize $E_{\rm pk}$. When the $E_{\rm pk}$ value exceeds the maximum obtained during the cw measurement, processing events can be seen. These are recognized by a fast ($< 5\ \mu$s) collapse of cavity stored energy. After a typical processing event, a higher field level in the cavity can be sustained at the same incident power and coupling. When $E_{\rm pk}$ can no longer be increased by adjusting Q_e, a higher incident power is required. Before the power is raised, however, the coupler is withdrawn to a safe, high Q_e position. The power can then be raised and the Q_e slowly lowered to carry out further processing. These steps are repeated until one of the following limits is encountered: maximum power, maximum coupling strength, or a reproducible thermal breakdown.

13.7.2 HPP Results

In the first stage of the investigation on HPP, tests were carried out on several 1-cell 3-GHz cavities. After proving the effectiveness of HPP and dissecting the cavities to find processed sites (described in Chapter 12), the method was applied to two 9-cell units. Both structures had their RRR improved to about 500 by solid state gettering with yttrium when the cavities were in the half-cell stage.

We postpone the presentation of the 1-cell results to the next section, moving immediately to the more important multicell cases. Figure 13.16 shows a typical

HIGH-POWER PULSED RF PROCESSING

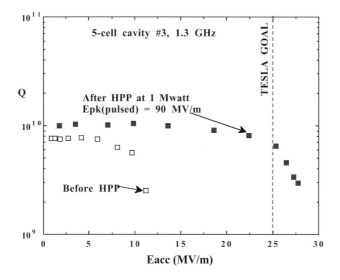

Figure 13.17: Before HPP the 5-cell, 1.3-GHz cavity was limited by heavy field emission to $E_{acc} = 12$ MV/m. After applying 1 MW of power and reaching 90 MV/m in the pulsed mode, the field emission was processed away and $E_{acc} = 28$ MV/m was possible. The RRR of this cavity was improved by solid-state gettering with titanium to about 500.

example of the benefits of HPP for a 9-cell, 3 GHz cavity. Before HPP, x rays were observed above 6 MV/m, and the Q_0 decreased with field from 2×10^{10} to 2×10^9 at $E_{acc} = 14$ MV/m. After measuring the initial performance, HPP was carried out in stages, to a maximum pulsed power of 100 kW. In the pulsed stage, the corresponding maximum E_{pk} reached was 68 MV/m. Following HPP, a significant improvement was seen in the cavity performance. The maximum E_{acc} was 20 MV/m and the associated Q_0 was 4.5×10^9. The Q_0 remained greater than 10^{10} for fields as high as $E_{acc} = 18$ MV/m. The lowest field at which x rays were detected increased from 6 MV/m before HPP to 13.5 MV/m after HPP.

In eight separate tests on two 9-cell structures the field levels improved from $E_{acc} = 8$ to 16 MV/m before HPP to 15 to 20 MV/m after HPP. In all tests, the maximum cw field was limited by field emission loading, so that higher power for processing might well have been further effective. The 9-cell cavities did not show thermal breakdown at the highest cw field, showing that the step taken to raise the RRR to 500 was an effective measure to avoid quench.

After the uniform success with 3-GHz single-cell and multicell cavities, the HPP technique was applied to 1.3 GHz structures, using 1 MW pulsed power and 150 μs pulse length. Two of the 5-cell cavities received titanium solid state gettering to improve RRR to about 500. *Before RRR improvement, these cavities were found to be limited by thermal breakdown at cw accelerating fields of 14 and 12.5 MV/m.* Of the three 5-cell units studied, one did not show a low field breakdown, and its RRR was, accordingly, not enhanced.

The gradient in three 5-cell structures was raised from 10–22 MV/m, before HPP, to 26–28 MV/m after HPP. One example "before-and-after" HPP result for a 5-cell unit is shown in Figure 13.17. Before HPP, field emission limited the maximum field to $E_{\text{acc}} = 13$ MV/m. Another example, Figure 13.18 shows how the maximum field was first limited by thermal breakdown, but after RRR improvement by titanium heat treatment it was possible to process field emission all the way up to $E_{\text{acc}} = 27$ MV/m. At the highest cw field level after HPP, one cavity was limited by available rf power, and one cavity was limited by the radiation level safety trip point. A third cavity was limited by thermal breakdown. Note that, with the MW level pulsed power available at 1.3 GHz, we were able to reach $E_{\text{pk}} \approx 90$ MV/m as compared to ≈ 70 MV/m with the 3-GHz cavities. As a result, we were able to more effectively process field emission and reach higher fields in the cw mode. We could not process much above $\approx$ MW because the input coupler of the test stand would break down quite frequently at the window.

The success of HPP in curing field emission carried over to larger structures for TTF at DESY [172]. To process 9-cell 1.3-GHz cavities, the power capability installed was 1 MW and a longer pulse length, ≈ 1 ms. Using this capability, the field emission limit of several 9-cell structures was raised from $E_{\text{acc}} = 15$ MV/m to 26 MV/m, cw. One example result for HPP on a 9-cell, 1.3 GHz structure is shown in Figure 13.19. At an operational pulse length of 1 ms, $E_{\text{acc}} = 25$ to 30 MV/m was also achieved. The long pulse length gradients are particularly significant because a future linear collider (TESLA) based on superconducting cavities will operate at a duty cycle of about 1% and an rf pulse length of about 1 ms, rather than in the cw mode.

13.7.3 The Controlling Parameter for RF Processing

The overriding factor determining the success of HPP is the peak electric field reached during the pulsed rf stage. If this field is not raised, no benefits are observed for the cw low-power performance. This correlation is made clear by Figure 13.20, which shows the results from all tests on 1-cell, 2-cell and 9-cell 3-GHz cavities. We see that the higher the field reached in the pulsed mode, the higher the maximum field for the cw mode.

A similar correlation is evident from the results of processing 1.3-GHz, 5-cell units, as shown in Figure 13.21. For both 1.3-GHz and 3-GHz cavities we observe that

$$E_{\text{pk}} \text{ (cw)} = (0.6 - 0.7) \times E_{\text{pk}} \text{ (pulsed)}. \tag{13.6}$$

At the maximum field given by the above relation, there is still some field emission present. To obtain field emission *free* behavior at a certain operating field level it is necessary to condition cavities to even higher values — i.e., approximately to twice the operating field.

$$E_{\text{pk}} \text{ (cw)} = 0.5 \times E_{\text{pk}} \text{ (pulsed)}. \tag{13.7}$$

HIGH-POWER PULSED RF PROCESSING

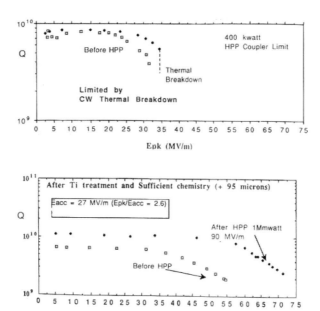

Figure 13.18: Improvement in performance of cavity due to post-purification and HPP. Note that the abscissa is $E_{\rm pk}$. (Upper) Before post-purification, thermal breakdown limited the maximum field to $E_{\rm acc} = 13.5$ MV/m. (Lower) After RRR improvement, the field could be raised to $E_{\rm acc} = 21$ MV/m, but in the presence of heavy field emission. This was successfully processed using HPP. During the processing stage, 1 MW of power applied raised $E_{\rm pk}$ to 90 MV/m in the pulsed mode. After processing, the CW field reached $E_{\rm acc} = 27$ MV/m.

The key to effective HPP processing is therefore to force the peak fields during processing to the highest possible value. The klystron power, pulse length, and coupling need to be arranged accordingly. It was found that conditioning for longer times at the same field level, or with longer pulses at the same field level, did not help to reduce field emission or to reach higher fields. This experience is consistent with the finding that emitter processing takes place only when the emitter current density and the total current exceed certain threshold values, which we discussed in the Chapter 12. Both emission current density and total current depend on $E_{\rm pk}$.

We see that the maximum fields reached in the cw mode (after HPP) are never as high as those reached during the pulsed stage. The statistical model, discussed earlier, suggests a plausible reason for Equation 13.7. A Monte Carlo study was carried out for a one-cell 3-GHz cavity, using the $\beta_{\rm FN}$ distribution, $A_{\rm e}$ distribution, emitter density ≈ 0.3 emitter/cm^2, and processing threshold (e.g., 100 W) described in Section 13.3. Table 13.1 gives the results for 10 simulated rf tests.

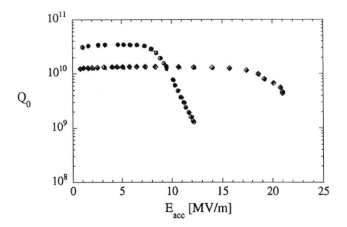

Figure 13.19: Improvement in gradient of a 9-cell cavity by HPP. The Q_0 drop could be recovered by warming up to room temperature. (Courtesy of DESY.)

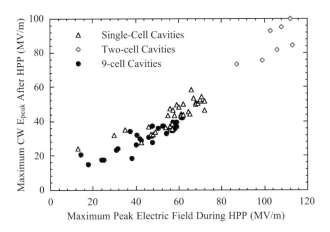

Figure 13.20: Maximum cw peak electric field plotted as a function of the maximum peak electric field during the HPP processing. One-cell, 2-cell and 9-cell S-band cavity results are shown.

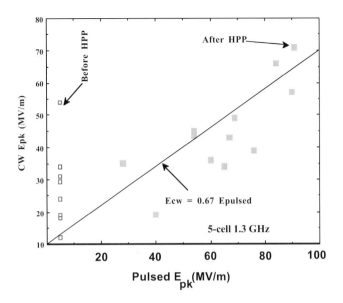

Figure 13.21: A summary of the benefits of HPP on 1.3-GHz cavities. The open squares are the cw results before HPP, and are plotted offset from the vertical axis. The filled squares are results after HPP. The higher the field achieved in the pulsed conditioning stage, the higher the cw operating field.

The performance of a typical 3-GHz single-cell cavity at $E_{\rm pk} = 40$ MV/m is characterized by the average power deposited by the active emitters. The simulated average power per test is ≈ 18 W, corresponding to a Q_0 of $\approx 5.5\times 10^8$, a situation considered as heavy field emission loading. In a single-cell, 3-GHz cavity at 40 MV/m, the quantity of power is due to field emission from an average of 1.7 emitters per test. If, during a simulated HPP run, the field level is raised to 50 MV/m, many emitters "process," because the individual emitter-deposited power becomes larger than the 100-W processing threshold. In the example, six emitters processed. After the simulated HPP at 50 MV/m, only about 1.1 emitters/test remain. The average field emission related power for one run at 40 MV/m falls to 2.5 W, equivalent to a $Q_0 = 4 \times 10^9$.

If the simulated HPP field is further raised to 60, 70, and 80 MV/m in separate runs, more emitters exceed the 100-W processing threshold and are turned off. The emitters that process and the surviving emitters are listed in Table 13.2, along with the power they deposit at 40 MV/m. After 60 MV/m, the Q_0 at 40 MV/m rises to 8.3×10^9. After 70 MV/m the Q_0 is above 10^{11}. Finally, only 0.1 emitters/test remain after processing at 80 MV/m. We may essentially call this field emission free behavior at 40 MV/m.

The statistical model confirms that for field emission free behavior at 40 MV/m, it is necessary to process at 80 MV/m, — i.e., the processing field needs to be twice as high. Similar results were obtained from Monte Carlo

Table 13.1: Monte Carlo simulation for HPP of 1-cell 3-GHz cavities. The individual results from 10 runs are shown, along with the averages

Run No.	Watts From Each Emitter After Processing to 40 MV/m		After Processing at 50 MV/m Measured Watts at 40 MV/m	
1	5.8	25.9	5.8	0 (processed)
2	8.5	0.6	8.5	0.6
3	23.3	1.6	0 (processed)	1.6
4	none	none	0	0
5	78	none	0 (processed)	0
6	1.7	3.6	1.7	3.6
7	3.9	0.3	0 (processed)	0.3
8	1.5	0.1	1.5	0.1
9	18	0.18	0 (processed)	0.18
10	7.5	0.6	0 (processed)	0.6
Average Power for One Run	18.1 watt		2.5 watt	
Q_0 at 40 MV/m	5.5×10^8		4×10^9	

Table 13.2: Continuation of Monte Carlo Simulations of Table 13.1

Processing Field $\longrightarrow$	60 MV/m	70 MV/m	80 MV/m
Power at 40 MV/m $\longrightarrow$ Run No.	Watt	Watt	Watt
1	5.8	0	0
2	0	0	0
3	1.6	0	0
4	0	0	0
5	0	0	0
6	1.7	0	0
7	0.3	0.3	0.3
8	1.5, 0.1	0	0
9	0.18	0	0
10	0.6	0.6	0
Average Power for One Run (watt)	1.2	0.063	0.03
Q_0 at 40 MV/m	8.3×10^9	1.6×10^{11}	3.3×10^{11}

studies on multicell structures. The relationship Equation 13.7 between the cw field and the processing field was preserved for the multicells. The model further shows that if the emitter density is reduced by better cleaning methods, the ratio may be more favorable than Equation 13.7; i.e., the pulsed field and pulsed power levels required in the processing stage will be lower.

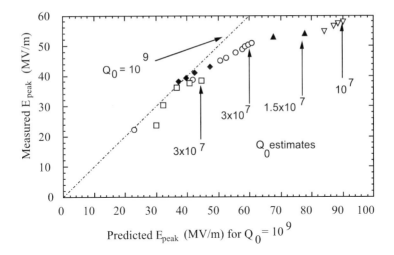

Figure 13.22: Graphical demonstration of the method of Q_0 estimation during HPP processing of a 9-cell 3-GHz cavity. The scatter plot is the measured value of E_{pk} during processing. The line plot is the predicted values for E_{pk} from the experimental conditions for $(P_g, Q_e, t_{\text{rf}})$ and a constant $Q_0 = 10^9$. The Q_0 estimates explain why the observed E_{pk} did not equal the predicted E_{pk}.

13.7.4 Limitations to HPP

One limit to the maximum achievable field during HPP is the rapidly falling Q_0 with increasing E_{pk}. Field emission is the obvious source of Q_0 decline. It would be interesting to know the Q_0 during the pulsed mode; but this is difficult to measure directly because Q_0 is difficult to separate from Q_L, when $Q_0 \gg Q_L$. However, an estimate for Q_0 at E_{pk} can be made, based on Equation 13.4 and Equation 13.5, together with the underlying simplifying assumption that Q_0 is constant through the pulse. Note that E_{pk} reached depends on Q_0.

We show an example of the method of Q_0 estimation from the experimental conditions $(P_g, Q_e, t_{\text{rf}})$ of a typical HPP processing session on a 9-cell 3-GHz cavity. Figure 13.22 compares the peak electric field attained to the predicted peak electric field. Above 40 MV/m, the measured E_{pk} value starts to depart from the predicted value. The Q_0 estimates indicate how much Q_0 must fall to bring the measured value into agreement with the predicted value. In the experiment under examination, the Q_0 must have dropped to 10^7 during the pulse in order to account for the fact that E_{pk} only reached ≈ 60 MV/m rather than the predicted 90 MV/m expected from the chosen conditions for $(P_g, Q_e, t_{\text{rf}})$. Using this estimation technique, a survey of HPP experiments reveals that Q_0 can become as low as 10^6 during the processing stage.

An even more severe Q_0 decline occurs if thermal breakdown is reached during the rf pulse. After this stage is reached, further gains in E_{pk} can come only with very large increases in pulsed power, because the growing normal conduct-

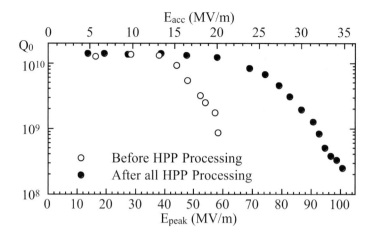

Figure 13.23: Field emission reduction by HPP for a special 2-cell cavity with a reduced $H_{\rm pk}/E_{\rm pk}$ ratio, designed to reach higher $E_{\rm pk}$ by avoiding thermal breakdown. The upper curve shows results of the cw measurement after HPP with 130 kW, and $E_{\rm pk} = 113$ MV/m in the pulsed mode.

ing region absorbs the available rf power very rapidly. However, it has been experimentally shown that the quench field may be exceeded for short periods. Even when thermal breakdown starts, $E_{\rm pk}$ can be pushed up by increasing the incident power, and decreasing $Q_{\rm e}$, so that the cavity fills more rapidly. Under these conditions, the electric field continues to increase in the time it takes for the quench to propagate. Thus it is possible to extend processing *beyond* the thermal breakdown limit, at the expense of higher power.

If thermal breakdown is the hard limit to HPP, it may in the future be worthwhile to circumvent it by altering the cavity geometry so as to lower $H_{\rm pk}/E_{\rm pk}$. There has been a proof-of-principle test for this idea [248]. By using a special geometry 2-cell, 3-GHz cavity, it was found possible to extend processing in the pulsed mode to $E_{\rm pk} \approx 110$ MV/m. The cavities used for the single-cell and 9-cell experiments described above have $H_{\rm pk}/E_{\rm pk}$ ratios of 20-23 Oe/(MV/m), respectively. The special 2-cell cavity had a larger rounding of the equator region to reduce $H_{\rm pk}/E_{\rm pk}$ to 14.2 Oe/(MV/m).

The results of the experiment with the 2-cell cavity are shown in Figure 13.23. On the initial increase of power, the cavity performance was similar to that of other single-cell cavities, dominated by field emission. After HPP, the improvement was not only substantial, but the maximum attained cw field was $E_{\rm pk} = 100.6$ MV/m. Once again, thermal breakdown at $H_{\rm pk} = 1430$ Oe was the limit, close to the expected threshold of the global thermal instability. The gradient at $E_{\rm pk} = 100.6$ MV/m was $E_{\rm acc} = 34.8$ MV/m.

A possible concern about the HPP method is the degree to which the Q_0 may degrade due to the presence of a large number of destroyed emitters. An-

HIGH-POWER PULSED RF PROCESSING

other concern is whether the craters may become defects that cause thermal breakdown. Some of the Q_0 vs. E curves after HPP presented above do indeed show a Q_0 degradation due to HPP. Generally, however, the Q_0 is not degraded below 10^{10}. A single-cell 3-GHz cavity, which was processed by HPP to $E_{\rm pk} \approx 70$ MV/m, still showed a $Q_0 = 10^{10}$ at $E_{\rm pk} = 40$ MV/m cw. Furthermore, on dissection, more than 40 starburst-crater areas were found in this cavity. Therefore the large numbers of processed sites do not seriously degrade Q_0 nor create a thermal breakdown problem. This is not very surprising, since the molten regions are generally less than 10 μm in size so that even a large number of craters do not pose a serious threat. One may, of course, wonder whether processing at higher fields could prove more dangerous. Figure 13.23 shows that even after processing at 110 MV/m, the Q_0 remains greater than 10^{10} at $E_{\rm acc} = 20$ MV/m.

Nevertheless, it is important to bear in mind the early experience with HPP which showed the importance of raising the field in small steps to avoid premature thermal breakdown. It is also important to condition the high-power coupler as much as possible in the absence of the superconducting cavity; sparking in the coupler regions could deposit foreign material into the cavity. For an adjustable coupler, this conditioning can be carried out by withdrawing the antenna. For a fixed coupler, it may be best to condition it in a separate high power test stand.

Experience with HPP has shown that, at the high processing fields, thermal breakdown in the pulsed mode presents a limit to HPP. The specially designed 2-cell cavity showed that higher fields in both the pulsed and cw modes can be reached by avoiding thermal breakdown. Therefore high thermal conductivity for niobium also proves beneficial to processing emission by HPP. By raising the thermal conductivity and avoiding thermal breakdown, it is possible to reach higher fields and thus to process more emitters. Moreover, high thermal conductivity cavities will better tolerate the heat deposited by the higher emission currents that will arise when the HPP field is raised. It is important to recognize that efforts to avoid quench by preparing defect-free cavities will not eliminate the need for high thermal conductivity for effective field emission processing.

13.7.5 Stability of Processing Benefits and Recovery from Vacuum Accidents

An important benefit of HPP is that the technique can be applied to recover cavities which may be accidentally contaminated, as for example in a vacuum mishap. This has been demonstrated using 9-cell structures at 3 GHz [248]. With HPP, it was possible to recover high gradient performance after deliberately introducing field-emitting contaminants through vacuum accidents while the cavity was at room temperature, and further vacuum accidents when the cavity was at liquid helium temperature.

A 9-cell cavity was processed using HPP to $E_{\rm acc} = 18$ MV/m. After warming to room temperature, the cavity was exposed to filtered air (0.3 micron HEPA filter) for 24 hours, and then reevacuated. When recooled to liquid helium

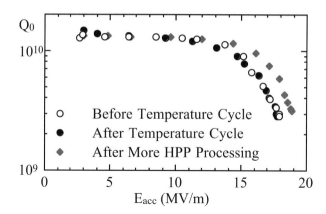

Figure 13.24: Change in FE loading in a 9-cell, 3-GHz cavity following a temperature cycle with exposure to filtered air, and also following further HPP processing.

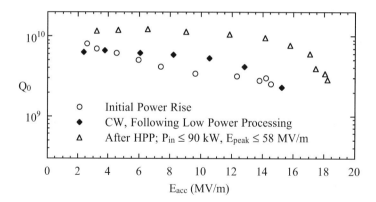

Figure 13.25: Deterioration of a 9-cell 3-GHz cavity following a vacuum accident. There was a slight improvement by cw rf processing, and substantial improvement by subsequent HPP.

temperature, the emission behavior was remeasured. Figure 13.24 shows the effect of the exposure. There was no significant increase in emission. Indeed, following further HPP processing, the cavity reached an even higher field. This test confirms the previously reported finding [213] that exposure to filtered air does not degrade the field emission behavior of a superconducting cavity.

Vacuum accidents are always a danger in accelerator systems, and the contamination due to such accidents can cause significant degradation in the performance of an accelerator cavity. We present here the results of three tests involving the exposure of 9-cell cavities to *unfiltered* air. The circumstances

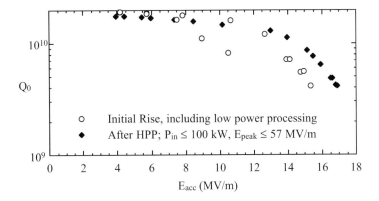

Figure 13.26: Strong degradation of a 9-cell cavity following a second vacuum accident, significant recovery by cw rf processing, and further recovery by HPP.

of the first "accident" were as follows. The cavity had been cooled to 4.2 K, and, while cold, it was exposed to the plumbing connected to the rough vacuum pumps, normally used to reduce the helium bath temperature to 1.5 K. Because of the accident, the Q_0 at low power dropped from 2×10^{10} to 7×10^9 (Figure 13.25). Following reevacuation of the cavity (but no warmup), the initial increase of power was characterized by very heavy emission, some of which was processable with low level cw power. The cavity was then HPP processed with 90 kW, at a field of $E_{pk} = 58$ MV/m. HPP not only reduced the field emission, but also improved the low field Q_0, possibly through removal of dust particles and/or gases condensed on the rf surface.

The second accident occurred on the same cavity, following the above experiment. After the cavity was warmed to room temperature, it was exposed to one atmosphere of *unfiltered* air. Again, the cavity was re-evacuated, and cooled down for experiments. As in the previous test, the initial power increase was characterized (Figure 13.26) by very heavy emission, accompanied by significant low power cw processing. After HPP with 100 kW and $E_{pk} = 57$ MV/m, the cavity substantially regained its previous performance.

The final experiment was a severe cold vacuum "accident" with the same cavity. Following the above test, the cavity was cycled to room temperature, recooled, and retested. While the cavity was at liquid helium temperature, its interior was exposed to unfiltered atmosphere. When the cavity was pumped out and remeasured, it showed heavy field emission, as well as a low field Q_0 degradation (Figure 13.27). Following a room temperature cycle and pump out, the cavity was HPP processed with 105 kW and $E_{pk} = 42$ MV/m. This time there was significant, but not total recovery.

Based on these results, we conclude that, after damage by vacuum accidents,

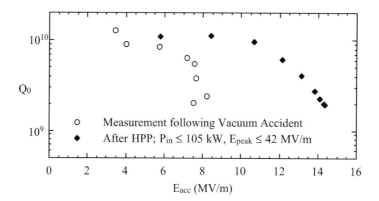

Figure 13.27: Q_0 vs. E_{acc} curves describing the behavior of a 9-cell cavity following the third vacuum accident, and partial recovery by HPP.

the performance of a superconducting cavity may be substantially regained through low power and HPP processing. These results are very encouraging for in situ performance recovery.

13.8 CLOSING REMARKS ON THE GRADIENT QUEST

In order to reach the highest accelerating fields, the highest thermal conductivity niobium is essential to avoid thermal breakdown both from defects as well as from intense field emission current that must be sustained for effective processing of emitters. High-pressure rinsing and high-power processing are essential to avoid field emission. Figure 13.28 summarizes the quest for high gradients in multicell accelerating structures by presenting results from many laboratories. Single-cell results are excluded.

In the early days of multipactor-free spherical and elliptical cavities, when the RRR was about 30 to 40, the maximum accelerating field was limited by quench to ≈ 5 MV/m. When the RRR improved to 250, the average quench field improved to ≈ 13 MV/m, but field emission dominated the results. HPR helped to eliminate field emission, but there was no improvement in gradient, as we discussed in connection with Figures 13.10 and 13.11. The experience with HPP was similar, as we mentioned; field emission could be overcome, but quench remained limiting. Finally, with the combined effort of improving the RRR to 500, coupled with HPR and/or HPP, gradients increased to ≈ 25 MV/m. At this level, both field emission and quench are about equally prevalent.

For the future, we have discussed the ideas under exploration to further improve RRR. HPR continues to appear promising as a technique to provide clean surfaces. HPP continues to appear promising for destroying emitters that enter the cavity despite all precautions or due to occasional vacuum accidents. A

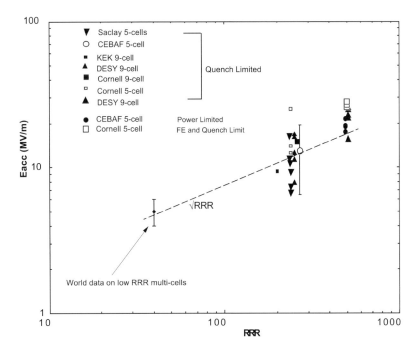

Figure 13.28: A summary of the results of multicell cavities showing the importance of high RRR coupled with emission reduction techniques such as HPP and HPR. The line shows a $\sqrt{\mathrm{RRR}}$ dependence expected from the simple theory of thermal breakdown.

processing field of 110 MV/m has been reached in a cavity where thermal breakdown was avoided by designing for low H_{pk}. There was no serious performance degradation even as a result of 40 crater.

However, HPP has its limits. To achieve emission-free performance at a desired E_{acc}, processing must be carried out at $\approx 2 \times E_{\mathrm{acc}}$. A comparison with the statistical model confirms this relation, at least for a density of 0.3 emitter/cm^2, typical of a surface prepared by chemical etching. Because the critical magnetic field presents a limit of $E_{\mathrm{acc}} \approx 60$ MV/m, the processing relation Equation 13.7 suggests that HPP may not be able to provide field emission free behavior above $E_{\mathrm{acc}} \approx 30$ MV/m. However, it is conceivable that surfaces with a lower density of emission sites may not need to be processed at a factor of two higher field. Although the needed simulations and tests have yet to be carried out, we expect that the processing field for a cleaner surface may not have to be as high as two times the field emission free operating field. We know that HPR reduces the emitter density significantly, so that HPP on an HPR surface is likely to yield better results.

CHAPTER 14

Alternate Materials to Solid Niobium

14.1 INTRODUCTION

Based on the fundamental aspects of superconductivity, discussed the Chapter 3, for a material to be useful in accelerators, the primary requirements are a high transition temperature, T_c, and a high superheating critical magnetic field, H_{sh}, which is of the order of the thermodynamic critical field, H_c. Among the elemental superconductors, niobium has the highest T_c. While lead, coated on to a copper cavity, has been very useful in early studies and heavy-ion accelerator applications, the higher T_c and H_c has made niobium the more attractive choice. Technical considerations, such as ease of fabrication, and the ability to achieve uniformly good material properties over a large surface area, have also proven favorable for niobium. For example, as we discussed in Chapter 6, high-purity niobium is readily available, has a reasonably high thermal conductivity, is mechanically workable to form cavity shapes, and can be electron beam welded without introducing problems. There are certain aspects at each stage of fabrication that deserve close attention; for example, the electron beam welding of niobium must be carried out with a defocused, preferably a rastered, beam. Niobium can also be used as a superconducting thin film on a copper substrate cavity. After some intensive development effort [98], this method has also proved successful, and is one of the main topics of this chapter.

The realm of superconducting compounds has been much less explored because of technical complexities that govern compound formation. In looking at candidates, such as Nb_3Sn, NbN, and the new high-temperature superconductors (HTS), such as $YBa_2Cu_3O_7$, it is important to select a material for which the desired compound phase is stable over a broad composition range. A comparison of the phase diagram of Nb_3Sn with Nb_3Ge reveals that the A-15 phase is stable for the former between 18 and 25 atomic % Sn, but for the latter, the A-15 phase is stable only between 18 and 19 atomic % Ge. With this criterion, formation of the compound may prove more tolerant to variations in experimental conditions, which in turn would make it possible to achieve the desired single phase over a large surface area. With a T_c of 18 K, Nb_3Sn is the most successful compound explored to date [252]. At low fields, residual resistance values comparable to niobium have been achieved. However, the maximum fields reached to date are far lower than for sheet niobium cavities. NbN ($T_c = 16.2$ K for the

δ phase) is another compound that has received some attention [155], but the rf results to date are rather disappointing, and this may be due to the narrower composition range in the phase diagram. The new HTS are even further from the performance level desired for application to accelerators. There have been several review papers about the rf properties of HTS as well as some of their nonaccelerator applications [253, 254, 255, 256].

14.2 SPUTTERED NIOBIUM ON COPPER

The chief motivation for using niobium-coated copper (Nb/Cu) cavities is to provide increased stability against thermal breakdown, by virtue of the higher thermal conductivity of copper. The thermal conductivity of copper is between 300 to 2000 W/m-K, depending on the purity and annealing conditions, as compared to the thermal conductivity of 300 RRR niobium, which is 75 W/m-K at 4.2 K. The cost saving of niobium material is another potential advantage, significant for large-size (350 MHz) cavities, as for LEP-II. At present, the base copper cavity is made by the same methods as the sheet metal niobium cavities — i.e., forming half-cells by spinning, trim machining, cleaning, electropolishing, electron beam welding and so on, as discussed in Chapter 6. For the future, techniques for hydroforming the base copper multicell cavity from a single tube are also being explored [111]. This would avoid the expensive electron-beam welding procedures, and thus lead to further cost reduction. Copper lends itself more easily than niobium to hydroforming from a single tube because copper is easier to work mechanically and because it is easier to carry out the needed interstage anneals, which can be at low temperature ($< 800\ °C$) and in air.

Another advantage of thin niobium films on copper is the relative insensitivity of the residual resistance to dc magnetic fields, as compared to niobium. It is typically 0.15 nΩ/mOe at 350 MHz, as compared to 0.6 nΩ/mOe for niobium. In Chapter 9 we mentioned that the reason for this effect is the high H_{c2} of the sputtered niobium films (see Equation 9.4). The upper critical field of Nb/Cu films has been measured to be 2.5 to 3.5 Tesla at 4.2 K [140]. This effect reduces the amount of magnetic shielding needed in the cryostat. The insensitivity was a fortuitous discovery in the development process of Nb/Cu cavities.

In the most successful coating method to date [257], thin film deposition is carried out by cylindrical magnetron sputtering, as shown in Figure 14.1. Before the coating stage, the copper cavity is degreased, chemically polished ($\approx 20\ \mu$m) rinsed with high-purity, dust-free water and alcohol, and dried under clean laminar air flow. After bakeout of the copper cavity at 150 °C for 24 hours to reach a good vacuum, a typical coating time is 4 hours. The coating thickness is a few microns at a substrate temperature of 180–200 °C. The RRR of the deposited niobium serves as one of the monitors of film quality. The sputtering rate and substrate temperature should be optimized to reach a RRR greater than 20. Note that the low RRR relative to bulk niobium is not a problem because the film is very thin.

The rod-like grains of the niobium film are up to 1 μm long and 10–150 nm in

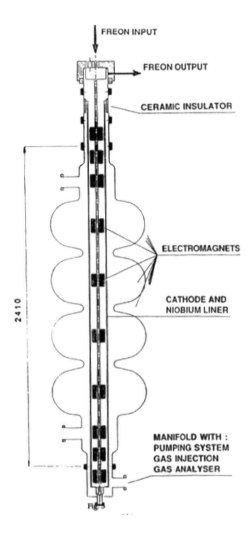

Figure 14.1: Cylindrical magnetron sputtering setup for depositing niobium on a copper cavity. Electromagnets are placed inside a tube of niobium, which is the sputtering target. Cells are coated one at a time. (Courtesy of CERN.)

diameter. When studied with transmission electron microscopy, the individual grains show a high density of defects, consisting of dislocations and point defect agglomerates [105]. The distance between two defects varies from 2 to 20 nm. The onset T_c of the films when removed from the copper substrate is 9.2 K, but the transition width is larger than for bulk niobium, stretching to 5 K in some cases. While on the substrate the T_c of the films is higher (as much as 9.6 K) due to the compressive stress from the different expansion coefficient of niobium and copper.

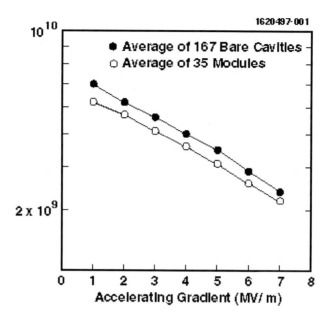

Figure 14.2: Average performance of 167 Nb/Cu cavities for LEP-II measured in the vertical dewar acceptance test. The performance of the same cavities is remeasured when they are placed inside the horizontal cryomodule (lower data set). The difference in Q_0 between the horizontal and vertical tests is about 8%. The reason for the uniform slope in the Q_0 vs. E curves is discussed in the text. (Courtesy of CERN.)

After an extensive program to determine the best copper cleaning and niobium coating methods at CERN, and at industrial vendors, more than two hundred 4-cell cavities have been successfully produced [258] and tested to reach $E_{\rm acc} > 6$ MV/m at $Q_0 > 4 \times 10^9$. Figure 14.2 gives the average performance under different conditions of a large number of units. A maximum gradient of 10 MV/m at a Q_0 of 4×10^9 has been achieved.

At low fields, the Q_0 of the Nb/Cu cavities is somewhat higher than the bulk niobium cavities at the same rf frequency. Due to the small grain size and high oxygen content (RRR = 20) of the films, the mean free path of the sputtered layer is low. The lower mean free path implies a lower BCS surface resistance, as we discussed in Chapter 4. Another interesting aspect is that the surface resistance starts to increase with $E_{\rm acc}$, as seen by the exponential decrease of Q_0 in Figure 14.2. The Q_0 slope is attributed to intergrain losses, a basic effect for fine-grained superconductors, which we will discuss later in connection with the performance of high-temperature superconductors.

More than 140 of these cavities are now installed to upgrade the energy of LEP from 50 to 80 GeV/beam. The remainder of the installation is continuing in order to upgrade the energy to 90 GeV/beam. A total of 256 Nb/Cu cavities

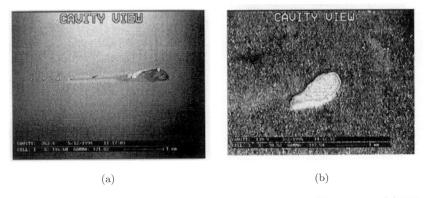

Figure 14.3: Typical defects occasionally found in a Nb/Cu cavity. (a) Blister of length about 2 cm. (b) Defect of size about 1.5 mm. For both types of defects, the niobium layer is stripped and the cavity recoated. (Courtesy of CERN.)

will be installed in 64 cryostat-modules.

Some of the problems encountered with early coatings were poor adhesion (blisters) over regions of the copper substrate contaminated by particles or stains or chemical retention in pits. Typical defects of this nature are shown in Figure 14.3. Another problem was that the damage layer (80–120 μm) in the copper sheet extended deeper than the material removed by etching the copper cavity prior to coating. All these problems have now been solved by improved chemistry and cleanliness as well as by deeper etching. It is important to note that the success represents a technological feat of sputter deposition of good quality niobium over an area of six square meters per 4-cell unit.

A disadvantage of Nb/Cu films is the slope in the $\log Q_0$ vs. E curve. The slope increases with rf frequency as shown in Figure 14.4. At present this makes sputtered niobium on Cu unattractive for high frequency (e.g., 1300 MHz) applications, such as TESLA. The highest accelerating field reached at 1500 MHz is 15 MV/m [105], as compared with niobium cavities at the same frequency which have now reached $E_{\text{acc}} = 40$ MV/m.

14.3 Nb$_3$Sn

The basic properties of Nb$_3$Sn are given in Table 14.1 This is a promising material because of the higher T_c and the higher superheating field. In principle, the higher T_c should lead to a high Q_0 at 4.5 K, allowing a future high-frequency (e.g., 1.3 GHz) machine, such as a linear collider, to operate near the boiling temperature of liquid helium, rather than at 2 K. This would lead to a large savings in both the capital and operating cost of the refrigerator provided that the absence of superfluid helium cooling does not prove to be a problem at the

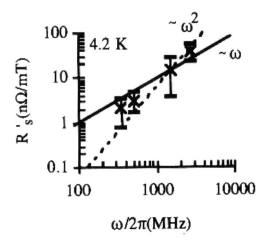

Figure 14.4: Dependence of the Q_0 slope on rf frequency. At each frequency the data are averaged over many tests. (Courtesy of CERN.)

Table 14.1: Fundamental properties of Nb$_3$Sn

T_c	18.2
H_c (Oe)	5350
H_{sh} (Oe)	4000
$\Delta(0)/kT_c$	2 – 2.2
ξ_0 (nm)	6
$\lambda_L(0)$ (nm)	60
Mean free path (nm)	1
H_{c1} (Oe)	200
H_{c2} (Oe)	20,000 – 22,000
κ	20

higher fields. However, the high Q must be achieved at the operating field. The higher superheating field opens up the possibility of accelerating gradients higher than fundamentally allowed for niobium cavities.

Below $T < 0.7\,T_c$, the BCS resistance of Nb$_3$Sn can be conveniently described by

$$R_{\rm BCS} = 9.4 \times 10^{-5} \frac{f^2}{T} \exp\left(\frac{-2.2T_c}{T}\right). \tag{14.1}$$

With a thermodynamic critical field equal to 5350 Oe and κ of 20, $H_{sh} = 4,000$ Oe, according to Equation 5.16. The theoretical maximum accelerating field is therefore near 100 MV/m. However, the highest rf magnetic field reached to date with a Nb$_3$Sn cavity was 890 Oe [259], still far below the ultimate potential. Note that $H_{\rm rf} > H_{c1}$ has been established for this extreme Type II material, confirming that the field at which dc magnetic flux will start to penetrate does not pose any problems to the rf performance; i.e., the rf surface resistance continues to correspond to the ideal superconducting behavior.

The only fabrication technique pursued so far has been vapor diffusion of tin into niobium, at a reaction temperature of 1200 °C. A practical advantage of this method for the far future is that it will permit existing niobium cavities to be converted to Nb_3Sn. When a procedure is developed to realize the potentially higher gradient at the higher operating temperature it would be an attractive upgrade path for a machine already built.

The solid state reaction procedure is outlined in Figure 14.5. The tin source sits in a tungsten crucible underneath the niobium cavity and inside a long niobium tube. The temperature of the tin source, and therefore the tin vapor pressure, is controlled independently of the temperature of cavity. This feature allows good nucleation of the right phase, and prevents excess tin at the surface. The RRR of the niobium cavity is protected with titanium on the outside of the cavity, as discussed in Chapter 13.

In the bakeout part of the heating cycle, the cavity and source are heated to 200 °C for outgassing, after which the cavity is closed in situ with a niobium lid, making a sealed reaction chamber. In the first stage of compound formation, both source and cavity are heated for 5 hours to 500 °C for the nucleation of the Nb_3Sn layer, which is promoted by the addition of $SnCl_2$ in the source crucible. In the reaction stage, the source is held at 1100 °C and the cavity at 1200 °C for 3 hours. The tin vapor pressure is estimated to be 10^{-3} mbar. Finally, the source heater is switched off 30 minutes prior to the cavity heater to avoid excess tin deposition at the rf surface. The thickness of the resulting compound layer is about 2 μm, and the average grain size is also about 2 μm (Figure 14.6).

The best reaction procedure used today has evolved over several stages of improvement [260]. In earlier attempts, it was difficult to uniformly nucleate the correct phase on a high RRR niobium substrate. One way to promote nucleation was to start with a thick oxide (Nb_2O_5) layer on the niobium surface, by first electrolytically anodizing the niobium surface. The disadvantage here was that the RRR of the underlying niobium was lowered at the reaction temperature because the oxygen from the thick oxide film diffused into the bulk. Another early difficulty was the presence of excess tin on the surface, making it necessary to chemically remove the first reacted layers by a mild etching procedure, called oxipolishing. This involves first growing an oxide layer on the Nb_3Sn by electrolytic anodization, followed by dissolution of the oxide layer in HF. By independently controlling the temperature of the tin source during the reaction, it was possible to ameliorate both the difficulties of nucleation and excess tin. Now good cavity performance results are obtained on an as-deposited surface.

The surface resistance obtained in a number of cavities by a variety of laboratories is shown in Figure 14.7. Most of the earlier results for residual rf surface resistance can be fit by

$$R_0 \propto 1.6 \times 10^{-8} f^2, \tag{14.2}$$

where f is the rf frequency (GHz). Note that the residual resistance increases as the square of the frequency as is characteristic of intergrain losses (see next

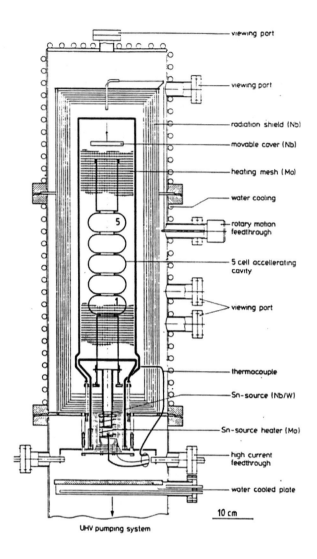

Figure 14.5: Schematic of the furnace used to produce Nb$_3$Sn in a niobium cavity by reaction with Sn vapor. (Courtesy of Wuppertal.)

section). Note also how the latest procedure gives a low-field residual resistance value that is considerably improved over past results. Residual resistances of 2–10 nΩ are possible at 1.5 GHz.

A characteristic feature of Nb$_3$Sn-on-niobium cavities is that the Q_0 values are sensitive to cool-down rate, presumably because of trapped magnetic flux due to thermoelectric currents generated from the bimetallic Nb/Nb$_3$Sn layer [252]. Therefore it is necessary to cool slowly through the transition tempera-

HIGH-TEMPERATURE SUPERCONDUCTORS

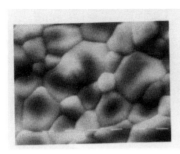

Figure 14.6: SEM photograph showing the structure of the Nb_3Sn layer. (a) Surface, (b) deliberately cracked layer with thickness of 5 μm. In both figures, the scale is 1 μm/div. (Courtesy of Wuppertal.)

ture, typically 10 K per hour. The low-field Q_0 value has also been observed to drop irreversibly if there is a local thermal breakdown. This effect is again attributed to thermocurrents and the associated magnetic flux generated at breakdown.

The performance of Nb_3Sn cavities is not yet on par with niobium cavities, although it must be emphasized that the relative effort devoted to Nb_3Sn is far less than for niobium cavities. The lowest residual loss achieved in a cavity (1.5-GHz, single-cell) was 2.2 nΩ, which corresponds to $Q_0 > 10^{11}$. Figure 14.8 shows the Q_0 vs. E curves for a recently prepared 1.5-GHz single-cell cavity at 4.2 and at 2 K. The maximum peak surface electric field that was achieved by the latest reaction method was 33 MV/m, which corresponds to $E_{\mathrm{acc}} = 16$ MV/m for typical structure parameters. There was no field emission in these tests. The increase of surface resistance with field is possibly due to intergrain losses. However, it is too early to rule out regions of imperfect stoichiometry, especially since the excess losses appear above $E_{\mathrm{pk}} = 5$ MV/m and not throughout the Q_0 vs. E curve, as is usually seen with fine-grain material, such as sputtered niobium.

14.4 HIGH-TEMPERATURE SUPERCONDUCTORS

Since the high T_c revolution, more than 90 superconducting compounds have been found with T_c above 23 K. High-temperature superconductors (HTS) belong to three families: cuprates, bismuthates, and fullerites. For a recent review of HTS see [261]. The cuprates include superconductors with T_c above 77 K. The maximum T_c reached in this family is 134 K for a HgBaCaCuO compound. The most commonly studied material is $YBa_2Cu_3O_7$ with a T_c of about 90 K. The cuprates have an anisotropic layered structure consisting of copper-oxygen planes which are considered to be the active planes for superconductivity. There are also intervening metal or metal-oxygen layers which serve as the charge-

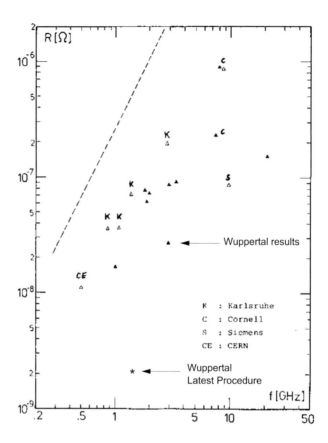

Figure 14.7: Residual resistance of Nb_3Sn measured by various labs. Solid triangles are results from Wuppertal. Open triangles are results from various labs as indicated. The dashed line shows the BCS resistance of niobium at 4.2 K for reference. Results from the latest procedure are labeled by *. (Courtesy of Wuppertal.)

carrier reservoir. The number of copper-oxygen planes is represented in the chemical formula by the index, n. The mechanism for superconductivity in the cuprates is still not resolved. The second family of HTS are the bismuthates. The formula is $Ba_{1-x}A_xBiO_3$, where A = K or Rb. The bismuthates have a cubic, isotropic structure with a maximum T_c of 30 K. There is good evidence that the bismuthates are BCS type superconductors. Relative to the 90 K superconductors, the bismuthates ($T_c \approx 30$ K) have been much less studied. The third family, the fullerites, have composition A_3C_{60}, where A = K, Rb, or Cs. Their crystal structure is also cubic and isotropic, but the chemical stability of these materials is an open issue.

Early enthusiasm over the remarkable strides made in critical temperature

HIGH-TEMPERATURE SUPERCONDUCTORS

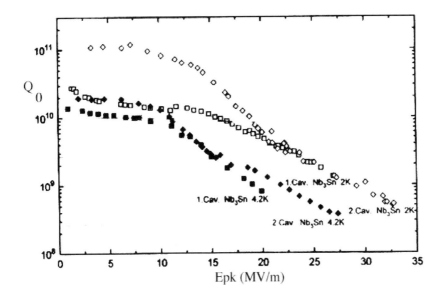

Figure 14.8: Performance of a Nb$_3$Sn cavity before coating (i.e., the niobium base cavity) and after coating. Results are given for 4.2 and 2 K. (Courtesy of Wuppertal and TJNAF.)

Table 14.2: Fundamental properties of high temperature superconductors compared to other superconductors interesting for accelerator application

Material	T_c (K)	Δ/kT_c	ξ_0 (nm)	λ_L (nm)	H_c (Oe)	H_{sh} (Oe)
Pb	7.2	2.17	83	48	800	1200
Nb	9.2	1.97	39	40	2000	2300
Nb$_3$Sn	18	2.2	5.7	110	5350	4000
YBaCuO	93	$1.5 - 4$	$ab < 2$	140	$10,000 -$	$7,500 -$
					$14,000$	$10,500$
			$c < 0.4$	770		

are now tempered with difficulties in achieving useful properties, such as a high critical current density. The intrinsic properties of HTS are very different from the familiar superconductors. Table 14.2 compares the fundamental properties of the superconductors of interest to accelerator applications. We restrict the discussion on HTS to the more frequently studied cuprates. For the HTS, the coherence lengths are very short (17 Å within the copper–oxygen planes and 3 Å perpendicular to the planes, respectively). There is also a large anisotropy of the magnetic and electrical properties between the c-axis and the ab-planes, with superior behavior when the current flow is in the ab-plane.

The HTS are extreme Type II materials with a GL parameter of ≈ 100.

There is a related intrinsic disadvantage that stems from the large penetration depth. In a simple two-fluid model R_s scales as λ_L^3, (see Equation 4.38). For cuprates, the penetration depths at $T = 0$ K are very large, ≈ 200 nm in the ab-direction and ≈ 1000 nm in the c-direction. Therefore the ideal part of the rf surface resistance is expected to be high.

To produce good-quality HTS films, it is therefore necessary to orient the grains so that the c-axis is normal to the rf surface everywhere. This restriction will be a serious disadvantage for realizing HTS in existing accelerating cavity shapes.

Besides proper orientation, fabrication of good material poses many challenges. For example, it is essential to have the right stoichiometry and oxygen content. Because of the short coherence length, transport properties are extremely sensitive to minute defects, such as grain boundaries and their associated imperfections. Decoupling of superconducting grains is believed to occur because the coherence lengths approach the scale of the grain boundary thickness, forming only weak links between individual grains. As a result, the intergrain critical current is 2 to 3 orders of magnitude lower than intragrain critical current. Even at a clean grain boundary, the scale of the disorder that exists from breaking up of a unit cell can exceed the coherence length, especially in the direction of the c-axis.

A phenomenological theory of granular superconductors has been developed based on a network of grains coupled weakly, so that a supercurrent flows from one to the next by tunneling. (For a review, see [253].) In the weak coupling limit and for low magnetic fields, the model leads to a simple correlation between the residual resistance, the rf frequency (f), the energy gap (Δ), and penetration depth λ_L

$$R_0 = f^2 \frac{\mu_0 h}{\Delta(0)} \lambda_L. \tag{14.3}$$

Accordingly, the lowest possible residual resistance is of the order

$$R_0 \approx 0.5 \; \mu\Omega f \; (\text{GHz})^2 \quad \text{in the c direction} \tag{14.4}$$

and

$$R_0 \approx 60 \; \text{n}\Omega f \; (\text{GHz})^2 \quad \text{in the ab plane}. \tag{14.5}$$

Figure 14.9 shows the best values of R_s achieved at low fields and compares them to Nb, Nb$_3$Sn, and Cu.

Above a threshold surface field, which corresponds to the critical supercurrent across the junction, decoupling of the grains leads to additional residual losses. The simple theory of weakly coupled junctions predicts that losses increase linearly with f and with magnetic field according to

$$R_0 \; (\Omega) = \frac{4\mu_0 f}{3 J_c} H = \frac{0.17 f \; (\text{GHz})}{J_c \; (\text{A}/\text{cm}^2)} H \; (\text{A}/\text{m}). \tag{14.6}$$

Therefore it is important that the films have a high current-carrying capability, characterized by J_c, the critical current density, introduced in Chapter 3. The

HIGH-TEMPERATURE SUPERCONDUCTORS

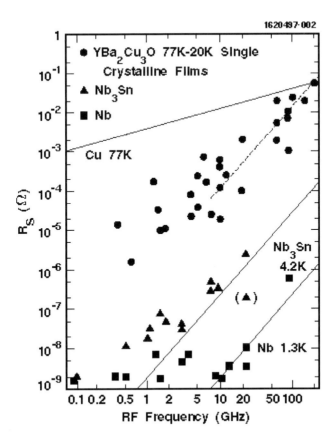

Figure 14.9: Residual resistance of HTS compared to the same for Nb, Nb$_3$Sn, and Cu. In the case of HTS, the resistance is quoted at 77 K if it is already residual; if the resistance is still decreasing with temperature the residual is obtained at 20 K. (Courtesy of Wuppertal.)

values of J_c for some of the best films are in the range of 2×10^6 A/cm^2. Even for these best values, Equation 14.6 implies a disappointing R_s of 14 mΩ at 100 Oe and 21.5 GHz. On the other hand, the best observed R_s for a 2.5-cm -diameter disk are $R_s \approx 10^{-4}$ Ω at 100 Oe and 21.5 GHz, suggesting that the predictions of Equation 14.6 are too pessimistic. Measurements [262] of good-quality strip lines also show that the rf field at the edges of the lines approach 1000 Oe without significant increase of the R_s. These results are encouraging for the future, as they suggest that the problem is not intrinsic but related to achieving good-quality films over a large area.

At the present performance levels, HTS films are attractive for low-field, passive electronic applications involving planar devices, such as multipole band-pass filters or compact delay lines, but not for high-field accelerator applications [256].

Of the three classes of the new HTS, the cuprates have received the most attention because of their high T_c. In the future, it may be appropriate to study the medium temperature superconductors further, because of their cubic isotropic properties. The bismuthates are also reported to have a larger coherence length, 60 Å [263], which is comparable to that of Nb_3Sn, raising the prospects of obtaining Q_0 values in the 10^{10} range.

Part III

Couplers and Tuners

CHAPTER 15

Mode Excitation and Its Consequences

15.1 INTRODUCTION

In previous chapters we saw how a monopole cavity mode can accelerate charge. We will now illustrate that the converse is true as well: a charge passing through a cavity can excite modes. Beam-induced modes have far reaching consequences for power dissipation and beam stability.

The charge-induced fields provide a retarding force. The energy loss must be taken into account when calculating the energy received by the charges from the accelerating mode. The modes excited by bunches can seriously affect subsequent charges passing through the cavity. If not sufficiently damped, they can lead to beam instabilities and beam loss. This is especially true in storage rings, where a bunch will pass through the same cavity many times. Even without beam breakup, higher-order modes (HOMs) can degrade the beam's quality, leading to loss of luminosity for colliders or loss of brightness for synchrotron radiation and FEL light sources.

In the case of superconducting cavities, HOMs also increase the cryogenic losses due to the additional power dissipation in the cavity walls. Even though the HOM power may only be of the order of watts, it is dissipated in the helium bath. Unless the modes are sufficiently damped, the additional refrigeration load can be expensive.

In this chapter, we will derive expressions for the excitation of HOMs using quantities which we already examined in Chapter 2. We will be able to define criteria which determine whether a particular higher-order mode is problematic. Beam stability issues related to HOM excitation will also be briefly examined.

15.2 MONOPOLE MODE EXCITATION BY A POINT CHARGE

Since the modes of a cavity form a complete, orthogonal set, we can treat the excitation of the modes individually and sum over our results in the end. To begin, we only consider monopole modes.

Consider a point charge moving on axis through a cavity. To determine the induced voltage, we will use the fact that energy is conserved in the cavity–charge system and that the cavity fields obey the principle of superposition;

that is, we can add the fields excited by the charge to any fields already present in the cavity.

Only monopole modes can be excited by charges moving on axis because other modes have no longitudinal electric field on axis. Let the charge-induced voltage in one mode of the cavity be

$$\tilde{V}_q = V_q e^{i\alpha} e^{i\omega_n t}, \qquad (15.1)$$

where V_q is the magnitude of the complex quantity (phasor) $\tilde{V}_q$ and thus *is always positive*. The angle α is the phase (yet to be determined) between the charge traversing the cavity and $\tilde{V}_q$, and ω_n is the eigenfrequency of the mode.

We postulate that the induced voltage acts on the charge itself. A priori, it is not obvious that the charge "sees" the full voltage. Thus we write the effective voltage acting on the charge ($\tilde{V}_{\text{eff}}$) as some fraction f of the total voltage; that is,

$$\tilde{V}_{\text{eff}} = f\tilde{V}_q. \qquad (15.2)$$

To calculate the induced voltage, we send a particle through a lossless cavity which has previously been excited to a voltage $\tilde{V}_c$. We let the phase of $\tilde{V}_c$ at the time of passage be some arbitrary angle φ so that

$$\tilde{V}_c = V_c e^{i\varphi} e^{i\omega_n t}, \qquad (15.3)$$

where V_c is a positive real quantity.

The stored energy (U_i) in the cavity before the passage is

$$U_i = \frac{V_c^2}{\omega_n \dfrac{R_a}{Q_0}}, \qquad (15.4)$$

where R_a and Q_0 are the mode's shunt impedance and unloaded quality, respectively. Once the charge has left the cavity, the voltage within is given by superposition, so that the final stored energy (U_f) is

$$U_f = \frac{|V_c e^{i\varphi} + V_q e^{i\alpha}|^2}{\omega_n \dfrac{R_a}{Q_0}}. \qquad (15.5)$$

The difference between the initial and final energy is

$$\Delta U_c = U_f - U_i = \frac{2 V_c V_q \cos(\varphi - \alpha)}{\omega_n \dfrac{R_a}{Q_0}} + \frac{V_q^2}{\omega_n \dfrac{R_a}{Q_0}}. \qquad (15.6)$$

Note that the charge itself has changed its energy by an amount ΔU_q:

$$\Delta U_q = q(V_c \cos\varphi + f V_q \cos\alpha). \qquad (15.7)$$

Due to energy conservation we have

$$\Delta U_q + \Delta U_c = 0, \qquad (15.8)$$

that is,
$$-q(V_c \cos\varphi + fV_q \cos\alpha) = \frac{2V_c V_q \cos(\varphi - \alpha)}{\omega_n \frac{R_a}{Q_0}} + \frac{V_q^2}{\omega_n \frac{R_a}{Q_0}}. \tag{15.9}$$

It is important to note that the superposition principle requires
$$V_q \propto q, \tag{15.10}$$
so that Equation 15.9 has terms in both q and q^2. If (15.9) is to hold for arbitrary q we can equate terms with the same powers of q. Thus

$$-qV_c \cos\varphi = \frac{2V_c V_q \cos(\varphi - \alpha)}{\omega_n \frac{R_a}{Q_0}} \tag{15.11}$$

$$-qfV_q \cos\alpha = \frac{V_q^2}{\omega_n \frac{R_a}{Q_0}}. \tag{15.12}$$

Since the phase φ is arbitrary, Equation 15.11 can only hold if α is 2π times an integer if $q < 0$, and π times an odd integer if $q > 0$. Thus we find that

$$\tilde{V}_q = \mp \frac{\omega_n R_a}{2Q_0} |q| e^{i\omega_n t} \tag{15.13}$$

and, using (15.12),
$$f = \frac{1}{2} \quad \text{or} \quad \tilde{V}_{\text{eff}} = \frac{\tilde{V}_q}{2}. \tag{15.14}$$

The upper sign in (15.13) applies when $q > 0$, and the lower sign applies when $q < 0$.

Only half of the induced voltage acts back on the charge itself and Equation 15.14 is known as the *fundamental theorem of beam loading* [264]. Note also that the phase α is always such that the induced voltage acts to retard the charges. Intuitively this is reasonable, since a charge cannot accelerate itself.

From (15.13) we see that the beam-induced voltage is related to the shunt impedance, which we have shown previously to be only dependent on the geometry of the cavity (see Chapter 2). The quantity

$$k_n = \frac{\omega_n}{4} \frac{R_a}{Q_0} \tag{15.15}$$

is called the mode *loss factor*. It can be calculated for a given geometry with numerical codes via the R_a/Q_0.

Using (15.4) for the energy stored in a cavity when the cavity voltage is V_c, we find that the energy deposited by the charge is

$$U_q = k_n q^2, \tag{15.16}$$

and hence the loss factor can also expressed as

$$k_n = \frac{V_c^2}{4U}. \tag{15.17}$$

Equation 15.13 applies to a point charge moving through a lossless cavity. When losses are included, (15.13) still applies, provided that the charge exits the cavity before the fields have a chance to decay substantially. The decay time of the fields in a copper cavity at 500 MHz is typically a few microseconds, which indeed is much longer than the nanosecond transit time of a relativistic charge. In the case of a superconducting cavity, the decay times are even longer.

15.3 MONOPOLE MODE EXCITATION BY A BUNCH

The mode excitation by a point charge is an idealized case. In an accelerator, numerous charges are grouped together in a bunch, and it is impossible to make the bunch point-like. Rather, it has a distribution that in many cases can be approximated by a Gaussian with a characteristic length σ_z along its direction of motion. The mode excitation by such a charge distribution can be calculated by superimposing the voltages

$$d\tilde{V} = 2k_n dq e^{i\omega_n t_q}$$

induced by the individual charges (dq) as they pass through the cavity at time t_q. In the process, we need to take into account the phase advance of the cavity rf fields. If we choose the time $t = 0$ to coincide with the passage of the bunch's center through the cavity, then we can write

$$\tilde{V} = 2k_n \int e^{i\omega_n t_q} dq = 2k_n \int_{-\infty}^{+\infty} I(t) e^{i\omega_n t} dt \qquad (15.18)$$

for the voltage induced by the entire bunch, where $I(t)$ is the current through the cavity as a function of time. For a Gaussian profile with total charge Q_b, we can write

$$I(t) = \frac{Q_b}{\sqrt{2\pi}\sigma_t} \exp\left(-\frac{t^2}{2\sigma_t^2}\right), \qquad (15.19)$$

where σ_t is the length (standard deviation) of the bunch in time. Equation 15.18 then evaluates to

$$\tilde{V} = 2k_n Q_b \exp\left(\frac{-\omega_n^2 \sigma_t^2}{2}\right) = 2k_n Q_b \exp\left(-\frac{\omega_n^2 \sigma_z^2}{2c^2}\right), \qquad (15.20)$$

where we have taken the velocity of the bunch to be c so that $\sigma_z = c\sigma_t$. We see that a bunch is only effective at exciting modes for which $1/\sigma_t \gtrsim \omega_n$. If $1/\sigma_t \gg \omega_n$, the bunch can be treated as a point charge. If, on the other hand, $1/\sigma_t \ll \omega_n$ then the induced voltage is much less than for a point charge. Hence most HOMs with angular frequencies greater than $1/\sigma_t$ do not need to be considered. The inability of a bunch to couple to modes above $1/\sigma_t$ is due to the fact that the beam current lacks Fourier frequency components above this frequency.

15.4 MONOPOLE MODE EXCITATION BY A TRAIN OF BUNCHES

We expand our discussion to consider an infinitely long "train" of bunches (each of charge q), spaced in time by T_b. We will assume that the bunches are short enough to be considered point charges, although the correction term in (15.20) for Gaussian bunches could easily be included in the following discussion. Analogous to the previous section, we will obtain the steady-state voltage excited by the beam by using the superposition principle. However, we can no longer neglect the power dissipation in the cavity. The fields in a given mode will decay exponentially with a time constant $T_d = 2Q_L/\omega_n$.

Following the first bunch's passage, the cavity voltage evolves as

$$\mp V_q \exp(i\omega_n t) \exp\left(-\frac{t}{T_d}\right),$$

where t is the elapsed time since the bunch's passage. Immediately following the second bunch's passage the total voltage in the cavity therefore is

$$\mp V_q \left\{1 + \exp(i\omega_n T_b) \exp\left(-\frac{T_b}{T_d}\right)\right\}.$$

The same argument is applied to yield the voltage after all the subsequent bunches. In the steady state, we thus obtain the beam-induced voltage ($\tilde{V}_b$) by summing an infinite number of terms:

$$\tilde{V}_b = \mp V_q \sum_{n=0}^{\infty} \exp(in\omega_n T_b) \exp\left(-n\frac{T_b}{T_d}\right) \qquad (15.21)$$

where we have omitted the $\exp(i\omega_n t)$ factor common to all terms in the summation. To simplify (15.21), we write

$$\omega_n = \omega + \Delta\omega, \qquad (15.22)$$

where

$$\omega = \frac{2\pi h}{T_b}, \qquad h = \text{integer}. \qquad (15.23)$$

The integer h is the harmonic number of the beam and ω is the frequency that governs the bunch spacing. Normally this is the frequency of the generator used to drive the accelerating mode. Summing (15.21) yields the beam-induced voltage:

$$\tilde{V}_b = \mp \frac{V_q}{1 - \exp\left(-\frac{T_b}{T_d}\right) \exp(iT_b \Delta\omega)}. \qquad (15.24)$$

To calculate the voltage seen by the beam ($\tilde{V}_b'$), we need to subtract $\mp V_q/2$ from (15.24) to take into account that a given bunch only sees half of the voltage it induces itself. Thus,

$$\tilde{V}_b' = \mp \frac{V_q}{1 - \exp\left(-\frac{T_b}{T_d}\right) \exp(iT_b \Delta\omega)} \pm \frac{V_q}{2}. \qquad (15.25)$$

We can separate the real and imaginary parts of $\tilde{V}_b'$ by writing

$$\tilde{V}_b' = \pm V_q(F_r + iF_i), \tag{15.26}$$

where

$$F_r = \frac{1 - \exp\left(-2\frac{T_b}{T_d}\right)}{2\left[1 - 2\exp\left(-\frac{T_b}{T_d}\right)\cos(\Delta\omega T_b) + \exp\left(-2\frac{T_b}{T_d}\right)\right]} \tag{15.27}$$

$$F_i = \frac{\exp\left(-\frac{T_b}{T_d}\right)\sin(\Delta\omega T_b)}{\left[1 - 2\exp\left(-\frac{T_b}{T_d}\right)\cos(\Delta\omega T_b) + \exp\left(-2\frac{T_b}{T_d}\right)\right]}. \tag{15.28}$$

The power P_b dissipated by the beam in exciting this mode is

$$P_b = I_0 \Re(\tilde{V}_b') = \frac{I_0^2 R_a}{(1+\beta)} \frac{T_b}{T_d} F_r, \tag{15.29}$$

where $I_0 = |q|/T_b$ is the time-averaged beam current and $\Re(\tilde{x})$ is the real part of $\tilde{x}$. In (15.29) we have allowed for external coupling with coupling strength β between the mode and a load. (Recall that $T_d = 2Q_L/\omega_n = 2Q_0/[1+\beta]\omega_n$.)

Since the modes of a cavity are orthogonal, we can sum the power absorbed by each mode to obtain the total power dissipated in HOMs (P_{total}). Similarly, the total voltage in the cavity due to HOM excitation (V_{total}) can be obtained by summing over each mode:

$$P_{\text{total}} = \sum_n^\infty P_{b,n} \tag{15.30}$$

$$V_{\text{total}} = \sum_n^\infty V_{b,n}, \tag{15.31}$$

where the summation is over all higher-order modes. Since the power lost by the beam to HOMs must be compensated for by an increase in the accelerating voltage, it is important to understand the extent of HOM excitation for a given cavity design. Beyond that, there are beam stability issues which need to be taken into account.

The values of V_{total} and P_{total} depend on T_b, T_d, and $\Delta\omega$. In particular, the beam-induced voltage is critically dependent on where the lines of the beam spectrum lie in relation to the cavity modes. Figure 15.1 depicts F_r as a function of $\theta = \Delta\omega T_b$ for two different ratios of T_b to T_d. For $T_b \ll T_d$ we see that a substantial resonant buildup can occur if θ is small. This corresponds to a close match between a Fourier component of the beam current and a mode's eigenfrequency. For convenience, we set the reference point at which resonant buildup becomes significant at $F_r = 1/2$ — that is, when $\cos\theta = \exp(-T_b/T_d)$. Clearly, for $T_b \ll T_d$ the probability of resonant buildup is small, but if it does happen, the resulting mode excitation is severe. Hence it is important that

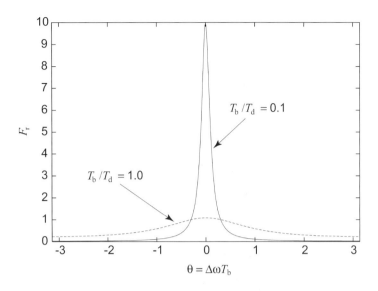

Figure 15.1: Functional dependence of F_r on θ.

all potentially dangerous modes are carefully examined to ensure that resonant excitation is avoided.

In general, the HOM spectrum is not known with sufficient accuracy and a stochastic approach is needed. The mode spacing is not regular and we may assume that the phase θ varies randomly between $-\pi$ and π for all HOMs. Furthermore, let us assume for simplicity that T_b/T_d is the same for all modes. In that case we can find the *average* mode excitation by integrating F_r over the phase θ; that is,

$$\langle F_r \rangle = \frac{1}{2\pi} \int_{-\pi}^{\pi} \frac{w}{u+v\cos\theta}\, d\theta = \frac{1}{\pi} \int_{0}^{\pi} \frac{w}{u+v\cos\theta}\, d\theta, \qquad (15.32)$$

where

$$u = 1 + \exp\left(-2\frac{T_b}{T_d}\right),$$
$$v = -2\exp\left(-\frac{T_b}{T_d}\right),$$

and

$$w = \frac{1 - \exp\left(-2\frac{T_b}{T_d}\right)}{2}.$$

This integral can be evaluated using the fact that

$$\int \frac{d\theta}{u+v\cos\theta} = \frac{2}{\sqrt{u^2-v^2}} \tan^{-1}\left(\frac{\sqrt{u^2-v^2}\tan(\theta/2)}{u+v}\right)$$

to give the simple result
$$\langle F_r \rangle = 1/2. \tag{15.33}$$
Thus, if we do not know the exact mode spectrum of a cavity, an estimate of the power dissipated by the beam into HOMs is given by using $\langle F_r \rangle$ instead of F_r in (15.29) and (15.30). Equation 15.33 simply states that excessive power loss to resonantly excited modes ($F_r > 1/2$) is compensated for by other modes for which $F_r < 1/2$, so that on *average* the power loss to each mode is given by the single-pass expression (15.16) divided by T_b.

However, T_b/T_d is generally not the same for all modes, and a few modes may dominate the cavity behavior. Furthermore, since the above evaluation is a stochastic approach, it does not include the unlucky case of excessive resonant losses which can occur due to a single mode, that is when $\theta = 0$. In that case, when $T_b \ll T_d$, we find
$$V_b = V_q \frac{T_d}{T_b} = \frac{\omega_n}{2} \frac{R_a}{Q_0} I_0 T_d. \tag{15.34}$$
To be safe, we have to ensure that (15.34) is within acceptable limits.

15.4.1 Cryogenic Losses

When using superconducting cavities in accelerators, one must take into account cryogenic losses due to HOM excitation. Each excited mode n dissipates a power $P_{c,n}$ in the cavity walls, where
$$P_{c,n} = \frac{V_{b,n}^2}{R_a}. \tag{15.35}$$
In the worst case ($\theta = 0$, $T_b \ll T_d$) Equation 15.34 applies, and we obtain
$$P_{c,n} = \frac{\omega_n^2 R_a I_0^2 T_d^2}{4 Q_0^2}. \tag{15.36}$$
Substituting $T_d = 2Q_L/\omega_n = 2Q_0/(1+\beta)\omega_n$ we obtain
$$P_{c,n} = \frac{R_a I_0^2}{(1+\beta)^2}. \tag{15.37}$$
The power dissipated into a load coupled to the nth mode with coupling strength β is $\beta P_{c,n}$ so that, by energy conservation, the power dissipated by the beam is
$$P_{b,n} = (1+\beta)P_{c,n} = \frac{R_a I_0^2}{1+\beta}. \tag{15.38}$$
Equation 15.38 can just as easily be derived from (15.29) by noting that $F_r = T_d/T_b$ in the resonant case (with $T_d \gg T_b$).

The "antiresonant" case ($\theta = \pi$) is the most favorable situation and one can show that for $T_d \gg T_b$
$$P_{c,n} = \frac{V_q^2}{4R_a}. \tag{15.39}$$

As an example, consider the Cornell 500-MHz superconducting cavity to be used in the CESR luminosity upgrade. One of the HOMs has a resonant frequency of 1120 MHz, an R_a/Q_0 value of 6.85 Ω, and a Q_0 value of 1.44×10^8 (undamped cavity). If a current of 0.1 A is to be accelerated, the power dissipated in the cavity walls (and hence in the helium bath) in the resonant case is

$$P_{c,n} = \frac{R_a I_0^2}{(1+\beta)^2} = \frac{9.86 \text{ MW}}{(1+\beta)^2} \qquad (\theta = 0)$$

which, of course, is unacceptable if no coupling to that mode is provided ($\beta = 0$). For comparison, we quote the result for the antiresonant case, where HOM power is negligible:

$$P_{c,n} = 0.009 \text{ W} \qquad (\theta = \pi).$$

Here we assume $1/T_b = 400$ kHz.

Due to the low Carnot efficiency of helium refrigerators, even a few watts dissipated in the helium bath can be a significant load. A typical refrigerator has an efficiency of $\approx 1/750$, when the technical efficiency is also taken into account. Thus we need to reduce the HOM power below one watt. This can be achieved either by ensuring that the HOM is excited antiresonantly, or by increasing the coupling to the load, thereby increasing β in (15.37). In damping the HOMs we must ensure that the coupling to the accelerating mode is minimized by placing HOM couplers in a low-field region of the accelerating mode or by using rejection filters to prevent the fundamental from reaching the HOM absorber. These approaches will be discussed in more detail in Chapter 16.

Continuing with our example, if we want to limit the losses in the resonant case to one watt, we need $\beta = 3140$, yielding a loaded Q for that mode of 4.6×10^4. The beam-induced HOM voltage then is

$$V_{c,n} = \frac{I_0 R_a}{1+\beta} = 3.14 \times 10^4 \text{ V}.$$

Besides the issue of cryogenic losses due to HOMs, we also need to consider beam instabilities induced by HOMs. Not only must the accelerating mode supply the additional energy lost by the beam to HOMs, but problems such as coupled-bunch instabilities can rapidly be amplified by cavity HOMs in storage rings, if the modes are not sufficiently damped. We will cover this topic briefly in later sections.

The important conclusion to be drawn from the above example is that a single HOM in a cavity can be dangerous if it is not damped sufficiently. Thus, it is imperative to examine the mode spectrum at least up to the beam pipe cutoff frequency. Even above cutoff, care must be taken to identify trapped modes, which also require damping. As we progress to higher frequencies the R_a/Q_0 values decrease since

$$\frac{R_a}{Q_0} = \frac{V_c^2}{\omega_n U}$$

decreases, and the decay times $T_d = 2Q_L/\omega_n$ also diminish.[1] Thus mode excitation is less severe. This is further aided by the fact that Gaussian bunches of length σ_z do not contribute to HOM excitation for modes above $\omega = c/\sigma_z$ (as discussed in Section 15.3).

15.5 DIPOLE MODE EXCITATION

A charge moving on axis is only able to excite monopole TM modes because the longitudinal electric field vanishes on axis for all other modes. However charges in an accelerator execute transverse oscillations about the design beam trajectory and such charges can excite dipole, quadrupole, and other HOMs. Due to their strong deflecting field, dipole modes are especially important. We briefly illustrate in this section how dipole mode excitation can be determined in the case of a point charge.

In Chapter 2 we showed that in a simple pill-box cavity of length d and radius R, the longitudinal electric field for TM_{mnp} modes is

$$E_z = E_0 J_m\left(\frac{u_{mn}\rho}{R}\right) \cos\left(\frac{p\pi z}{d}\right) \cos m\phi. \quad (15.40)$$

For $|x| \ll 1$, $J_1(x) \approx x/2$. Hence, for a dipole mode, E_z increases linearly with ρ for small distances from the axis.

The field distribution is affected by the addition of beam tubes. Nevertheless, it is possible to show that the longitudinal (accelerating) electric field, when averaged along a path parallel to the axis, varies as [265]

$$E_z = E_a \left(\frac{\rho}{a}\right)^m \cos m\phi \quad (15.41)$$

for a TM_{mnp} mode in an axisymmetric structure. Here E_a is the average field at the radius of the beam tube (a) and ρ is the distance off axis.[2] Equation 15.41 is in agreement with our treatment of the pill-box cavity in the case of a dipole mode ($m = 1$) and an infinitesimal beam tube radius.

The derivation of the voltage excited by a charge moving off axis is very similar to the treatment of monopole mode excitation. We begin by defining the dipole loss factor (k_d) analogous to the monopole loss factor in (15.17). Since the accelerating voltage in a dipole mode increases linearly with distance from the axis, we need to choose a suitable reference distance. For this purpose we define V_a to be the accelerating voltage at the beam tube radius. Hence, an appropriate choice for the dipole loss factor is

$$k_d = \frac{V_a^2}{4U}. \quad (15.42)$$

---

[1] We are assuming that the ratio of the accelerating voltage squared to the stored energy as well as the cavity quality do not increase significantly for HOMs.

[2] The average is taken in the same manner as when calculating the accelerating field of monopole modes (see Chapter 2).

DIPOLE MODE EXCITATION

Again we let a charge q travel through a previously excited cavity. The stored energy prior to the charge passing through the cavity is

$$U_\mathrm{i} = \frac{V_a^2}{4k_\mathrm{d}}, \tag{15.43}$$

whereas after the passage it is

$$U_\mathrm{f} = \frac{|V_a e^{i\varphi} + V_q e^{i\alpha}|^2}{4k_\mathrm{d}}. \tag{15.44}$$

The phases have the same meaning as in the discussion of monopole modes, and V_q is the accelerating voltage (at the beam tube radius) induced by the charge. The total energy change of the charge moving a distance ρ off axis on the other hand is

$$\Delta U_q = q(V_a \cos\varphi + f V_q \cos\alpha)\frac{\rho}{a}, \tag{15.45}$$

where we have taken into account the linear field variation with distance from the axis. (Note that the direction from the axis to the charge defines $\phi = 0$.) Again, we equate the energy gained by the charge with the energy lost by the cavity. Superposition requires that $V_q \propto q$ and we find, analogous to the monopole case, that $\alpha = 2\pi k$ (k = integer) for $q < 0$ and $\alpha = \pi l$ (l = odd integer) for $q > 0$. Furthermore, $f = 1/2$. Thus

$$\tilde{V}_q = \mp 2k_\mathrm{d}|q|\frac{\rho}{a} e^{i\omega_n t}, \qquad \rho \leq a. \tag{15.46}$$

Again, the upper case applies when $q > 0$ and the lower case applies when $q < 0$.

For a given stored energy in a dipole mode, the quantity $V_\parallel/\rho$ is constant for $\rho \leq a$, where $V_\parallel$ is the accelerating voltage a distance ρ off axis. Hence we can define the dipole shunt impedance R_d by

$$R_\mathrm{d} = \frac{V_\parallel^2}{(\omega_n/c)^2 \rho^2 P_\mathrm{c}} = \frac{V_a^2}{(\omega_n/c)^2 a^2 P_\mathrm{c}}, \tag{15.47}$$

where, as always, P_c is the power dissipated in the cavity walls. The factor $(\omega_n/c)^2$ is included in (15.47) so that the shunt impedance is in units of ohms, as in the case of monopole modes. Given the definition (15.47), the dipole loss factor can be written as

$$k_\mathrm{d} = a^2 \left(\frac{\omega_n}{c}\right)^2 \frac{\omega_n}{4} \frac{R_\mathrm{d}}{Q_0}, \tag{15.48}$$

and the energy lost by a charge to the dipole mode is

$$U_q = k_\mathrm{d} q^2 \left(\frac{\rho}{a}\right)^2. \tag{15.49}$$

Due to their deflecting nature, dipole modes also pose a threat to beam stability. The deflecting force ($\mathbf{F}_\perp$) experienced by a unit charge traveling off axis can be related to the longitudinal electric field [which we know from (15.46)] by the Panofsky–Wenzel theorem [266]. It states that

$$i\omega \int_0^d \mathbf{F}_\perp e^{i\omega z/v} \frac{dz}{v} = q\left[\mathbf{E}_\perp e^{i\omega z/v}\right]_0^d - q \int_0^d \boldsymbol{\nabla}_\perp E_z e^{i\omega z/v} dz, \quad (15.50)$$

where v is the longitudinal velocity of the charge, the $\perp$ subscript refers to the transverse components of the subscripted vector, and $\boldsymbol{\nabla}_\perp = \boldsymbol{\nabla} - \hat{z}\,\partial/\partial z$. The first term on the right side is zero provided that the electric field vanishes at $z = 0$ and $z = d$, which usually is the case if the path starts and ends some distance into the beam tubes. The left side of (15.50) is related to the *effective deflecting voltage* $V_\perp$ experienced by the charge in moving from $z = 0$ to $z = d$, which is

$$V_\perp = \frac{c}{qv} \int_0^d \mathbf{F}_\perp e^{i\omega_n z/v}\,dz. \quad (15.51)$$

Hence

$$V_\perp = \frac{ic \int_0^d \boldsymbol{\nabla}_\perp E_z e^{i\omega_n z/v}\,dz}{\omega_n}. \quad (15.52)$$

If we can neglect the variation of the charge's transverse coordinates during its transit through the cavity then

$$V_\perp = \frac{i \int_0^d E_z(\rho = a) e^{i\omega_n z/v} c\,dz}{\omega_n a} \quad (15.53)$$

and hence

$$\frac{|V_\perp|^2}{\omega_n U} = \frac{\left|\int_0^d E_z(\rho = a) e^{i\omega_n z/v} dz\right|^2}{(k_n a)^2 \omega_n U}, \quad (15.54)$$

where $k_n = \omega_n/c$. Comparing (15.54) with (15.47) we find that

$$\frac{R_\mathrm{d}}{Q_0} = \frac{|V_\perp|^2}{\omega_n U}, \quad (15.55)$$

analogous to the monopole-mode shunt impedance.

15.6 INSTABILITIES FROM BEAM CAVITY INTERACTIONS

To understand the need for damping HOMs and the degree to which they should be damped, we describe how beam-cavity interactions limit the beam quality in linear accelerators, and how they limit the total current in circular and multipass accelerators. This is an extensive subject covered elsewhere [264, 267, 268]. We present only a brief, qualitative treatment to appreciate coupler requirements.

At one extreme of the time scale there are the single-bunch, single-pass effects which are driven by the short time behavior of wakefields. The feature that is most important here is the large beam hole of a superconducting cavity which reduces the wakefields far below those of a normal conducting cavity.

The longitudinal wakefields decrease with aperture as $1/a^2$, and the transverse wakes decrease as $1/a^3$. Furthermore, since superconducting cavities do not waste energy in wall dissipation, one can afford the large stored energy associated with low-frequency structures. The low rf frequency again means large apertures and stronger reductions in wakefields. We will see in Chapter 21 how these factors make rf superconductivity attractive for linear colliders. For high-current, factory-like storage rings (Chapter 20), substitution of superconducting cavities for normal conducting cavities is expected to greatly improve current limits not only because of the larger aperture but also because fewer superconducting cavities will be required to provide the needed accelerating voltage. As we mentioned, superconducting cavities provide higher cw gradient than normal conducting cavities. Due to the shorter length of the rf system, there is an overall reduction in short-range wakes.

At the other extreme of the time scale are the coupled bunch effects. These become important if wakefields linger for times long compared to the bunch separation. In superconducting cavities, the unloaded Qs of the HOMs are typically 10^9 and must be appropriately damped by HOM couplers not only to keep the HOM-induced cryogenic load acceptable, but also to avoid multibunch beam instabilities.

Instabilities are classified as either longitudinal or transverse. The former are induced when the beam interacts with monopole higher-order modes, and the latter are induced when the beam interacts with the deflecting (chiefly dipole) modes.

15.6.1 Single-Bunch Effects

Single-bunch effects are driven by the short-time behavior of wakefields which arise when there are geometric or material discontinuities along the accelerator beam pipe. The short-range wakes act over the time period of the bunch length. Since the bunch is very short (in the picosecond to nanosecond range) these wakes do not depend on the Q, but only on the R/Q_0 of the modes — that is, on the cavity geometry.

The short-range wakes generated by particles at the head of the bunch act on the particles at the tail. Wakefields arising from interaction of the beam with monopole modes cause energy loss, as discussed earlier. We showed how the loss factor of individual modes can be calculated from the individual R/Q_0. For the total loss factor ($k_\parallel$) of an axially symmetric structure one can use wakefield calculation programs such as TBCI [269] and ABCI [270].

A variation of the longitudinal wakefield across the bunch causes varying energy loss, thereby introducing harmful energy spread. In a circular e^+e^- collider, the accompanying bunch lengthening compromises the ability to focus the beams at the interaction point for high luminosity. In a linear collider, the energy spread of the beam degrades the final focus of the beam, making it difficult to achieve a small spot size. A rough estimate for the wakefield-induced energy spread is

$$\frac{\delta E_{\rm b}}{E_{\rm b}} = \frac{2qk_{\|}}{E_{\rm acc}} \qquad (15.56)$$

Here $E_{\rm b}$ is the beam energy and $k_{\|}$ is the total structure loss factor in V/(C-m).[3]

The monopole wakes decrease as the square of the aperture. Besides using larger apertures as permissible with superconducting cavities, energy spread can also be reduced by arranging for the bunch to arrive earlier than the crest of the rf wave. Under this condition the particles which arrive later achieve a larger energy gain from the accelerating field. This partly compensates for the deceleration by the wakefield.

We now turn to the short-range transverse wakes. These deflect the tail of the bunch, causing emittance growth and beam halo. In analogy to the longitudinal loss factor, the total transverse loss factor ($k_\perp$) can also be determined from codes such as ABCI. The overall effect of transverse wakes is to limit the amount of charge in a bunch. The transverse field witnessed at the tail is proportional to the displacement of the head. As the bunch proceeds down the linac, executing betatron oscillations, the displacement of the tail (Δx) increases with length. If the tail displacement exceeds the displacement of the head, there will be unstable growth. If the head and the tail in a simple two-particle model of the beam are separated by $2\sigma_z$, then the relative displacement of the tail with respect to the head is roughly

$$\frac{\Delta x}{x} = \frac{eNk_\perp}{2E_{\rm acc}} \langle \beta \rangle \ln\left(\frac{E_{\rm bf}}{E_{\rm b0}}\right) \qquad (15.57)$$

Here N is the number of particles per bunch, $k_\perp$ is the total dipole loss factor in V/(C-m^2), $E_{\rm b0}$ is the injection energy, $E_{\rm bf}$ is the final energy, and $\langle \beta \rangle$ is the average focusing strength along the linac.

Transverse wakes decrease as the cube of the aperture which makes large-aperture superconducting cavities favorable. Transverse wakes can also be reduced by stronger transverse focusing forces of quadrupole magnets.

For the case of a superconducting linear collider, TESLA (see Chapter 21), the relevant accelerator and rf structure parameters are given in Table 15.1 [192]. Using these, we make a rough estimate that the HOM-induced power is 3.7 W/m, the energy spread due to longitudinal wakefields is about 0.5%, and the emittance growth through the linac is about 50%. The HOM power is high enough compared to the rf dissipation in the fundamental mode (1.5 W/m) and therefore must be extracted efficiently by the HOM couplers. Both the energy spread and emittance growth can be substantially improved over these estimates by choosing the proper arrival phase of the bunch as well as by choosing a good focusing lattice for the linac. We used $\langle \beta \rangle = 50$ m here for illustration only. A more realistic design and estimate for TESLA yields an emittance growth of less than 5% [271]. Note that the design of TESLA (as for other linear collider options) is still under evolution, so that the parameters quoted here are likely

---

[3]Note that this loss factor is *per unit length* of a structure. We therefore use the symbol $k_{\|}$ to distinguish it from k_n.

INSTABILITIES FROM BEAM CAVITY INTERACTIONS

Table 15.1: RF structure and TESLA parameters used to estimate the effect of short-range wakes

f:	RF frequency	MHz	1300
R/Q_0:	Geometric shunt impedance	Ω/m	830
E_{acc}:	Gradient	MV/m	25
Q_0:	Unloaded Q		5×10^9
	Duty factor	%	1
N:	Number of particles/bunch		5×10^{10}
f_c:	Beam collision frequency	Hz	8000
$k_\parallel$:	Total longitudinal loss factor	V/pC-m	9
$k_\perp$:	Transverse loss factor	V/pC-m^2	16
σ:	Bunch length	mm	1
E_0:	Injection energy	GeV	3
E_f:	Final energy	GeV	250
$\langle \beta \rangle$:	Average focusing strength	m	50

to change in the near future. But the example is meant to be representative of the effects.

15.6.2 Coupled-Bunch Instabilities

If the wakefields linger for at least the duration of the bunch spacing there can be coupled bunch instabilities. A long train of bunches may also coherently excite a mode for a time on the order of Q_L/ω. While the R/Q_0 of a mode describes the single pass excitation of a cavity mode by a beam bunch, the quantity, $R/Q \times Q_L$, describes the coherent summation of multiple excitations over the filling time of the mode. The primary approach to avoiding coupled-bunch instabilities is to lower the Q_L of the HOMs.

Single-Pass Effects

Regenerative Beam Breakup When a train of bunches passes slightly off axis through an accelerating structure, dipole HOMs are excited and deflect the beam into regions of higher longitudinal field. The longitudinal electric field off axis decelerates the beam, and feeds more energy into the mode. The transverse magnetic field further deflects the beam. This type of instability can arise even within a single structure. The threshold current is given by

$$I_{\text{th}} \propto \frac{E_b}{(r_d/Q_0)Q_L L^2}, \quad r_d = R_d \left(\frac{\omega_n}{c}\right)^2, \tag{15.58}$$

where r_d/Q_0 is the transverse dipole shunt impedance (Ω/m^2), E_b is the beam energy, and L is the length of the structure. Since superconducting structures are only a few cells long, the L^2 dependence makes this limit less important. However, modes with high Q_L could mean trouble for low-energy (MeV) beams.

Cumulative Beam Breakup Suppose the first bunch enters the first cavity of an accelerating system offset from the beam axis. It excites a dipole mode. But the first bunch experiences no additional deflection. Subsequent bunches experience deflection due to HOM excitation by the first bunch. The level of excitation varies with time depending on the relative frequencies of the bunch train and the excited HOM. A series of cavities produces an amplification of an initial offset at the first cavity. This is called *cumulative beam breakup*. Reduction of initial offset, or of cavity misalignments, reduces the maximum beam displacement. HOMs which are near harmonics (within a half width) of the bunch frequency are of the most concern.

Multipass Effects

Longitudinal Instability High-Q modes mediate energy transfer between bunches. Consider the example of a storage ring. When a bunch passes through a cavity, it induces a voltage in a monopole HOM. As the bunch goes around the ring, the vector increment will undergo a phase shift relative to the synchrotron oscillation phase of the bunch. If the vector increment drives the oscillation amplitude of the bunch to larger values on subsequent bunch passages, an instability can grow. Note that the bunch also experiences a phase shift as it goes around the ring. The resonance condition is fulfilled if

$$mf_r + f_s = f_n, \tag{15.59}$$

where f_r is the revolution frequency, f_s is the synchrotron frequency, f_n is the HOM frequency, and m is an integer [272]. When the phase is such that wakes from successive bunches add constructively, an instability may occur. The growth rate is a sensitive function of

$$\epsilon = \frac{f_n}{f_r} - (m + \nu), \tag{15.60}$$

where ν is the synchrotron tune — that is, the ratio of the synchrotron oscillation frequency to the bunch revolution frequency. For ϵ near one, the impedance of the HOM peaks near a negative frequency line in the beam power spectrum leading to instability. If ϵ is near zero, the impedance peaks near a positive frequency line, which leads to damping.

Therefore, whether a particular mode causes an instability depends very much on its exact frequency. In the most pessimistic case, the longitudinal instability growth rate due to a single HOM is roughly given by

$$\frac{1}{\tau_{gl}} = \frac{\Omega_s \left(\frac{R}{Q_0}\right) Q_L I_b e^{-\omega_n^2 \sigma_z^2/c^2}}{2V_c \cos\phi_s} D\left(\frac{\omega_n T_b}{Q_L}\right) \tag{15.61}$$

Here Ω_s is the synchrotron angular frequency, V_c the cavity voltage, ϕ_s the synchronous phase, and R/Q_0 the geometric shunt impedance, ω_n the angular frequency of the HOM. D is a function which takes into account the decrease of

INSTABILITIES FROM BEAM CAVITY INTERACTIONS   347

Table 15.2: Parameters for the superconducting cavity for the KEK-B factory and for the high-energy ring (HER) used to evaluate the typical growth rate of longitudinal multibunch instability

R/Q :	Ω	6.6
f_{HOM} :	MHz	1018
Q_L :		106
I_b :	A	1.1
Q_s :	KHz	1.6
V_c :	MV	1.6
$\cos\phi_s$:		0.22
σ :	cm	0.4

the induced voltage during the time T_b between two bunches [273]. To guarantee stability, the rise time must be longer than the synchrotron radiation damping time or the damping time of any longitudinal multibunch feedback system.

As an illustration, we consider the superconducting cavity under development for the high-energy ring (HER) of KEK-B [274]. Since the design beam current is 1.1 A, low values for both the R/Q_0 and Q_L are desired. The bunches will be spaced 2 ns apart leaving little damping time between bunch passages even for Q_L values of 100. Table 15.2 gives the relevant cavity and accelerator parameters for the KEK-B high-energy ring. Using Equation 15.61, we obtain the instability rise time for the highest R/Q modes of 0.6 sec, which is significantly longer than the radiation damping time of 23 ms, and the damping time (≈ 10 ms) of the planned bunch-by-bunch feedback system.

For low-current (< 100 mA) storage rings now in operation, such as HERA or LEP, the damping requirements are not so stringent, (e.g., $Q_L \approx 10^3$ to 10^5). The time between bunch passages is also larger (microseconds). In calculating the realistic instability growth rate, the R/Q_0 and Q_L from *all modes* of all cavities must be taken into account. Both the damping and the antidamping contributions of all the modes have to be taken into account. This is usually done by a Monte Carlo method once the R/Q_0 and Q_L values for monopole modes are determined [272].

In studying the effect of a large ensemble of cavities, an important consideration is that manufacturing tolerances lead to a statistical distribution of higher-order mode resonant frequencies. The spread in the frequencies is equivalent to lowering the Q_L of the mode, if there are a large number of cavities in the accelerator. This by no means implies that HOM damping can be circumvented. There is always a significant probability that a single high-Q_L mode will be located at a dangerous frequency and will dominate the onset of a multibunch instability. If the accelerator operating conditions are changed slightly (as is usually the case from time to time), a different high-Q_L, high-R/Q_0 mode can take its place. Therefore the safest strategy is to identify all the high-R/Q_0 modes and provide sufficient damping for all.

Table 15.3: Parameters for the superconducting cavity for KEK-B factory and for the high-energy ring to evaluate the growth time of transverse multibunch instabilities

R/Q :	Ω	15.3
f_{HOM} :	MHz	705
Q_L :		94
I_b :	A	1.1
E_0 :	GeV	8
b :	cm	5
ν :	Betatron tune	43

Transverse Instability When a bunch passes through a cavity, it induces a voltage proportional to its displacement from the cavity beam axis. It also receives a deflection proportional to the vector sum of previously induced voltages. The sum includes the attenuation of the voltages due to the Q_L of the mode, and the phase shift of the HOM voltage during the time between bunch passages. When the beam recirculates through the accelerator structure there is a feedback between the cavities and the beam. Exponential growth can occur above a threshold beam current. Many cavities can participate in the instability. This is the multipass beam breakup in recirculating linacs and is the multibunch instability in storage rings. In the former, only transverse modes are important because the recirculating arcs are made isochronous; that is, the energy is independent of path length so that there are no synchrotron oscillations. In storage rings both longitudinal and transverse modes can drive instabilities. The resonance condition is

$$mf_r \pm f_{\text{H,V}} = f_n. \tag{15.62}$$

Here $f_{\text{H,V}}$ is the horizontal or vertical betatron oscillation frequency. Again we see that the instability depends very much on the exact mode frequency.

In the worst case, the transverse instability growth rate [273] from the excitation of a single mode can be roughly expressed as

$$\frac{1}{\tau_{\text{gt}}} = \frac{\left(\frac{R}{Q_0}\right) Q_L I_b c^2 e^{-\omega_n^2 \sigma_z^2/c^2}}{b^2 4\pi \nu \omega_n E_b} D\left(\frac{\omega_n T_b}{Q_L}\right) \tag{15.63}$$

Here E_b is the beam energy in eV, which should be taken as the injection energy for the most unfavorable case, ν is the betatron tune, σ_z is the bunch length, and R/Q_0 is shunt impedance of the transverse mode measured at a distance b off axis. Again, using KEK-B HER as an illustration, we obtain from Equation 15.63 that the instability rise time for the highest R/Q_0 mode is 0.6 sec, much longer than the radiation damping time.

A realistic instability growth rate must take into account contributions of all the modes by a Monte Carlo method. For high-Q_L modes, statistical or deliberately built-in spreads in the frequency of the HOMs of different cavities will help because cavities will not act coherently. A higher injection energy will

CODE EXAMPLES FOR HOM STUDIES

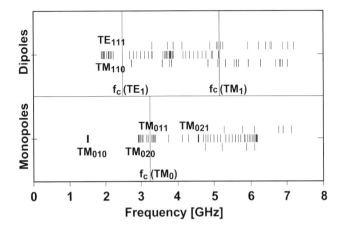

Figure 15.2: URMEL computed frequencies for monopole and dipole modes in a 6-cell accelerating cavity. The first 61 monopoles and the first 50 dipole modes are plotted.

increase the beam rigidity and reduce the amount of deflection. To guarantee stability in a storage ring, the rise times must be longer than the synchrotron radiation transverse damping time or the transverse feedback system damping time. Tighter focusing also reduces the transverse displacement resulting from the transverse kick.

15.7 CODE EXAMPLES FOR HOM STUDIES

We have already discussed the fundamental concepts of HOMs and why they need to be damped. In this discussion we showed that in order to understand the effects of an HOM one requires knowledge of its frequency, R/Q_0, and Q_L. To know these values in real structures, we must resort to computer codes. This application is very similar to using the codes to study the fundamental mode. The differences are as follows: We want to examine many modes instead of the unique fundamental, we may need to extend this analysis to modes which are above the cutoff frequency of the beam tubes, and we are interested in exploring ways of damping the modes.

In Section 2.3 of Chapter 2 we described how the code URMEL [71] could be used to understand the cavity's fundamental mode. URMEL can also be used to calculate higher-order monopole modes and dipole modes. Figure 15.2 shows a map of the frequencies of the lowest monopole and dipole modes of a 6-cell accelerating structure computed by URMEL. We are especially interested in modes with a high R/Q_0.

An example of a monopole higher-order mode with one of the highest R/Q_0 values is the TM_{011} mode. The field pattern for one of the modes in the TM_{011}

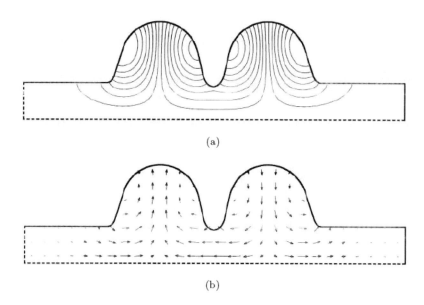

Figure 15.3: An example nonpropagating monopole HOM in a 2-cell cavity as calculated by URMEL. The cavity is excited in one of the modes of the TM_{011} passband. This HOM has one of the highest values for R/Q_0. (a) Equipotential lines. (b) Electric field plot.

passband for a 2-cell cavity is shown in Figure 15.3. The frequency of this mode is below the cutoff frequency of the beam tube; it is nonpropagating. Figure 15.4 shows an example calculated by URMEL of a monopole HOM in a 6-cell cavity. The mode shown is a propagating mode since it is 40% above the TM_{01} cutoff frequency of the beam tube.

As can be seen in the electric field plot in Figure 15.4b, although small, the fields are not attenuated along the distance of the beam tube. This poses a problem that did not exist for nonpropagating modes, such as the fundamental mode. Since the fields reach the edge of the computer modeled beam tube, the choice of boundary condition becomes significant. Since URMEL cannot present a true waveguide boundary condition, one has to ensure that the overall HOM solution is not very sensitive to the choice of boundary condition at the end of the modeled beam tube. If the solution is sensitive to the chosen boundary condition, then other methods may be needed. These methods include (a) using "lossy" codes such as SEAFISH or CLANS to be discussed later in this section or (b) applying an analytic method in postprocessing, discussed in [275].

In addition to monopole modes, dipole modes may be studied using URMEL. Figure 15.5 shows the fields of two such dipole modes. One mode (a) just does not propagate in the beam tube cutoff, and the other mode (b) is propagating. Discovering the presence of non-propagating modes below cutoff and trapped

CODE EXAMPLES FOR HOM STUDIES 351

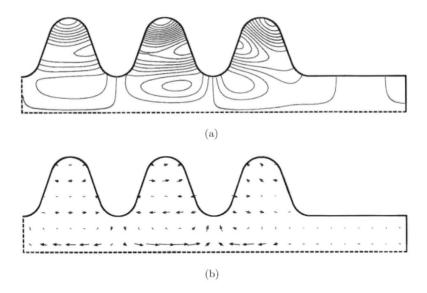

Figure 15.4: An example HOM in a six-cell cavity as calculated by URMEL. The frequency of this mode is well above the beam tube's cutoff frequency. (a) Equipotential lines. (b) Electric field plot.

modes (above cutoff) is very important. In a multicell cavity, the couplers designed to remove HOM energy are placed at the ends of the structure as discussed in the next chapter. The HOMs need to be able to propagate from the inner cells out to the HOM couplers where they can be damped. If either a nonpropagating or trapped HOM is found, the structure must be redesigned to eliminate it. This can be accomplished by reshaping the cell.

One example of a trapped mode is shown in Figure 15.6. For comparison, we also show a propagating mode. Both modes are above the cutoff frequency of the beam tube. The trapped mode even has a higher frequency than the propagating mode chosen for comparison.

As we will discuss in the next chapter, HOMs can be damped in two ways: dedicated coaxial or waveguide couplers to transmit the HOM power to a load, or HOM dampers where a segment of the beam pipe is used as the HOM coupler. In order to model the latter case with a numerical code, it is necessary to model lossy materials — that is, those with complex μ and ϵ. SEAFISH [276], a variant of SUPERFISH [70] is an example of a code that allows for strongly lossy materials such as rf damping ferrites. Similarly CLANS [75], which is a variation of SUPERLANS [72], can also be used to calculate the R/Q_0 and Q_0 of a mode with lossy materials on the beam pipe. The solutions for the cavity shown in Figure 15.6 were computed by CLANS using a band of ferrite in the beam tube to serve as an HOM damper.

Figure 15.7 shows an example from SEAFISH of four HOMs in the Cornell

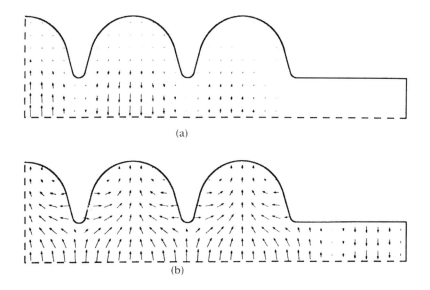

Figure 15.5: Examples of dipole modes in an accelerating cavity. (a) Nonpropagating dipole mode. (b) Propagating dipole mode.

B-cell structure before and after the introduction of a ferrite HOM damper in the beam pipe. Without the ferrite load, the HOMs are not damped and may contribute to beam instability. With the ferrite load, calculations show that most of the HOMs are damped to have a Q_0 less than 100.

Codes such as SEAFISH and CLANS are still limited to dealing with "electric" or "magnetic" boundary conditions, but lossy materials can be used to create the effect of a terminated waveguide boundary. By cleverly modeling a matched load in the beam tube of a cavity, the HOMs can be properly modeled even though they are propagating modes.

CODE EXAMPLES FOR HOM STUDIES

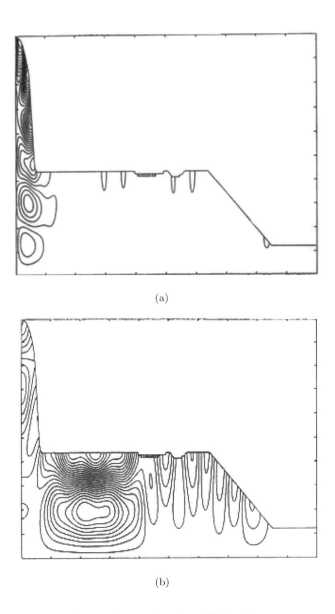

Figure 15.6: Higher-order modes in the Cornell B-cell structure computed by CLANS. The cell is at the far left and the beam tube occupies most of the length of the plot. A lossy band of ferrite is modeled at the center of the plot. (a) Trapped mode at 3314 MHz. (b) Propagating mode at 2209 MHz.

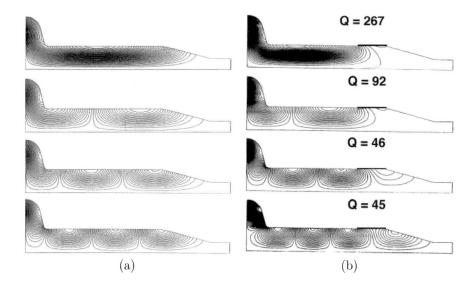

Figure 15.7: SEAFISH computations of HOMs in the Cornell B-cell. (a) Without ferrite load, Q_0 values are in the range of 10^9–10^{10}. (b) With ferrite load (designated by black bar on the cavity beam tube). The presence of the lossy ferrite lowers the HOM Q values and affects the field patterns considerably.

CHAPTER 16

Higher Order Mode Couplers

16.1 INTRODUCTION

In the previous chapter, we discussed HOM damping requirements for removing beam induced power from the cavity in order to avoid resonant buildup of beam-induced voltage, and in order to avoid beam instabilities. Here we present some of the considerations that come into play during the process of designing HOM couplers to provide the damping desired for operation in accelerators. We will discuss three major varieties of HOM couplers: waveguide, coaxial, and beam tube. After presenting some of the design aspects, we give a brief report on the performance of each coupler type. Some of the couplers we discuss are not the final ones used for the intended application. But the early versions are nevertheless useful to study as illustrations of the important concepts in HOM coupler development.

16.2 PRELIMINARY DESIGN CONSIDERATIONS

In the very early stages of the development of superconducting cavities, HOM couplers were located in the cell, where large HOM fields are present. Even though the fields are not as strong as inside the cell, it is preferable to put HOM couplers on the beam tube, outside the end cells; this ensures that the couplers do not disturb the cell geometry. We also wish to avoid reopening the possibility of low field multipacting in the favorably shaped superconducting cells. Techniques have now advanced to the point of reaching the desired damping with couplers attached to the beam pipe.

The first step in designing HOM couplers is to identify the monopole and dipole HOM modes and to carry out field calculations using codes such as URMEL, as we discussed in the previous chapter. The focus is on modes with high R/Q_0. The field intensity of the modes in the end cells should be examined to ensure that there are no high R/Q_0 modes with unusually low stored energy in the end cell. This can happen if there is a HOM with low propagation velocity, a situation often encountered for a mode near the edge of a mode family (commonly referred to as the mode passband). In this case, a possible solution is to incorporate a small, but designed, asymmetry in one end cell to "tilt" the

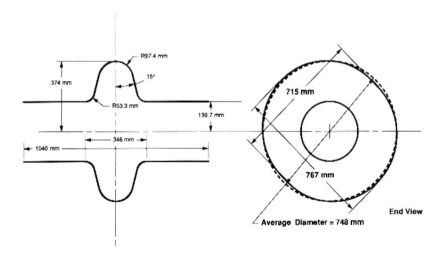

Figure 16.1: Providing azimuthal flats in the cell geometry to polarize the dipole modes. For the degree of asymmetry shown, the measured splitting of the TM$_{110}$ deflecting mode at 500 MHz was 14 MHz [278].

field profile in the HOM, and thereby provide the needed stored energy at one of the two end cells [277]. It is important to verify that the intentional asymmetry does not compromise the ability to make the field profile flat in the accelerating mode.

The highest frequency of concern depends on the bunch spectrum, which, in turn, is determined by the bunch length. Generally, one studies modes up to the cutoff frequency of the beam pipe. It is important, however, to also examine modes above the cutoff frequency, to make sure that there are no modes that are "trapped"—i.e., modes without stored energy in the end cells. If high R/Q_0 trapped modes are found, it is once again necessary to modify the cell shapes to make stored energy available in the end cells.

A minimum of two HOM couplers per cavity are desirable so that both polarizations of the dipole modes can be adequately damped. Each multipole mode of an ideal cavity is doubly degenerate, having fields which vary azimuthally as $\cos(m\phi - \phi_0)$ or $\sin(m\phi - \phi_0)$ about an axis of symmetry ϕ_0, determined by the coupler or by accidental asymmetries. Here the $m = 0, 1, 2..$ modes are known as monopole, dipole, and quadrupole modes, respectively. The presence of a single coupler will break the multipole mode degeneracy, aligning the fields of one polarization along the coupler axis and the other polarization at an angle $\phi = \pi/2m$. The external coupling will be proportional to $\cos 2(m\phi)$ or $\sin 2(m\phi)$ and will always be zero for one of the two polarizations of each multipole mode.

To reduce the cost of HOM couplers it may prove useful to polarize the modes [279] by introducing an azimuthal asymmetry in the cell shape (Figure 16.1). To disrupt the cell symmetry, the azimuthal boundary of the cavity can be made up of two chords joined by circular arcs which meet the chords tangentially. It is desirable to keep the cavity axisymmetric at the iris so that the accelerating field will be uniform in the beam hole. To make the iris axisymmetric, the width of each chord is reduced linearly with the wall radius, eventually vanishing at the start of the iris curvature. The modes are oriented 90° from the coupler axis. When the modes are so aligned, only one waveguide coupling port is necessary, as confirmed by rf measurements.

The cells can be fabricated with the usual deep drawing technique using "polarized" dies. Therefore the polarized cavity is no more difficult to fabricate than the cylindrical cavity. One concern is that the disruption in the azimuthal geometry will give rise to multipacting. Cold tests on niobium polarized cavities have proven the absence of multipactor [280].

If beam instability from quadrupole modes is a concern, a 90° angle between the two couplers should be avoided so that quadrupole modes can still be intercepted without additional couplers. However, the quadrupole modes have lower R/Q_0 than the dipole modes and are not generally a concern for beam instabilities.

HOM couplers have been developed in both the waveguide and the coaxial varieties. Waveguide couplers are more often found in higher-frequency (i.e., 1000 and 1500 MHz) cavities. The coaxial variety is preferred for the lower frequency (350–500 MHz) cavities because these couplers are more compact. On the other hand, the filter needed to reject the fundamental mode in a coaxial coupler must be designed and tuned carefully to allow $Q_e > 10^{10}$ for the accelerating mode. Moreover, the design must allow the inner parts of the coaxial coupler to be properly cooled to avoid quench. For high current, B-factory-type applications where strong damping and high beam-induced power are expected, the entire beam tubes are used as the HOM couplers. We will discuss the development and performance of each of the three types of couplers.

16.3 WAVEGUIDE COUPLERS

The basic idea of a waveguide HOM coupler is illustrated in Figure 16.2. The electric field of the HOM couples to the field of the waveguide propagating in the TE_{01} mode. The length of the shorted waveguide stub on the other side of the beam tube is empirically adjusted to optimize the coupling to the most dangerous HOMs. The cutoff feature of a waveguide makes rejection of the fundamental mode conceptually simple, as compared to coaxial couplers to be discussed later. There is no need for tuning a rejection filter and there are no stop bands to interfere with HOM propagation. The waveguide width and length are selected so that the guide will propagate the lowest-frequency HOM, and yet provide sufficient attenuation of the fundamental accelerating mode. The first waveguide flange should be located far enough away so that the fundamental mode dissipated power at the joint is negligible.

Waveguide HOM couplers were developed at Cornell for the muffin-tin cavities and the elliptical 5-cell, 1500-MHz cavities and were used in CESR beam tests [281, 282]. The couplers for the muffin-tin cavity were on the cell and

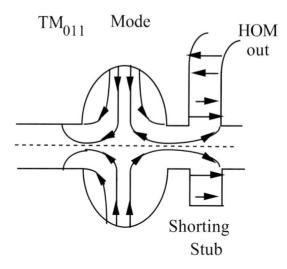

Figure 16.2: Coupling of the cavity HOM electric field to waveguide HOM port. The length of the shorting stub is chosen to maximize the damping of the most dangerous modes.

will not be further discussed. We will discuss the waveguide couplers for the elliptical cavity in some detail because these are now in use on a large scale at TJNAF.

The development of beam tube couplers for the Cornell elliptical cavities was subject to several mechanical constraints imposed by the desire to reuse the existing cryostat for the muffin-tin cavity. The slightly shorter waveguide width selected as a result did not allow propagation of the lowest frequency dipole mode. To alleviate this problem, adjustments were made to the waveguide input coupler (see the introduction of the stub-on-stub discussed in Chapter 18).

In its first incarnation, the waveguide coupler was a four-arm, symmetric waveguide cross, centered on the cavity beam tube, with one arm parallel to the input coupler waveguide. Two orthogonal arms were selected for HOM power extraction while their corresponding guides on the opposite side of the beam tube were shorted. The lengths of the shorted waveguides were optimized empirically to achieve good coupling to the dangerous HOMs. Techniques to calculate damping by waveguide couplers were not available at the time, but are available now [283, 275].

An important difficulty arose in that the waveguide input coupler pinned the polarization of some of the deflecting modes (TM_{110} and TE_{111}), leaving one member of a troublesome mode pair well coupled to the coupler and the other (orthogonally polarized) member of the same pair weakly coupled. To solve this problem [282], the HOM coupler was reshaped into the final Y-shape configuration shown in Figure 16.3. Finally, the lowest-frequency dipole mode was coupled through the input coupler waveguide by adding a stub-on-stub, also

WAVEGUIDE COUPLERS

Figure 16.3: Waveguide HOM couplers developed at Cornell, now in use at TJ-NAF. The entire coupler assembly is made from niobium. The two waveguides at right angles to each other are the coupler ports. The auxiliary waveguide stub at the bottom is empirically adjusted to reach the desired degree of damping. There is an extra sideport on one of HOM waveguides for a field monitor probe.

seen in Figure 18.4.

The waveguide HOM couplers developed at Cornell remain essentially unchanged for use in TJNAF. Since the HOM-induced power in TJNAF is very low (a few tenths of a watt) the loads are placed in the helium vessel. The output arms are lengthened to provide sufficient cutoff for the accelerating mode, and the arms are curved to fit inside the He vessel (Figure 1.23). The absorber (Figure 16.4) is fabricated from an aluminum nitride ceramic containing glassy carbon and is brazed to a stainless steel flange [284]. The absorber material shows sufficient losses at 2 K over the frequency range of 2 to 20 GHz. There are no windows between the loads and cavity since the absorbed power is low and the absorber material is compatible with UHV and dust-free requirements.

In an early effort at DESY [285], 9-cell 1-GHz cavities with waveguide HOM couplers (Figure 16.5) were explored for upgrading the energy of PETRA. To avoid the mechanical and electrical complexity of bellows in the rectangular waveguide, a broadband transition from waveguide to coax was inserted near the exit of the vacuum vessel. The HOM power is absorbed at room temperature by a standard coaxial termination. The cavities and couplers were tested in the PETRA beam.

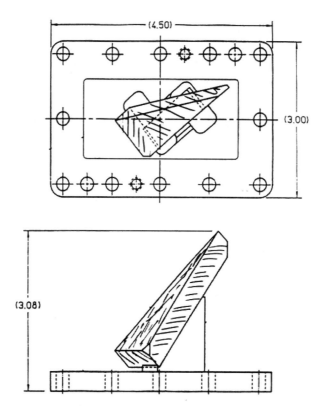

Figure 16.4: Configuration of the HOM absorber relative to the HOM waveguide for the TJNAF load. The lossy material is glassy carbon in an aluminum nitride matrix. The rectangular flange bearing the absorber is mounted on the waveguide inside the helium vessel. Dimensions are in inches. (Courtesy of TJNAF.)

16.3.1 Performance of Waveguide HOM Couplers

The damping achieved for the prominent (high R/Q_0) modes for the Cornell and DESY couplers is given in Table 16.1. Waveguide HOM couplers for the elliptical 1500-MHz cavities were used in CESR beam tests to extract a total of 280 W of HOM power induced by a 22 mA beam. The measured threshold currents for beam instabilities agreed within a factor of 2 with Monte Carlo simulations of beam instabilities in CESR, indicating that no important modes were missed [26].

In the PETRA test of the DESY cavity, the waveguide HOM couplers removed 280 W induced by 8 mA of beam in good agreement with the power expected. Because only one HOM coupler was present, one transverse mode of the TE_{111} family was insufficiently damped ($Q_e \approx 10^7$) and caused a trans-

COAXIAL COUPLERS

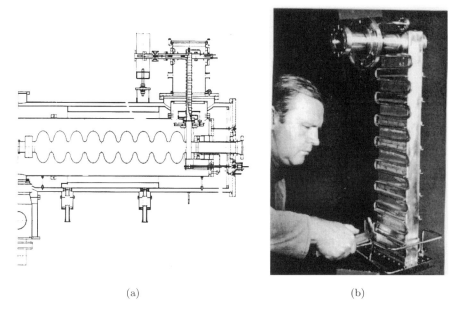

(a)   (b)

Figure 16.5: (a) Waveguide HOM couplers for the PETRA 1-GHz niobium cavity showing the configuration from the cavity through the cryostat to the standard coaxial termination. (b) Vacuum vessel segment of PETRA HOM waveguide coupler showing heat exchanger pipes and transition to coax. (Courtesy of DESY.)

Table 16.1: HOM-loaded Q's for waveguide couplers used in the CESR and PETRA beam tests. Multiple Q values are for the two polarizations of dipole modes

Mode	1.5 GHz, 5-cell	1 GHz, 9-cell
TM_{011}	7×10^2	5×10^3
TM_{020}	2×10^3	1×10^4
TE_{111}	3×10^4, 1.6×10^4	3×10^3, 10^7
TM_{110}	1.3×10^4, 1×10^4	4×10^4

verse instability during the ramping of the beam energy. This instability was overcome by increasing the beam energy ramp rate [286].

16.4 COAXIAL COUPLERS

The design and performance of coaxial couplers are reviewed in References [287, 288]. The coaxial HOM coupler is compact and suitable for low frequency (350–500 MHz) cavities, and medium HOM power extraction. Figure 16.6 shows

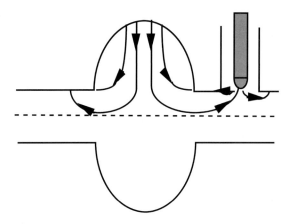

Figure 16.6: Coupling of HOM electric field to a coaxial antenna coupler.

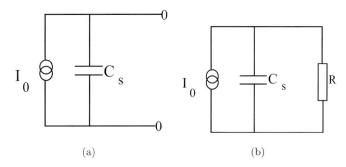

Figure 16.7: (a) Current generator, equivalent circuit model for probe type coupler and (b) with terminating load.

how the electric field lines in the high impedance TM_{011} HOM terminate on the antenna of a coaxial HOM coupler. One can also use a loop at the end of the center conductor to couple to the magnetic field. For brevity, we will focus on the case of the antenna coupler. The analogous treatment for loop couplers can be found in references [287, 288]. Illustrative examples will cover both antenna and loop couplers.

The challenges in designing coaxial couplers are twofold: (a) to maximize the coupling to the fields over the broad frequency band of the dangerous HOMs, and (b) to provide excellent rejection for the fundamental mode.

Circuit Model We view the coupler port as the output port of a generator, as for example a current source, in parallel with an internal impedance, C_s. If E is the electric field in the HOM, I_0 is the displacement current picked up by the probe with a short circuit across the port. I_0 can be calculated by integrating

COAXIAL COUPLERS

the electric field over the area (ds) of the coupling antenna.

$$I_0 = \omega \epsilon_0 \int \mathbf{E} \cdot d\mathbf{s}. \tag{16.1}$$

Field calculation codes are used to give the stored energy in the HOM, and to first order, the corresponding field E at the antenna. We assume that the field is not changed by the presence of the coupler. If we connect a real load R across the coupler (Figure 16.7(b)), the maximum dissipated power P that can be obtained is

$$P = \frac{1}{4} \frac{I_0^2}{\omega C_s}, \tag{16.2}$$

when the load R is at its optimum value, $1/\omega C_s$, i.e., equal to the internal impedance of the generator. We can design the probe so as to minimize the stray capacitance, i.e., avoid all features that add to the capacitance, but do not collect electric field lines from the HOM. The corresponding Q_e of the coupler is:

$$Q_e = \frac{\omega U}{P} = \frac{4 U C_s}{\epsilon_0^2 \left(\int \mathbf{E} \cdot d\mathbf{s} \right)^2} = \frac{4 U}{\omega R \epsilon_0^2 \left(\int \mathbf{E} \cdot d\mathbf{s} \right)^2}, \tag{16.3}$$

when R is optimal.

To make Q_e as low as possible we must design the probe to pick up as much of the cavity field as possible — i.e., maximize the integral. However, a simple increase of antenna size does not necessarily produce the expected benefit in damping. The short-circuit current will rise, but so will the probe's self-capacitance. The reactance in the coupling device diminishes the extracted power, and hence the damping effectiveness of the coupler. Therefore, for a given antenna area, the probe capacitance and the stray capacitance should be made as small as possible; i.e., C_s should be minimized.

To suppress the coupling to the fundamental mode a parallel LC filter can be added in series to the load as shown in Figure 16.8. The niobium version of such a coupler, realized by KEK (Figure 16.8(b)) was used for TRISTAN [51]. It consists of a capacitively loaded $\lambda/4$ resonator on the outer conductor with a T-stub at the exit to allow cooling of the inner conductor. The coupler is fully demountable, because the current is zero at the HOM coupler flange. A disadvantage of this approach is that, besides the sharp notch to reject the fundamental mode, there are other stop bands at higher frequencies, as shown in Figure 16.9. The frequency of the second stop band is arranged to be different for the two HOM couplers that are used. This is accomplished by adjusting the diameter of both the inductive and capacitive parts as well as the length of the T-stub.

Compensation In further developments of HOM couplers to provide stronger damping, a problem was realized with the elementary approach of Figure 16.8(a). Suppose that it is desired to use a standard (50 Ω impedance) coaxial transmission line to bring the HOM power out of the cryostat. For couplers with antenna area large enough to satisfy the damping requirements, the

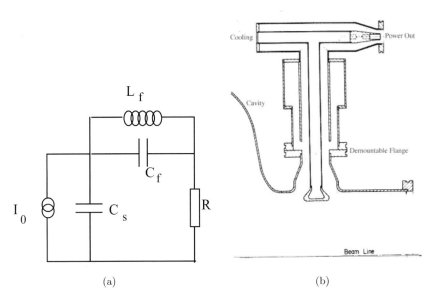

Figure 16.8: (a) Equivalent circuit for TRISTAN HOM coupler with fundamental mode rejection filter and (b) hardware version of the coupler with rejection filter. (Courtesy of KEK.)

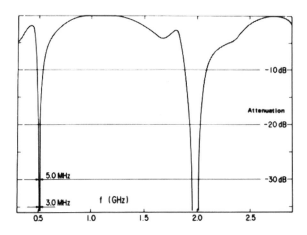

Figure 16.9: RF transmission characteristics of the TRISTAN HOM coupler, showing the rejection at the fundamental frequency at 0.5 GHz and a second stop band at 2 GHz. (Courtesy of KEK.)

COAXIAL COUPLERS

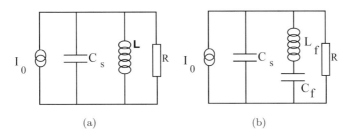

(a)   (b)

Figure 16.10: (a) Equivalent circuit diagram for a coupler with compensating inductor and (b) circuit with addition of fundamental mode filter.

optimum value of R tends to be far away from the conventional 50 Ω load resistor. To circumvent the problem created by the competing requirments of a large antenna and a low stray capacitance, the new idea was to compensate for the stray capacitance C_s by using an inductor L in parallel with C_s (Figure 16.10).

Exact compensation can be achieved in this simple way, but only at a single frequency

$$f_c = \frac{1}{\sqrt{LC_s}}. \tag{16.4}$$

The inductance L is chosen so that f_c matches the frequency of the mode with the highest R/Q_0, so as to provide the strongest damping. The key idea of the compensation is that by choosing the appropriate inductance L we can make all the current I_0 to pass through the load resistor, R. The damping is maximal for one mode, but at the expense of other neighboring modes. This may also be called resonant coupling. We will see later how to broaden the approach to damp other modes.

Fundamental mode rejection can be integrated into the compensation scheme by adding a filtering capacitor in series with the compensating inductor as shown in Figure 16.10(b). The filter notch must be wide enough to allow pretuning of the suppression filter at room temperature. It must also be narrow enough to avoid hurting the coupling of the HOMs which are near the fundamental, such as the TE$_{111}$ mode. If we tune L_f and C_f to the fundamental mode frequency, then no power is dissipated in the load. At higher frequency, the series resonator develops an inductive term just as needed to compensate C_s.

It is important to determine the strength of the fundamental mode displacement current at the operating field (i.e., the fundamental field at the filter inductor) to ensure that it will not cause undue heating of coupler components. If the magnetic field is sufficiently high, the inductor must be made superconducting.

Broad-band Damping With the simple compensation scheme discussed above, one can achieve some compensation within a bandwidth Δf around the frequency f_c at which the compensation is optimal with

$$\Delta f \approx \frac{1}{RC_s}. \tag{16.5}$$

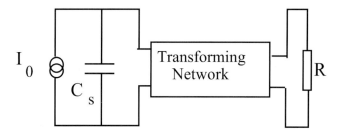

Figure 16.11: General scheme for compensating the probe capacitance at several frequencies to match several cavity HOM frequencies.

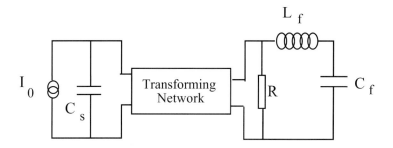

Figure 16.12: Circuit for a general compensating transformer network including the fundamental mode filter.

This approach provides some damping of nearby modes. There is obviously a trade-off between high damping and broad-band damping. According to Equation 16.3 the damping becomes stronger (lower Q_e) as R increases. As one makes R bigger to reach stronger coupling, the bandwidth also gets smaller, which means reduced coupling to other HOMs. Note that, according to Equation 16.5, it is again important to make C_s small to achieve a large bandwidth.

Figure 16.11 shows the principle of a more general compensation scheme to provide strong damping of several HOMs. We need a loss-free transforming network which has the shunt capacitor C_s as its first element and which transforms the load resistor R into an impedance with resonant character. The challenge is to make the real part of this impedance peak at several frequencies, selected to match the distribution of dangerous cavity modes.

To suppress coupling to the fundamental (accelerating) mode, it is also necessary to arrange for the impedance of the coupling network to have a zero in the real part. As we discussed earlier, this can be achieved with a series $L_f C_f$ filter resonator in parallel with the load (Figure 16.12).

A first step toward realization of such a coupler is the equivalent circuit shown in Figure 16.13(a). The coupling circuit is depicted more realistically in Figure 16.13(b) where the inductances are obtained via transmission line sections with impedance larger than 50 Ω and capacitances can be transmission line segments with impedance less than 50 Ω.

COAXIAL COUPLERS

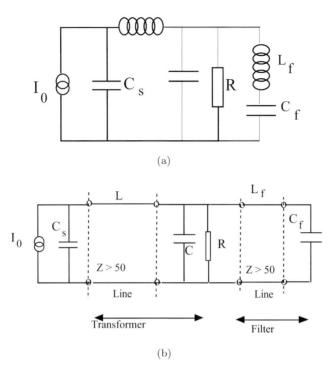

Figure 16.13: (a) Equivalent circuit of CERN Type I coupler with two coupler resonances to match two families of HOMs. (b) Realization of transformer inductances for the CERN Type I coupler with sections of coaxial line.

This is the idea behind the LEP Type I coupler which achieved resistance transformation around two compensation frequencies [287]. Figure 16.14(b) shows how to realize the circuit inductances with segments of coaxial lines of high impedance (compared to 50 Ω) and capacitances with coaxial line segments of low impedance. The fundamental mode filter forms a shunt across the load resistor. As we mentioned before, at the HOM frequency, the filter behaves as the compensating inductor. The resulting circuit is a transforming bandpass filter with two frequencies of compensation, as needed to damp two HOM families.

A complete CERN Type I coupler that satisfies other requirements such as mechanical support, cooling, and filter tuning requirements is shown in Figure 16.14(a). This coupler was used for the first series of LEP cavities made from sheet niobium. The small-diameter niobium tube which supports the inner conductor also admits liquid helium to the interior parts of the coupler. The rf window is in the cryostat insulation vacuum so as to reduce the risk to the cavity. The coupler is demountable, but it requires sapphire positioning pins to align the lateral rf output.

A clever method [289] to construct such a compensating transformer network

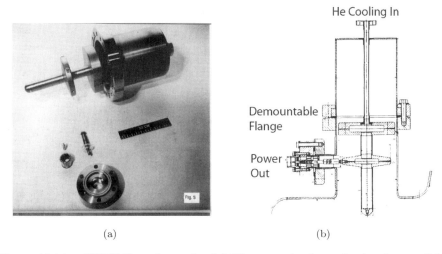

Figure 16.14: CERN Type I coupler (a) Photograph of coupler hardware. (b) Mechanical drawing. (Courtesy of CERN.)

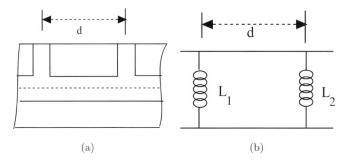

Figure 16.15: Realization of a shorted resonant line in a coax with two supporting stubs which also serve as the inlet and outlet ports for cooling fluid. The two stubs are represented as inductors in the lumped circuit model and spaced $\lambda/2$ apart.

COAXIAL COUPLERS

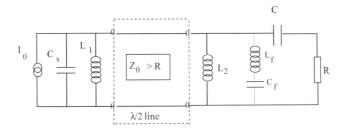

Figure 16.16: Addition of the fundamental mode filter section to an HOM coupler consisting of two inductive stubs.

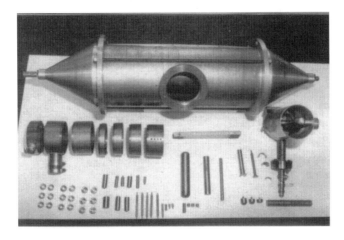

Figure 16.17: Elements used to model a two-stub HOM coupler. (Courtesy of DESY.)

is based on the use of two inductive stubs, as shown in Figure 16.15, along with the equivalent circuit for the two stubs. With two inductive hollow posts, this approach turns out to be an excellent method to provide cooling to the coupler antenna as well as stub supports for the center conductor. As usual, a series resonant filter, $L_f C_f$, can be added parallel to L_2 to suppress the fundamental mode (Figure 16.16).

Using circuit analysis procedures, the elements of such a network can be chosen to optimize the locations, peak heights, and bandwidths of the resonances. Constructing the physical structure corresponding to the network is finally a matter of cut and try experience, as well as intuition. Figure 16.17 displays a set of subelements used for construction of coupler models to realize a two-stub coupler [289].

The HERA 4-cell cavity uses a set of three two-stub coaxial couplers [289]; one coupler is mainly for the high-impedance TM_{011} mode family, and the other two are for the TE_{111}, TM_{110}, and TM_{012} mode families. The approach of multiple couplers was necessary to provide the needed heavy damping for the

Figure 16.18: HOM couplers for HERA based on the idea of two inductive stubs. The resonances are arranged to intercept HOMs at different frequencies. A total of three such couplers is used for each 4-cell HERA cavity. (Courtesy of DESY.)

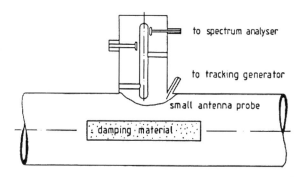

Figure 16.19: Measurement setup for evaluating the HOM coupler sensitivity curves. (Courtesy of CERN.)

high beam current (60 mA) anticipated in HERA. Example HOM couplers for HERA are shown in Figure 16.18.

As part of the modeling process, the frequency-dependent transmission properties of the evolving HOM coupler can be measured with an arrangement such as shown in Figure 16.19. Here the coupler is mounted onto a piece of beam tube with damping material inside. A network analyzer is used to measure the transfer of power from a short probe inserted near the coupling port and the output port, where the load would normally be connected. As an example result, the sensitivity curve of Figure 16.20 shows the broad resonances due to the transforming network successfully aligned with the dangerous HOM families in

COAXIAL COUPLERS 371

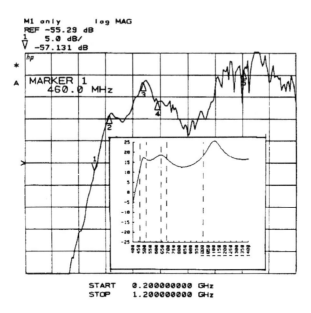

Figure 16.20: Power transmission curve for one version of the CERN HOM coupler. There are three coupler resonances to damp three dangerous mode families in the LEP cavity. The inset shows the response calculated from a circuit model. (Courtesy of CERN.)

the LEP-II cavity [287].

More recently, the code HFSS [290] has been used to facilitate the transition from the circuit model to the rf structure via calculation of the reflection and transmission properties in 3-D rf structures. One calculates the real part of the input impedance as a function of the frequency and identifies the location of the coupler resonances to compare these with the dangerous HOM frequencies.

A variation of the two-stub design of Figure 16.21 was developed to achieve a demountable coupler for the LEP-II Nb/Cu cavity. A capacitor is added in series with L_1 in the circuit of Figure 16.21 to serve as the filter. If the capacitance is achieved via a piece of open-circuit transmission line, the cavity end of the coupler resembles a fishing hook. When properly oriented, this type of LC series resonator filter naturally forms a loop that couples to the magnetic field of the TE dipole mode, but not to the magnetic field of the fundamental mode. This feature was discovered during HOM coupler development at Saclay, as discussed in [291]. Another important aspect for the demountable HOM coupler is that the flange joining the coupler unit to the cavity port is kept free of fundamental mode current. Finally, coupling to the load is via the capacitive gap. The physical realization of a complete demountable coupler is shown in Figure 16.21 along with the final equivalent circuit.

In any coaxial HOM coupler design, the parts of the coupler that are exposed

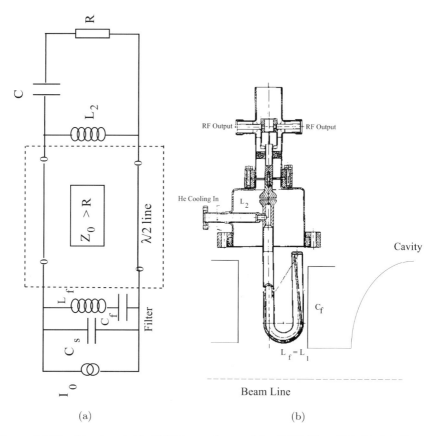

Figure 16.21: Demountable HOM coupler used for LEP-II at CERN. (a) Circuit diagram. (b) Hardware version showing circuit elements. (Courtesy of CERN.)

to high magnetic fields must be made from a superconducting material and be well cooled to avoid quench. The rejection filters must have the possibility of fine tuning the filter capacitor after the coupler is attached to the cavity to ensure that the rejection is sufficient. In a good design, the filter frequency needs to be set to within a few MHz of the fundamental cavity mode frequency to achieve $Q_e > 10^{10}$.

16.4.1 Performance of Coaxial Couplers

As part of the development process of any type of HOM coupler, damping measurements for individual HOMs are made with a model copper cavity at room temperature. Table 16.2 shows the damping achieved by the TRISTAN, HERA, and LEP coaxial couplers for the prominent HOMs. The performance of HOM couplers has been reviewed in Reference [292].

Table 16.2: Measured damping with coaxial couplers for dangerous HOMs. The R/Q_0 values in ohms are calculated by URMEL. For the transverse modes we give the longitudinal R/Q_0 at 5 cm off axis. The values given for LEP-II are for the Type I couplers.

Mode	LEP-II R/Q_0	LEP-II Q_e	TRISTAN R/Q_0	TRISTAN Q_e	HERA R/Q_0	HERA Q_e
TM_{011}			22	45,000	5.4	4,800
Family	46	8,000	56	15,000	48	600
	108	13,000	103	20,000	111	600
TE_{111}	8.3	27,000	25	55,000	17	6,200
Family	7.2	19,000	33	25,000	14.6	4,800
TM_{110}	11	58,000	13	55,000	4.3	5,100
Family	7.4	94,000	26	101,000	22.4	7,000
			7.2	216,000	15.4	15,000

The chosen HOM coupler design must also be tested in a superconducting cavity and must be operated at the design field to ensure that there is no unexpected heating in the coupler regions at the operating field. The vacuum feedthroughs, connectors and cables must be checked to ensure that they can handle the anticipated beam induced HOM power, plus any fundamental mode power expected from imperfections in setting the rejection filter. The cold test is also essential to ensure that any multipacting in the HOM coupler is well conditioned, without degrading the rejection filter. For the TRISTAN couplers, multipacting occurs at a low field level, but is conditioned in less than one hour [292]. In the early stages of operating experience with the TRISTAN coupler, it was reported that multipactor induced a quench and heated a part of the rejection filter, thereby changing its frequency and coupling large amounts of fundamental power. As much as 1 kW was observed due to 2 MHz detuning, along with a temperature rise to 300 K. The cure was to add an interlock that shut off the rf when the fundamental power leaking out exceeded a chosen threshold during the HOM coupler processing stage.

A cold test at high field also determines whether the performance of the HOM coupler is affected by field emission current emanating from the cavity, as was found for the LEP couplers [57]. This was cured by reducing the coupler penetration and by conditioning the cavity more thoroughly to reduce field emission.

Extensive experience with many HOM couplers in HERA showed that the two-stub coupler works stably without quenches in cw operation at 4–5 MV/m, and beam current up to 30 mA, even in the presence of strong field emission. The good cooling of the inner conductor by the two stubs makes the welded coupler less sensitive (than the demountable coupler) to x-ray and electron bombardment.

HOM power extracted during operation of superconducting cavities in TRISTAN and LEP was compared to the predictions from loss factor calculations.

At TRISTAN, for modes below the beam pipe cutoff frequency, the total cavity loss factor was 0.41 V/pC. Accordingly, the expected power lost by beams of 4 bunches×2 mA is 66 W per cavity, assuming no resonant buildup. The measured power was 70–100 W/cavity at the same current. At higher current (8 mA), the TM_{011} mode alone showed 50 W of power. No instabilities or resonant buildup caused by HOMs were observed up to 13 mA in TRISTAN [292]. For the LEP cavity, the loss factor was 0.22 V/pC, which gives a power loss of 10.9 W for a beam of 0.75 mA in reasonable agreement with the measured power of 12 W.

16.5 BEAM TUBE COUPLERS FOR HIGH-CURRENT APPLICATIONS

As we will discuss in Chapter 20, superconducting cavities are under development for future high current machines such as High Energy Ring of KEK-B and the luminosity upgrade of CESR. The primary need is to lower the narrow band impedances and thereby to increase the threshold for multibunch instabilities. To use superconducting cavities in these high current machines, it is necessary to damp HOMs to Q_e values < 100 and to remove about 10 kW of beam-induced HOM power. The strategy adopted to meet these demanding goals is to use large-diameter beam pipes so as to reduce the R/Q_0 values of the HOMs, as well as to propagate the HOMs freely out of the cavity to high-power loads located at room temperature [293]. The beam pipe serves as an excellent high power conduit, and is cutoff for the fundamental mode, providing a natural filter. For a 500-MHz fundamental mode frequency a beam pipe diameter of 24 cm is large enough to propagate all the monopole modes in a single-cell cavity. This was confirmed by using URMEL to examine the field patterns in the beam tubes, as well as using SEAFISH and CLANS to calculate the Q_e in the presence of beam pipe absorbers (as discussed in the previous chapter).

All but the two lowest-frequency dipole modes also propagate out of the 24 cm diameter beam pipe. In order to remove the two troublesome lowest-frequency dipole modes, the beam pipe on one side of the cavity is further enlarged (to 30 cm) as in the case for KEK-B, or is fluted in the case of CESR-III (Figure 16.22). MAFIA calculations show that the "rectangular" waveguides formed by the flutes guide out the TM_{110} and TE_{111} deflecting modes.

The HOM load is a 10-cm section of room temperature beam pipe lined on the inside with a ferrite material [294]. The ferrite, which is an excellent microwave absorber, is bonded to copper either by flux-free soldering (CESR-III) or by hot isostatic pressing (KEK-B). In the case of solder, the ferrite is first nickel plated. Latest studies show that nickel-vanadium is better [295]. Figure 16.23 shows the complete load for CESR-III. It consists of a stainless steel shell with 18 copper plates bolted along the inside. Each copper plate carries two ferrite tiles. Copper tubing is brazed to each plate for water cooling. More recently, the copper was replaced by a tungsten–copper composite called Elkonite [295]. This material has a better thermal expansion matching coefficient to ferrite.

BEAM TUBE COUPLERS FOR HIGH-CURRENT APPLICATIONS 375

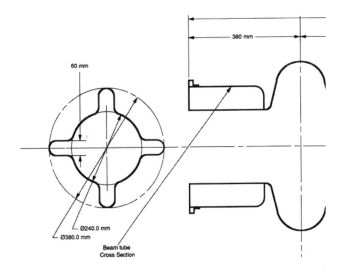

Figure 16.22: Flute-shaped beam pipe for guiding out the first two deflecting modes (TM_{110} and TE_{111}) from superconducting cavity for CESR-III.

In the case of the CESR-III cavity, Q_e and R_a/Q_0 values for a large number of monopole modes calculated by the program CLANS [75] are shown in Figure 16.24. The program allows inclusion of lossy materials, such as ferrites. The important quantity, $R/Q_0 \times Q_e$, is shown in Figure 16.24(b). The very low Q_e values obtained by the ferrite beam pipe load suggest that the ferrite layer behaves essentially as a matched beam pipe load. The complex dielectric permittivitty and magnetic permeability of the ferrite are required as input to CLANS. These were measured over the frequency range from 300 kHz to 20 GHz using the coaxial transmission line technique. Figure 16.25 shows example ferrite properties ϵ and μ [296].

Results are given in Table 16.3 for the *measured* HOM damping achieved with similar ferrite HOM loads for prominent monopoles and dipoles in the KEK-B cavity [274]. Similar data was measured for the CESR-III cavity with ferrite loads [296].

A possible concern about the ferrite-lined beam pipe is the broad-band impedance of the ferrite sections and the possible effect of short-range wakes on single-bunch stability. The impedance has been evaluated by two methods (a) using an analytical model of a conducting pipe with a material layer and (b) with the code AMOS [297], a program which calculates wakefields in the presence of absorbing materials. AMOS predicts a loss factor of 0.12 V/pC for a 10 cm long CESR-III HOM load at a bunch length of 1 cm [298]. This is a significant addition to the parasitic loss factor of a large beam pipe cavity (0.1 V/pC). Using the impedance of the ferrite beam pipe load, it was predicted that there should be no beam stability problems for a single-bunch charge of

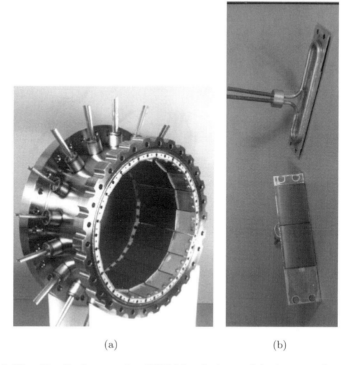

(a) (b)

Figure 16.23: Ferrite beam pipe HOM load shows (a) the complete load and (b) one of 18 individual panels with cooling tube on one side and ferrite on the side facing the beam.

Table 16.3: Measured values of Q_L for monopole and dipole HOMs in the superconducting cavity under development for the KEK-B High Energy Ring. The R/Q values are calculated by URMEL, but without ferrite present

Monopole Modes			Dipole Modes		
Frequency (MHz)	R_a/Q_0 (Ohms)	Q_L	Frequency (MHz)	R_d/Q_0 (Ohms)	Q_L
783	0.1	132			
834	0.3	72	609	0.15	92
1,018	6.6	106	648	2.96	120
1,027	6.4	95	688	11.8	145
1,065	1.6	76	705	15.3	94
1,076	3.2	65	825	0.36	60
1,134	1.7	54	888	0.19	97

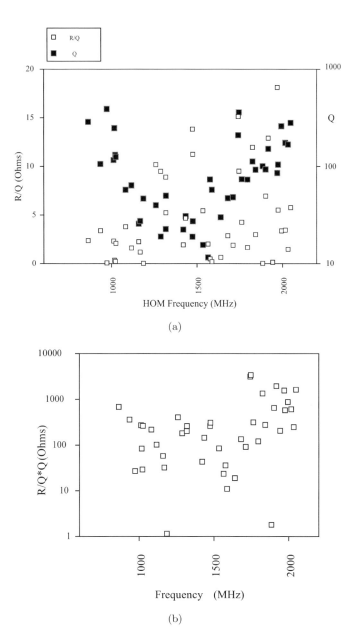

Figure 16.24: (a) Calculated R_a/Q_0 and Q_e values for monopole modes of the cavity intended for CESR-III. The CLANS code used for these calculations took into account the properties of the ferrite. (b) Same data as (a) except plotted as the total impedance, $R_a/Q_0 \times Q_e$, where Q_e is the damped Q.

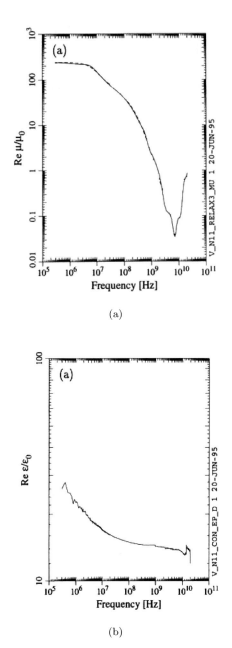

Figure 16.25: Measured properties of ferrite used for CESR-III HOM loads.

1.6 μC. For CESR-III, this corresponds to a *single* bunch current of 0.64 A as compared to the typical and anticipated single bunch currents of 10–15 mA. That the predicted instability threshold is so high is not surprising since the diameter of the ferrite beam pipes is very large.

16.5.1 Performance of Beam Pipe HOM Couplers

Ferrite-lined beam pipe absorbers have been tested up to 20 kW total dissipation (20 W/cm^2) in air and 7 kW in vacuum [299, 300]. These power levels are higher than anticipated for B-factory application. Even at 20 kW there was no cracking in the ferrite, but the temperature of the ferrite rose to 150°C.

Beam pipe ferrite couplers attached to 500-MHz superconducting cavities have been tested in the CESR [301] beam up to a beam current of 220 mA and in the Accumulation Ring at KEK [302] up to 600 mA (see also Chapter 20). The maximum beam-induced power extracted from the superconducting cavity was about 2 kW per load, limited by the maximum available beam current. No HOM-induced beam instabilities or resonant power excitation were encountered in either beam test, which indicates that all the dangerous cavity modes were well damped. During the beam tests, the frequencies of the HOMs were moved across dangerous harmonics to ensure that the damping is adequate for the future needs of high beam current accelerators.

To verify the broad-band impedance predictions for the ferrite-lined beam pipe HOM loads, a special beam test was conducted by placing a 60 cm long section of 9 cm diameter ferrite-lined pipe in CESR [303]. The smaller diameter accentuated the monopole impedance by about a factor of two above the predicted longitudinal impedance of the total of 8 HOM loads needed for CESR-III. The transverse impedance used was ten times larger the expected dipole impedance for a series of 8 HOM loads planned for CESR-III. As a measure of the total impedance added to the machine, the calculated loss factor of the test pipe was 2 V/pC at 1.5 cm bunch length. This should be compared to 9 V/pC, the total loss factor for CESR at the same bunch length. As expected from the calculated impedances, no problems were encountered up to the maximum single-bunch current (42 mA) that CESR could provide.

CHAPTER 17

Coupling Power to the Beam

17.1 INTRODUCTION

The primary role of the input coupler is to transfer radiofrequency (rf) power from the generator to the cavity and hence the beam. To do this, it must also provide a match between the generator impedance and the combined impedance of the cavity–beam system, so as to avoid unnecessary power reflections. In this chapter, we will explore this requirement. In the next chapter, we will discuss design considerations for couplers to fulfill this requirement and give some illustrative examples.

When a cavity is operated in a linear accelerator or a storage ring, energy is transferred from the cavity to the beam, resulting in additional energy loss over and above the dissipation in the cavity walls. Hence this situation is known as *beam loading*.

As the beam passes through the cavity, it excites fields in the accelerating mode, as well as in higher-order modes. We will study the interaction of the beam with the accelerating mode, and we will determine how the beam affects the operation of the cavity. The interaction of the beam with higher-order modes was covered in Chapter 15. We will also show that, in a storage ring, the cavity is used to provide longitudinal phase stability for the charge bunches, which are subject to energy perturbations as they travel around the ring. To be able to provide stability, certain criteria, which we will touch on, need to be satisfied. In particular, the cavity voltage cannot be in phase with the beam current, and the cavity frequency must be detuned with respect to the generator frequency.

We will use the equivalent circuit model we introduced in Chapter 8 to obtain expressions for the generator power requirements as a function of beam current, coupling, and generator frequency. We will also illustrate some of the situations that the input power coupler must be able to cope with. Some of this material has been covered in an excellent review by P. B. Wilson [264]. In these cases we will limit ourselves to summarizing the results presented there. The reader is encouraged to refer to Wilson's review for a more detailed discussion.

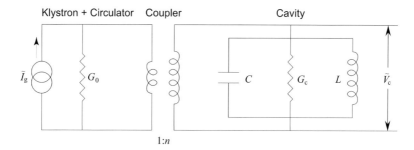

Figure 17.1: Circuit diagram used to model a cavity coupled to a generator.

17.2 THE EQUIVALENT CIRCUIT

We showed in Chapter 8 that a cavity connected to a generator via a circulator and matched load can be represented by the equivalent circuit in Figure 17.1. The circulator and load ensure that all reflected power is intercepted before it reaches the generator. We related the following circuit parameters to the cavity quantities

$$V_c = \text{cavity accelerating voltage} \tag{17.1}$$

$$\omega_0 = \frac{1}{\sqrt{LC}} \tag{17.2}$$

$$R_a = \frac{2}{G_c} \tag{17.3}$$

$$Q_0 = \frac{1}{G_c}\sqrt{\frac{C}{L}} \tag{17.4}$$

$$\frac{R_a}{Q_0} = 2\sqrt{\frac{L}{C}}. \tag{17.5}$$

Note that in this chapter, complex quantities (phasors) are denoted by a tilde, as, for example, in $\tilde{A} = |\tilde{A}|\exp(i\phi)$. Furthermore, we refer to the magnitude of the phasor by omitting the tilde; that is, $A = |\tilde{A}|$.

Both the cavity voltage $\tilde{V}_c$ and the generator current $\tilde{I}_g$ vary in time as $\exp(i\omega_g t)$, where ω_g is the angular frequency of the generator. Since the $\exp(i\omega_g t)$ term is common to both, we will omit it in the following discussion.

We now view the circuit as seen from the cavity end. The input circuit has an admittance G_0/n^2, and the time-averaged power dissipation in the cavity and in the input line are

$$P_c = \frac{G_c V_c^2}{2} \tag{17.6}$$

and

$$P_e = \frac{G_0 V_c^2}{2n^2}, \tag{17.7}$$

BEAM LOADING

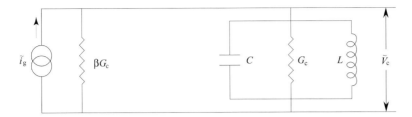

Figure 17.2: Simplified circuit diagram representing a cavity connected to a generator.

respectively. Hence the coupling is

$$\beta = \frac{P_\text{e}}{P_\text{c}} = \frac{G_0}{G_\text{c} n^2}, \tag{17.8}$$

and the external Q is

$$Q_\text{e} = \frac{Q_0}{\beta} = \sqrt{\frac{C}{L}} \frac{1}{\beta G_\text{c}}. \tag{17.9}$$

For simplification, we can remove the transformer from the equivalent circuit provided we replace G_0 by its transformed admittance $G_0/n^2 = \beta G_\text{c}$ and $\tilde{I}_\text{g}$ by its transformed current $\tilde{i}_\text{g} = \tilde{I}_\text{g}/n$ (Figure 17.2).

The power dissipation in the cavity is maximized when the cavity is matched to the input circuit. In this situation, the current flowing through the cavity is $\tilde{i}_\text{g}/2$ and the power dissipated is

$$P_\text{c}^\text{max} = \frac{\tilde{i}_\text{g}^2}{8\beta G_\text{c}} = P_\text{g}. \tag{17.10}$$

We identify this power as the available generator power P_g at a given generator current $\tilde{i}_\text{g}$.

17.3 BEAM LOADING

To determine the effect of the beam on the system, we need to include it in our equivalent circuit. To illustrate a possible approach we will restrict the discussion to electron or positron beams. In this case we make the following observations and assumptions:

1. In e^+e^- accelerators, the particles travel at close to the speed of light (c) and their velocity changes very little as they pass through a cavity. The beam is, to a good approximation, a current source.

2. The beam consists of a series of bunches that enter the cavity at intervals T_b, which is an integer multiple of the accelerating voltage's period. The bunches typically have a Gaussian profile along the direction

of motion with a characteristic length σ_z. By Fourier transforming the time-dependent current, we find that the beam has a frequency spectrum roughly given by a series of δ functions with a Gaussian envelope with a standard deviation $\sigma_\omega = c/\sigma_z$. The beam thus acts as an rf source which can excite the cavity's accelerating mode, provided the beam spectrum extends beyond the eigenfrequency of that mode. The principle of superposition allows us to add the beam-induced voltage and the voltage produced by the generator to obtain the total cavity voltage.

3. We postulate that the spacing of the beam's spectral lines is significantly larger than the bandwidth of the accelerating mode; that is, the cavity is excited by only one spectral line.

4. We assume that T_b is much less than the decay time of the accelerating mode's fields (T_d). This is usually a good assumption, and we will show later how it provides an important simplification.

5. We assume that any higher-order modes, which may be excited by other spectral lines of the beam, are damped to such an extent that we do not need to consider them here. Damping is achieved with higher-order mode couplers, which are discussed in Chapter 16.

6. Finally, we limit our discussion to steady-state operation. Our results will not apply to the initial injection of the beam.

Let us examine whether these conditions are typically satisfied. As an example, we consider a test of the Cornell 500-MHz superconducting cavity which is to be used in the Phase III upgrade of CESR. If only one bunch is stored in the ring, the following numbers apply:

$$
\begin{aligned}
E_b &= 5.3 \text{ GeV (electron beam energy)} \\
T_b &= 2.5 \text{ } \mu\text{s (single bunch operation)} \\
T_d &= 127 \text{ } \mu\text{s} \\
\sigma_z &= 1 \text{ cm} \\
\omega_0 &= 2\pi \times 500 \text{ MHz (frequency of the accelerating mode)} \\
\Delta f &= 2.5 \text{ kHz (bandwidth of the accelerating mode).}
\end{aligned}
$$

Clearly the beam is relativistic, thereby satisfying condition 1. We also find that the beam spectrum is significant out to an angular frequency of about $\omega_{\max} = c/\sigma_z = 2\pi \times 4.8$ GHz, implying that the beam can indeed excite the cavity's accelerating mode. However, we also have to ensure that all higher-order modes with strong coupling impedances up to at least $2\pi \times 4.8$ GHz are damped, to avoid their resonant excitation by the beam (see Chapter 15).

The ratio $T_b/T_d = 0.02$ clearly satisfies assumption 4. Increasing the number of bunches in the accelerator only reduces this ratio, so that single-bunch operation is a worst-case scenario. The beam's spectral line spacing is

BEAM LOADING

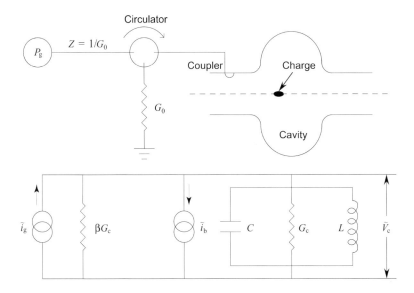

Figure 17.3: Equivalent circuit for a beam-loaded cavity.

$1/T_\mathrm{b} = 400$ kHz, which indeed is significantly larger than the bandwidth of the accelerating mode, thereby fulfilling assumption 3.

Having seen that our assumptions are valid, we model the beam-loaded cavity by the equivalent circuit. We simply add a current source representing the beam current in parallel with the generator (Figure 17.3).

We are only considering the Fourier component of the beam current that excites the cavity ($\tilde{i}_\mathrm{b}$). From the Fourier transform of the beam current, one finds that $\tilde{i}_\mathrm{b}$ has a magnitude $2I_0$, where I_0 is the time-averaged (dc) beam current. If the beam is to receive the maximum possible acceleration, it needs to be out of phase with the generator current by $180°$, as shown in Figure 17.3.

It is also important to note that successive bunches need to traverse the cavity at the same phase of the accelerating voltage. Thus

$$T_\mathrm{b} = \frac{2\pi h}{\omega_\mathrm{g}}, \tag{17.11}$$

where h is an integer, known as the harmonic number, and ω_g is the generator's angular frequency, which normally is tuned close to the eigenfrequency of the accelerating mode (ω_0). Hence the beam's spectral line closest to the cavity resonance is at the generator frequency, and $\tilde{i}_\mathrm{b}$ varies as $\exp(i\omega_\mathrm{g}t)$. However, the frequency ω_g is not necessarily the same as ω_0, because the cavity may need to be driven slightly off resonance, as we shall see later. Similar to $\tilde{V}_\mathrm{c}$ and $\tilde{i}_\mathrm{g}$, we will omit any explicit reference to the $\exp(i\omega_\mathrm{g}t)$ term in $\tilde{i}_\mathrm{b}$, since this is common to all the voltages and currents in Figure 17.3.

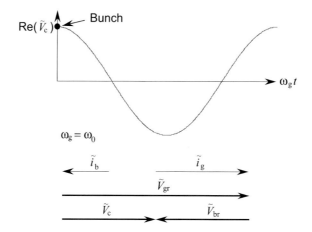

Figure 17.4: Relation between the various voltages and currents in Figure 17.3 when the cavity is excited on resonance and the bunches enter the cavity in phase with the cavity voltage.

17.4 RESONANT OPERATION

Let us consider the case $\omega_g = \omega_0$ — that is, where the generator is tuned exactly to the cavity resonance. Furthermore, we let the beam current and the generator voltage be 180° out of phase for maximum acceleration. We will use the phase of $-\tilde{i}_b$ as the reference phase.

On resonance, the cavity admittance is G_c so that the beam-induced voltage in Figure 17.3 is

$$\tilde{V}_{br} = \frac{\tilde{i}_b}{(1+\beta)G_c} = -\frac{I_0 R_a}{1+\beta}, \qquad (17.12)$$

and the generator voltage is

$$\tilde{V}_{gr} = \frac{\tilde{i}_g}{(1+\beta)G_c} = \frac{2\sqrt{R_a P_g \beta}}{1+\beta}, \qquad (17.13)$$

where we are restricting ourselves to the positive root of β. The net cavity voltage is given by superposition of the generator and beam-induced voltages; that is,

$$\tilde{V}_c = \tilde{V}_{gr} + \tilde{V}_{br} \qquad (17.14)$$

and therefore the voltage through which the beam is accelerated is

$$V_c = V_{gr} - V_{br}. \qquad (17.15)$$

This situation is depicted in Figure 17.4.

RESONANT OPERATION

By using (17.12) and (17.13) in (17.15) we find, for fixed P_g, that the accelerating voltage decreases linearly with increasing beam current as

$$V_\text{c} = \sqrt{R_\text{a} P_\text{g}} \frac{2\sqrt{\beta}}{1+\beta}\left(1 - \frac{I_0 \sqrt{R_\text{a}}}{2\sqrt{P_\text{g}\beta}}\right) = \sqrt{R_\text{a} P_\text{g}} \frac{2\sqrt{\beta}}{1+\beta}\left(1 - \frac{K}{\sqrt{\beta}}\right), \quad (17.16)$$

where

$$K = \frac{I_0}{2}\sqrt{\frac{R_\text{a}}{P_\text{g}}} \quad (17.17)$$

is the dimensionless *beam loading parameter*. K is a measure of how much the beam dominates the cavity behavior. This becomes clearer if we recast K in terms of powers. When $\tilde{V}_\text{c}$ and $-\tilde{i}_\text{b}$ are in phase, the power transferred to the beam is

$$P_\text{b} = V_\text{c} I_0 = \frac{V_\text{c} i_\text{b}}{2}. \quad (17.18)$$

Since the shunt impedance is

$$R_\text{a} = \frac{V_\text{c}^2}{P_\text{c}},$$

it is easy to show that

$$K^2 = \frac{P_\text{b}^2}{4 P_\text{g} P_\text{c}}. \quad (17.19)$$

The beam loading parameter thus relates the power received by the beam to the power dissipated in the cavity. Note that when $K = 0$ (i.e., $I_0 = 0$), we recover from (17.16) the expression for the power dissipated in the unloaded cavity, which we obtained in Chapter 8:

$$P_\text{c} = \frac{4\beta}{(1+\beta)^2} P_\text{g}. \quad (17.20)$$

The efficiency η_g with which power is transferred from the generator to the beam can be obtained by combining (17.16), (17.17), and (17.18) to yield

$$\eta_\text{g} = \frac{P_\text{b}}{P_\text{g}} = \frac{V_\text{c} I_0}{P_\text{g}} = \frac{4K\sqrt{\beta}}{1+\beta}\left(1 - \frac{K}{\sqrt{\beta}}\right). \quad (17.21)$$

Furthermore, the reflected power is obtained by invoking the conservation of energy:

$$\begin{aligned}
P_\text{r} &= P_\text{g} - P_\text{b} - P_\text{c} = P_\text{g} - \eta_\text{g} P_\text{g} - \frac{V_\text{c}^2}{R_\text{a}} \\
&= P_\text{g}\left[1 - \frac{4K\sqrt{\beta}}{1+\beta}\left(1 - \frac{K}{\sqrt{\beta}}\right) - \frac{4\beta}{(1+\beta)^2}\left(1 - \frac{K}{\sqrt{\beta}}\right)^2\right] \\
&= P_\text{g}\left(\frac{\beta - 1 - 2K\sqrt{\beta}}{\beta + 1}\right)^2. \quad (17.22)
\end{aligned}$$

If a cavity is matched to the generator ($P_r = 0$) when no beam is present ($\beta = 1$), it is no longer matched in the presence of a beam. Since the reflected power (along with the accelerating voltage and efficiency) is dependent on the beam current, we see that for fixed coupling reflectionless operation is only possible at one beam current. If there is a need to operate the accelerator at several different currents, it is important to either (a) fix the coupling at a value for which the reflected power never exceeds tolerable levels or (b) vary the coupling to always permit a match. The latter case is obviously the more advantageous, but the design of variable couplers is also an engineering challenge.

Another reason to provide variable coupling is due to unavoidable variations in cavities, couplers, and components of the rf distribution system. The external Q of input couplers will vary slightly from cavity to cavity because of fabrication tolerances and scatter in the field profiles of multicell cavities.

17.4.1 Optimal Coupling in the Presence of Beam Loading

We now determine the matching condition for a beam-loaded cavity operated on resonance. Note that the reflected power in (17.22) is zero when

$$K = \frac{\beta - 1}{2\sqrt{\beta}}. \tag{17.23}$$

It is also possible to show by differentiation of (17.16) and (17.21) that the accelerating voltage and the efficiency as a function of β for constant K are both maximized when (17.23) is satisfied.

Setting $\nu = \sqrt{\beta}$, we can solve for the optimum coupling. One finds

$$\nu_\pm = K \pm \sqrt{1 + K^2}. \tag{17.24}$$

Using (17.19) and recalling that $P_g = P_b + P_c$ in the case of zero reflection, one obtains

$$1 + K^2 = \frac{(P_g + P_c)^2}{4 P_g P_c}, \tag{17.25}$$

so that the solutions are

$$\nu_+ = \sqrt{\frac{P_g}{P_c}} \tag{17.26}$$

and

$$\nu_- = -\sqrt{\frac{P_c}{P_g}}. \tag{17.27}$$

Earlier we restricted ourselves to the positive root of β and hence only (17.26) is valid. Thus we have zero reflection when

$$\beta = \frac{P_g}{P_c} = 1 + \frac{P_b}{P_c}. \tag{17.28}$$

If the power transferred to the beam is significantly larger than the power dissipated in the cavity walls (i.e., $K \gg 1$), then strong coupling is required to achieve zero reflection.

RESONANT OPERATION

Expression (17.28) is not surprising if we recall that β was defined with respect to the unloaded cavity; that is,

$$\beta = \frac{Q_0}{Q_e} = \frac{P_e}{P_c} \qquad (17.29)$$

In the unloaded case, zero reflection is achieved for $\beta = 1$. Beam loading simply introduces additional losses which we could have lumped together with the wall losses by defining P_{cb} as

$$P_{cb} = P_c + P_b. \qquad (17.30)$$

Analogous to the unloaded case we would expect zero reflection with the beam present when

$$\frac{P_e}{P_{cb}} = 1 \qquad (17.31)$$

which requires that

$$\frac{P_e/P_c}{1 + P_b/P_c} = 1 \qquad (17.32)$$

or

$$\beta = 1 + \frac{P_b}{P_c} \qquad (17.33)$$

in agreement with (17.28).

Given the definition for the external Q in (17.29), Equation 17.28 yields

$$Q_e = \frac{Q_0}{1 + P_b/P_c}. \qquad (17.34)$$

Since in most cases for superconducting cavities $P_b \gg P_c$ (heavy beam loading), we have

$$Q_e \approx \frac{Q_0}{P_b/P_c} = \frac{V_c^2}{P_b(R_a/Q_0)}. \qquad (17.35)$$

This is the external coupling required for reflectionless operation of heavily beam loaded cavities.

17.4.2 Current and Frequency Fluctuations

In real situations, the current in an accelerator may differ from the design current so that the coupling is not optimal. Furthermore, mechanical oscillations of the cavity are commonplace; this can be the result of external vibrations, pressure variation in the helium bath, or even cavity deformation due to the Lorentz force exerted by the cavity electromagnetic fields. In each case the cavity is detuned with respect to the generator frequency. A fast-frequency tuner may be able to provide some compensation, but this is difficult for high-frequency oscillations. We therefore would like to determine how the power requirements change due to beam current variation and cavity detuning, *while keeping the cavity voltage constant in magnitude and phase with respect to* $\tilde{i}_b$ [287].

When the cavity is no longer operated on resonance, the cavity admittance ($\tilde{Y}_c$) acquires a reactive component. In the equivalent circuit of Figure 17.3, the cavity admittance is

$$\tilde{Y}_c = G_c + i\left(\omega_g C - \frac{1}{\omega_g L}\right). \tag{17.36}$$

Since $|\omega_g - \omega_0|/\omega_0 \ll 1$ and $G_c Q_0 = \sqrt{C/L}$ and $\omega_0 = 1/\sqrt{LC}$, we can write

$$\tilde{Y}_c \approx G_c\left(1 + 2iQ_0 \frac{\delta\omega}{\omega_0}\right), \tag{17.37}$$

where $\delta\omega = \omega_g - \omega_0$. By adding the loading of the cavity due to the external coupler (βG_c) we obtain the total admittance of the loaded cavity:

$$\tilde{Y}_L \approx G_c\left(\beta + 1 + 2iQ_0 \frac{\delta\omega}{\omega_0}\right) = G_c(1+\beta)\left(1 + 2iQ_L \frac{\delta\omega}{\omega_0}\right). \tag{17.38}$$

Some care now needs to be taken to ensure that the phases of the various voltages are tracked correctly. Using $-\tilde{i}_b$ for our reference phase, we see that the beam-induced voltage ($\tilde{V}_b$) is

$$\tilde{V}_b = \frac{\tilde{i}_b}{\tilde{Y}_L} = \frac{-i_b}{\tilde{Y}_L} e^{i\omega_g t}, \tag{17.39}$$

and that the generator-induced voltage ($\tilde{V}_g$) is

$$\tilde{V}_g = \frac{\tilde{i}_g}{\tilde{Y}_L} = \frac{i_g}{\tilde{Y}_L} e^{i(\omega_g t + \theta)}, \tag{17.40}$$

where θ is the required phase between $-\tilde{i}_b$ and $\tilde{i}_g$ to ensure that $\tilde{V}_c$ remains constant in phase and magnitude.

Since $\tilde{V}_c = V_c \exp(i\omega_g t) = \tilde{V}_g + \tilde{V}_b$ we find that

$$\tilde{i}_g = \left(i_b + V_c \tilde{Y}_L\right) e^{i\omega_g t}. \tag{17.41}$$

Note that the same result can also be obtained by summing the currents in Figure 17.3.

To obtain the generator power from (17.10), we need $|\tilde{i}_g|^2$:

$$|\tilde{i}_g|^2 = i_b^2 + V_c^2 G_c^2(1+\beta)^2 + 2i_b V_c G_c(1+\beta) \\ + 4\left[(1+\beta)Q_L G_c V_c \frac{\delta\omega}{\omega_0}\right]^2. \tag{17.42}$$

Since $P_g = |\tilde{i}_g|^2/8\beta G_c$, we obtain

$$P_g = \frac{i_b^2}{8\beta G_c} + \frac{V_c^2 G_c(1+\beta)^2}{8\beta} + \frac{i_b V_c(1+\beta)}{4\beta} + \frac{(1+\beta)^2 Q_L^2 G_c V_c^2 \delta\omega^2}{2\beta \omega_0^2}. \tag{17.43}$$

RESONANT OPERATION

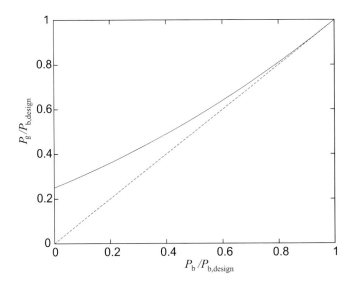

Figure 17.5: Generator power required to maintain a constant cavity voltage as a function of the power transferred to the beam for a cavity that is matched at $i_b = i_{b,\text{design}}$ and $P_b = P_{b,\text{design}}$. The dashed line corresponds to a cavity which is adjusted for a match at each current — that is, $P_g = P_b$. ($P_c \ll P_b$.)

Let us introduce a design current $i_{b,\text{design}}$. At this current, the power transferred to the beam is $P_{b,\text{design}} = V_c i_{b,\text{design}}/2$, and the power dissipated in the cavity walls is $P_c = V_c^2/R_a = G_c V_c^2/2$. From (17.28), we know that for reflectionless coupling at the design current we need

$$G_c = \frac{i_{b,\text{design}}}{V_c(\beta - 1)}. \tag{17.44}$$

Writing $i_b = f i_{b,\text{design}}$ permits us to express (17.43) as

$$P_g = \frac{P_{b,\text{design}}}{4\beta(\beta-1)} \left\{ [f(\beta-1) + (\beta+1)]^2 + \left[\frac{2(1+\beta)Q_L \delta\omega}{\omega_0}\right]^2 \right\}. \tag{17.45}$$

To simplify matters, we note that $f = P_b/P_{b,\text{design}}$, and for superconducting cavities $\beta \gg 1$ (when beam loading is heavy), so that

$$P_g \approx \frac{P_{b,\text{design}}}{4} \left[\left(1 + \frac{P_b}{P_{b,\text{design}}}\right)^2 + 4\left(\frac{\delta\omega}{\Delta\Omega}\right)^2 \right], \tag{17.46}$$

where $\Delta\Omega = 2\pi\Delta f = \omega_0/Q_L$ is the bandwidth of the cavity $\times 2\pi$.

The behavior of (17.46) is plotted in Figure 17.5 for $\delta\omega = 0$. We see that if no current is accelerated, we require a power $P_g = P_{b,\text{design}}/4$ to maintain the cavity voltage V_c. As the beam current is raised, P_g rapidly approaches

$P_{\text{b,design}}$ so that fairly little power is wasted relative to a perfect match. Not shown in Figure 17.5 is the fact that the additional power requirement due to cavity detuning scales quadratically with the frequency shift.

17.5 NONSYNCHRONOUS OPERATION

17.5.1 Phase Stability in the Presence of Little Beam Loading

We already alluded to the fact that in storage rings, the bunches cannot travel through the cavity at the phase for which the acceleration is maximized. This requirement is due to the fact that a bunch is subject to energy perturbations that produce variations in the arrival times of the bunch at the cavity. This is equivalent to a perturbation of the bunch's phase $\omega_g t$. To ensure stable operations, a mechanism must be employed to restore the phase to the design value. A detailed treatment of phase stability is beyond the scope of this book; we will restrict ourselves to the basics. The reader is referred to other texts for more details [304].

A bunch that has the design energy and momentum (E_b and p_b, respectively) is referred to as a *synchronous bunch*. It follows the design orbit of path length C_s. However, if this bunch is perturbed and its momentum changes by δp_b, then the path length of its orbit changes by δC_s, where for small perturbation

$$\frac{\delta C_s}{C_s} = \alpha_p \frac{\delta p_b}{p_b}. \tag{17.47}$$

The accelerator-dependent coefficient α_p is called the *momentum compaction*. For high-energy e^+e^- storage rings, α_p is positive, provided that the ring is operated above "transition." This is always the case during steady-state operation. Equation 17.47 thus implies that the time a relativistic bunch requires to complete an orbit *decreases* with *decreasing* energy.

In order to effectively stabilize energy oscillations, we have to ensure that the cavity provides *less* energy to a bunch which arrives at the cavity later than the synchronous bunch and *more* energy to a bunch which arrives too early.

When the beam loading is small (i.e., $K \ll 1$) this can be achieved by adjusting the phase (ϕ_s) between $\tilde{V}_c$ and $-\tilde{i}_b$, so that the voltage through which the beam is accelerated ($V_{c\phi}$) satisfies

$$\left.\frac{dV_{c\phi}}{dt}\right|_s < 0 \tag{17.48}$$

at the arrival time of a synchronous bunch. Condition (17.48) is known as the *first Robinson stability criterion*, although in this form it only applies to the case when beam loading is not significant. The cavity operation, as dictated by (17.48), is depicted in Figure 17.6 (top). The phase ϕ_s is known as the *synchronous phase*[1] and the accelerating voltage through which the beam is

---

[1] The reader should be aware of the fact that often ϕ_s is referenced with respect to the zero crossing of the cavity voltage, in which case $\phi_s \to \phi_s + \pi/2$.

NONSYNCHRONOUS OPERATION

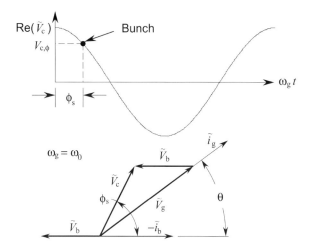

Figure 17.6: Phasor relation between the generator voltage ($\tilde{V}_g$), beam induced voltage ($\tilde{V}_b$), and cavity voltage ($\tilde{V}_c$) when $\phi_s \neq 0$. The cavity is being driven on resonance and the phasors rotate counterclockwise at the generator frequency.

accelerated is given by

$$V_{c\phi} = V_c \cos \phi_s. \tag{17.49}$$

In this case the power transferred to the beam is

$$P_b = I_0 V_c \cos \phi_s = I_0 V_{c\phi}. \tag{17.50}$$

By satisfying (17.48), we ensure that a bunch with a small energy or phase perturbation experiences a force acting to restore the synchronous phase. One can show that a perturbed bunch will oscillate at an angular frequency

$$\Omega_s = \sqrt{\frac{\alpha_p e}{E_b T_r} \left| \frac{dV_{c\phi}}{dt} \right|_s}, \tag{17.51}$$

where T_r is the revolution time for a synchronous bunch [304].

In order to have $\phi_s \neq 0$, we need to introduce a phase θ between $-\tilde{i}_b$ and $\tilde{i}_g$. With $-\tilde{i}_b$ as the reference phase, the phasor diagram is shown in Figure 17.6 (bottom), in which all phasors rotate counterclockwise at the rf frequency. As previously, the cavity voltage is given by the sum of the beam-induced voltage and the generator voltage, that is, $\tilde{V}_c = \tilde{V}_b + \tilde{V}_g$.

Figure 17.6 illustrates that the voltage induced by the beam ($\tilde{V}_b$) introduces a phase between the cavity voltage and the generator current whenever $\phi_s \neq 0$. In the case shown, the cavity, as seen by the generator, has acquired an inductance since $\tilde{V}_c$ leads $\tilde{i}_g$. Operating the cavity in this manner will cause unwanted reflections at the coupler.

17.5.2 Cavity Detuning

We can compensate for the inductance due to the beam by adding a capacitance to the system. This can be done by driving the cavity off resonance at a frequency $\omega_g = \delta\omega + \omega_0$, where

$$\left|\frac{\delta\omega}{\omega_0}\right| \ll 1. \tag{17.52}$$

To examine the response of the detuned system we revert to the equivalent circuit, taking into account that the LCR circuit representing the cavity alone is no longer a real impedance.

In the previous section we already determined the admittance of the cavity, loaded by the input coupler, when $\omega_g \neq \omega_0$. It is given by (17.38). The impedance ($\tilde{Z}_L$) is the reciprocal of the admittance,

$$\tilde{Z}_L = \frac{1}{\tilde{Y}_L} \approx \frac{R_a}{2(1+\beta)(1+2iQ_L\delta\omega/\omega_0)} = \frac{R_a}{2(1+\beta)} \cos\psi\, e^{i\psi}, \tag{17.53}$$

where the *detuning angle* ψ is defined by the relation

$$\tan\psi = -2Q_L\delta\omega/\omega_0. \tag{17.54}$$

The voltages produced by the beam and the generator are

$$\tilde{V}_b = \tilde{i}_b \tilde{Z}_L = \frac{-i_b R_a}{2(1+\beta)} \cos\psi\, e^{i\psi} = -V_{br} \cos\psi\, e^{i\psi} \tag{17.55}$$

and

$$\tilde{V}_g = \tilde{i}_g \tilde{Z}_L = \frac{i_g R_a}{2(1+\beta)} \cos\psi\, e^{i(\psi+\theta)} = V_{gr} \cos\psi\, e^{i(\psi+\theta)}. \tag{17.56}$$

Note that if ψ is negative (i.e., if $\omega_g > \omega_0$), then $\tilde{V}_b$ and $\tilde{V}_g$ lag $\tilde{i}_b$ and $\tilde{i}_g$, respectively, so we do indeed produce a capacitive reactance. This situation is depicted in Figure 17.7 (bottom).

The next step is to find the detuning required to cancel the inductance due to the beam. As seen by the generator, the effective beam impedance $\tilde{Z}_b$ for arbitrary ϕ_s is

$$\tilde{Z}_b = \frac{V_c e^{i\phi_s}}{i_b} = \frac{V_c e^{i\phi_s}}{2I_0}, \tag{17.57}$$

whereas the impedance of the coupler-loaded cavity is given by (17.53). Since $\tilde{Z}_b$ and $\tilde{Z}_L$ are in parallel, we require the sum of the admittances to be real; that is,

$$\Im\left(\frac{1}{\tilde{Z}_L} + \frac{1}{\tilde{Z}_b}\right) = 0, \tag{17.58}$$

where $\Im(\tilde{x})$ is the imaginary part of $\tilde{x}$. Thus

$$\frac{2(1+\beta)}{R_a}\tan\psi + \frac{2I_0}{V_c}\sin\phi_s = 0 \tag{17.59}$$

NONSYNCHRONOUS OPERATION

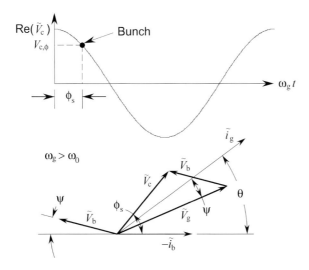

Figure 17.7: Phasor relation between the generator voltage, beam-induced voltage, and the cavity voltage when $\phi_s \neq 0$ and $\omega_g > \omega_0$.

and

$$\tan\psi = -\frac{V_{\text{br}}}{V_c}\sin\phi_s. \tag{17.60}$$

The amount of detuning required then is [see (17.54)]:

$$\frac{\omega_g - \omega_0}{\omega_0} = \frac{I_0 R_a \sin\phi_s}{2Q_0 V_c}. \tag{17.61}$$

Evidently the cavity frequency ω_0 needs to be tuned below the generator frequency ω_g to yield a real load. This is fortunate, since one can show not only that $dV_{c\phi}/dt < 0$ is required for stability, but also that ω_0 must be less than ω_g. The latter condition is known as the *second Robinson stability criterion* [305]. This requirement ensures that the amplitude of the bunches' phase oscillations are damped rather than amplified.

The same result in (17.60) can be obtained trigonometrically using Figure 17.7 and expressing ψ in terms of ϕ_s, θ, V_b, and V_c while satisfying the condition that the generator's load be real — that is, $\theta = \phi_s$ [264].

Given the detuning needed to make the cavity–beam system's impedance real, we are now in a position to calculate the required generator power. If (17.60) is satisfied, the total impedance seen by the generator is

$$\tilde{Z}_{\text{total}} = \left[\frac{1}{\tilde{Z}_L} + \frac{1}{\tilde{Z}_b}\right]^{-1} = \left[\frac{2(1+\beta)}{R_a} + \frac{2I_0\cos\phi_s}{V_c}\right]^{-1}. \tag{17.62}$$

Since the generator current is $i_g = V_c/Z_{\text{total}}$, (17.10) yields the required generator power:

$$P_\text{g} = \frac{R_\text{a} i_\text{g}^2}{16\beta} = \frac{P_\text{c}}{4\beta}\left[(1+\beta) + \frac{I_0 R_\text{a} \cos\phi_\text{s}}{V_\text{c}}\right]^2, \qquad (17.63)$$

which simplifies to

$$P_\text{g} = \frac{[(1+\beta)P_\text{c} + P_\text{b}]^2}{4\beta P_\text{c}}. \qquad (17.64)$$

We see that in the absence of the beam, (17.64) reverts to the expression we originally obtained for the dissipated power (Chapter 8), that is, Equation 17.20.

So far we have optimized the detuning to produce a real load. We also have to find the coupling (β_0) needed to avoid power reflections at the coupler. This is done by requiring that $P_\text{g} = P_\text{b} + P_\text{c}$ in (17.64). Hence

$$\beta_0 = \frac{P_\text{g}}{P_\text{c}} = 1 + \frac{P_\text{b}}{P_\text{c}}. \qquad (17.65)$$

Equation 17.65 is the same result as that obtained for a resonantly operated cavity with $\phi_\text{s} = 0$. Not surprisingly, this result can also be obtained by differentiating (17.64) with respect to β.

Since we can reexpress (17.60) as

$$\tan\psi = -\frac{P_\text{b}}{(1+\beta)P_\text{c}} \tan\phi_\text{s}, \qquad (17.66)$$

we can summarize our two criteria for reflectionless operation as follows:

$$\beta_0 = 1 + \frac{P_\text{b}}{P_\text{c}} \qquad (17.67)$$

$$\tan\psi = \frac{1-\beta_0}{1+\beta_0} \tan\phi_\text{s}. \qquad (17.68)$$

If the cavity is matched for a beam power $P_\text{b,design}$, then $\beta P_\text{c} \approx P_\text{b,design}$ (if $P_\text{b,design} \gg P_\text{c}$), and (17.64) can be rewritten as

$$\frac{P_\text{g}}{P_\text{b,design}} = \frac{1}{4}\left(1 + \frac{P_\text{b}}{P_\text{b,design}}\right)^2. \qquad (17.69)$$

This expression is identical to (17.46) when the cavity is not detuned. In fact, (17.46) applies whenever the beam-loaded cavity represents a real load, in which case $\delta\omega$ represents the detuning from the optimum frequency (see Problem 43).

17.5.3 Phase Stability in the Presence of Heavy Beam Loading

For superconducting cavities, condition (17.48) is often not quite correct to ensure phase stability. Normally $P_\text{b} \gg P_\text{c}$ (i.e., $K \gg 1$), and beam loading is

NONSYNCHRONOUS OPERATION

heavy. (This situation can also arise with normal conducting cavities.) Since the phase of the beam-induced voltage ($\tilde{V}_b$) is governed by the phase of the bunches, it is unable to contribute to phase stability. Only the generator voltage is able to provide a restoring force, and thus the condition for stability is

$$\left.\frac{d\Re(\tilde{V}_g)}{dt}\right|_s < 0, \tag{17.70}$$

rather than (17.48). Here $\Re(\tilde{x})$ is the real part of $\tilde{x}$. Referring to Figure 17.7, we see that (17.70) is equivalent to

$$0 < \theta + \psi < \pi \tag{17.71}$$

Note that counterclockwise rotations define positive angles and that $\omega_g \geq \omega_0$, that is, $-\pi/2 \leq \psi \leq 0$. Thus Figure 17.7 yields the requirement

$$V_c \sin \phi_s > -V_b \sin \psi \tag{17.72}$$

for stability. From the fact that $V_b = V_{br} \cos \psi$ [see (17.55)] it follows that (17.72) can be rewritten as

$$2V_c \sin \phi_s + \frac{I_0 R_a}{1+\beta} \sin(2\psi) > 0. \tag{17.73}$$

Expression (17.73) is the general form of the first Robinson stability criterion. Given (17.73), the beam current has to satisfy

$$I_0 < \frac{2V_c(1+\beta)\sin \phi_s}{R_a \sin|2\psi|}, \quad -\frac{\pi}{2} \leq \psi \leq 0 \tag{17.74}$$

to permit stable operation. Equation 17.74 can be recast as

$$I_0 < \frac{V_c(1+\beta)\sin(2\phi_s)}{R_a \cos \phi_s \sin|2\psi|}, \tag{17.75}$$

which in turn yields

$$P_b < (1+\beta) P_c \frac{\sin(2\phi_s)}{\sin|2\psi|}. \tag{17.76}$$

If we to want match a heavily beam loaded cavity, we have to satisfy $\beta = 1 + P_b/P_c \approx P_b/P_c$ and $\psi \approx -\phi_s$, although $-\psi/\phi_s < 1$ by a small amount. As a consequence, (17.76) demonstrates that the stability criterion is only barely satisfied for a matched superconducting cavity. This is also evident from (17.71), since $\theta = \phi_s$ for optimal detuning. Even a small phase perturbation of the bunches will result in an instability. To safeguard against such an occurrence, one can overcouple the cavity, or, alternatively, detune the cavity by an angle χ ($0 \leq \chi \leq \phi_s$) from the optimum, where χ is known as the *tuning offset*. By writing $\psi \approx -\phi_s + \chi$ we can express (17.74) as

$$I_0 < \frac{2V_c(1+\beta)\sin \phi_s}{R_a \sin[2(\phi_s - \chi)]} \tag{17.77}$$

to yield the maximum current that can be stored.

Note that cavity operation with $\chi \neq 0$ incurs power reflections at the coupler, since the cavity–beam system no longer represents a real load. Alternatively, the need for a tuning offset and the associated power wastage can be avoided by using a feedback or feedforward system [306].

17.6 REEXAMINATION OF THE CIRCUIT MODEL FOR BEAM LOADING

Now that we have calculated the effect of beam loading on the cavity operation, let us examine once more whether the voltage induced by a beam is indeed correctly modeled by the equivalent circuit. To this end, we use an expression derived in Chapter 15 for the voltage induced by an infinite series of (negatively) charged bunches. We found

$$\frac{\tilde{V}_b}{V_q} = \frac{-1}{1 - \exp\left(-\frac{T_b}{T_d}\right)\exp(-iT_b\delta\omega)}, \tag{17.78}$$

where

$$V_q = \frac{1}{2}\omega_0 T_b I_0 \frac{R_a}{Q_0} = \frac{I_0 R_a T_b}{T_d(1+\beta)} \tag{17.79}$$

is the cavity voltage induced by a single passage of a bunch through the cavity.[2] Note that

$$-T_b\delta\omega = -\frac{T_b}{T_d}\frac{2Q_L}{\omega_0}\delta\omega = \frac{T_b}{T_d}\tan\psi. \tag{17.80}$$

In the limiting case where $(T_b/T_d)\tan\psi \ll 1$ and $T_b/T_d \ll 1$, (17.78) becomes

$$\tilde{V}_b \approx -\frac{V_q}{\frac{T_b}{T_d}(1 - i\tan\psi)}, \tag{17.81}$$

which simplifies to

$$\tilde{V}_b \approx -\frac{I_0 R_a}{1+\beta}\cos\psi\, e^{i\psi}. \tag{17.82}$$

This is identical to (17.55), which we obtained with the equivalent circuit. We thus see that the circuit model is indeed applicable.

17.7 TYPICAL PARAMETERS

To illustrate some typical numbers for the parameters we just calculated, let us turn to the values proposed for the superconducting cavity to be used for the Phase III upgrade of CESR.

---

[2] Note that we used $\tilde{i}_b$ rather than $-\tilde{i}_b$ as our reference phase in Chapter 15. Hence a minus sign appears in (17.78).

TYPICAL PARAMETERS

$$
\begin{aligned}
\omega_g &= 2\pi \times 500 \text{ MHz} & &\text{Frequency of the generator} \\
R_a &= 89 \text{ }\Omega \times Q_0 & &\text{Shunt impedance of the accelerating mode} \\
Q_0 &= 10^9 & &\text{Quality factor of the accelerating mode} \\
E_{\text{acc}} &= 10 \text{ MV/m} & &\text{Design operating accelerating gradient} \\
I_0 &= 500 \text{ mA} & &\text{Beam current} \\
U_{\text{sr}} &= 1 \text{ MV} & &\text{Radiation losses per turn/per charge} \\
E_{\text{acc}}/\sqrt{U} &= 1.77 \text{ MV/m}\sqrt{J} & &\text{Accelerating gradient}/\sqrt{\text{stored energy}}
\end{aligned}
$$

These data enable us to calculate the conditions that must be satisfied to match the cavity to the generator. First we obtain the stored energy,

$$U = \left(\frac{E_{\text{acc}}}{1.77 \text{ MV/m}\sqrt{J}}\right)^2 = 31.9 \text{ J}.$$

Hence the power dissipated in the cavity walls is

$$P_c = \frac{\omega_0 U}{Q_0} = 100.3 \text{ W},$$

and the operating cavity voltage is

$$V_c = \sqrt{R_a P_c} = 2.99 \text{ MV}.$$

To replenish the synchrotron radiation losses incurred by the beam, the cavity has to accelerate each charge through $V_{c\phi} = U_{\text{sr}} = 1$ MV (assuming only one cavity is used). Thus the synchronous phase angle is

$$\phi_s = \cos^{-1}\left(\frac{U_{\text{sr}}}{V_c}\right) = 70.443°.$$

The power transferred to the beam is

$$P_b = I_0 V_{c\phi} = I_0 V_c \cos\phi_s = 500 \text{ kW}$$

so that the coupling for a match is

$$\beta_0 = 1 + \frac{P_b}{P_c} = 4987.$$

It is important to note that in the absence of the beam, the cavity is heavily overcoupled. This illustrates one situation a coupler must be designed to deal with. We will briefly discuss this case later. We also see by this example that most of the generator power is delivered to the beam and little is dissipated in the cavity itself. This is in stark contrast to normal conducting cavities. For example, if the same cavity were made of copper, we could expect $Q_0 = 4 \times 10^4$, in which case the power dissipated in the walls would be 2.5 MW! Hence $\beta = 1.2$ would suffice to match the cavity to the generator. In reality, copper cavities

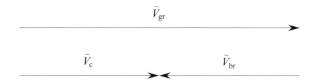

Figure 17.8: Relation between the various voltages in the equivalent circuit during normal operation of a heavily beam-loaded cavity ($\omega_g = \omega_0$, $\phi_s = 0$, $\psi = 0$).

have a different geometry and are operated at lower fields than superconducting ones, to reduce the dissipated power to acceptable levels. Nevertheless, we see that in superconducting cavities the behavior normally is dominated by the beam, whereas in copper cavities this is usually not the case.

Finally we need to know the amount of detuning required. Since we already know the optimum coupling, we find

$$\tan\psi = \frac{1-\beta_0}{1+\beta_0}\tan\phi_s \Rightarrow \psi = -70.436°.$$

If the generator frequency is fixed at $\omega_g = 2\pi \times 500$ MHz, the cavity frequency must satisfy

$$\frac{\omega_g - \omega_0}{\omega_0} = -\frac{\tan\psi}{2Q_L} = 7.02 \times 10^{-6}.$$

In other words, we need to tune the cavity 3509 Hz below the frequency of the generator, which is about 1.4 times the bandwidth of the cavity, this being 2.5 kHz ($= \omega_0/2\pi Q_L$).

17.8 SPECIAL CONSIDERATIONS

We have seen that β can be very large for a beam-loaded superconducting cavity. Due to this fact there are certain design considerations for the rf system. To illustrate some of these, consider a cavity operated on resonance with $\phi_s = 0$ and $\psi = 0$. In this case,

$$\frac{\tilde{V}_g}{\tilde{V}_b} = \frac{\tilde{V}_{gr}}{\tilde{V}_{br}} = -\frac{2\sqrt{P_g P_c \beta}}{P_b}. \tag{17.83}$$

The fact that the cavity is matched implies that $P_g = \beta P_c$ and $P_b = (\beta-1)P_c$ so that

$$\frac{\tilde{V}_{gr}}{\tilde{V}_{br}} = -\frac{2\beta}{\beta-1}. \tag{17.84}$$

If $\beta \gg 1$, the above ratio reduces to $\tilde{V}_{gr}/\tilde{V}_{br} \approx -2$. Figure 17.8 then represents the situation during normal operation.

SPECIAL CONSIDERATIONS

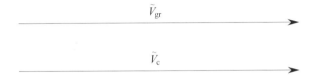

Figure 17.9: Relation between the various voltages in the equivalent circuit after a beam dump.

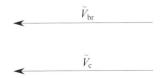

Figure 17.10: Relation between the various voltages in the equivalent circuit after the generator trips.

In the event that the beam suddenly is lost, the beam-induced voltage is absent and we are faced with the situation shown in Figure 17.9. Due to the absence of the beam, the cavity now is severely overcoupled and all the generator power is reflected at the coupler into the circulator and the load. These have to be able to handle the full generator power being delivered. In this scenario the cavity acts as an open circuit, producing a standing wave in the input line. The high fields of this standing wave can pose a problem for an rf window in the input line, so that the window location must be chosen with care.

In reality, the beam-induced voltage will only decay in a time $T_\mathrm{d} = 2Q_\mathrm{L}/\omega_0$, which, for the example given earlier, is about 127 μs. To avoid problems, the rf system has to be able to react to this situation within a time comparable to T_d and shut off the generator power.

A similar condition occurs if the generator trips off. In this case, V_gr decays rapidly and we are faced with the situation shown in Figure 17.10. Now $\tilde{V}_\mathrm{c} = \tilde{V}_\mathrm{br}$ instead of $\tilde{V}_\mathrm{c} = -\tilde{V}_\mathrm{br}$ during normal operation. The power emitted through the coupler is given by βP_c which equals the generator power before the trip. Again, the circulator and load in the input circuit need be able to handle the full generator power, and once more the beam must be dumped in a time comparable to the decay time of the cavity fields.

Finally, consider a doubling of the beam current while the generator power and coupling are kept constant. This scenario is unlikely to occur in reality, but we include it here for completeness. Since V_br doubles, we see that the cavity voltage reduces to zero so that, by conservation of energy, all power is once again reflected. In this case, the cavity acts as a short circuit.

CHAPTER 18

Input Power Couplers and Windows

18.1 INTRODUCTION

As discussed in the previous chapter, the primary role of the input coupler is to transfer rf power from the generator to the cavity and to the beam. The input coupler must also provide a match between the generator impedance and the combined impedance of the cavity-beam system, so as to minimize the wasted reflected power. For some applications, the input coupling strength may need to be adjustable. For others, it must provide sufficient damping of the undesired modes in the fundamental mode passband (see Chapter 7). In select cases, it has also been used to provide coupling to a few low-frequency HOMs. Finally, it is desirable for the input coupler to provide high peak power pulses to process away field emission.

In this chapter, we discuss the design issues and trade-offs for input couplers and windows, illustrating the choices made with couplers developed for superconducting cavities in various applications. Finally we present the performance of the couplers and windows. Much has been adopted from the technology of couplers and windows for normal conducting cavities. To some extent, the science and technology of high-power input couplers is still evolving and there remain uncertainties in many areas. Reviews for input coupler developments can be found in References [307, 308, 309].

18.2 COUPLERS

18.2.1 Design Issues

The presence of a coupling hole near the equator of a cell enhances the magnetic field and may cause premature thermal breakdown. If the coupling hole is placed near the iris, it enhances the electric field in the cell and can give rise to field emission. Disturbing the symmetry of the cell fields with the coupling hole may also give rise to multipacting within the cell. To avoid all these dangers, it is preferable to locate the input coupler for a superconducting cavity just outside the end cell, on the beam tube, where the fields are lower than in the cell. Locating the coupler near the end cell also damps the nonaccelerating modes of the fundamental passband of a multicell cavity. These modes usually need to damped to Q_L values of 10^5 to 10^6 for storage ring application.

The design and development of input couplers and windows involves trade-offs among many options. One of the first design choices is between the coaxial or waveguide variety. The coaxial variety is more compact and therefore suitable for lower frequencies (e.g., 350–500 MHz) where the dimensions of the waveguide become very large. If a variable coupler with a large range is desired, the coaxial variety is more suitable than the waveguide coupler because the penetration of the probe can be varied, for example, by introduction of a bellows on the inner conductor.

The waveguide coupler is attractive for higher-frequency cavities where it can be smaller, and for very high power applications because the cooling is simpler. Only the outer waveguide wall needs cooling, in contrast to coaxial couplers, which also need cooling for the center conductor. However, the waveguide adds mechanical complexity to the cryostat. Due to its large rectangular joints, it is more difficult to fabricate and assemble. Some reduction in waveguide size is possible by using reduced-height waveguide, but the width must remain large for the waveguide to propagate at the desired frequency. To make the waveguide somewhat more compact, putting a ridge in the broad wall of the waveguide is an option frequently discussed, but this adds further mechanical complexity at the joints and creates the need for the development of special transitions to match to standard microwave components. In spite of the large dimension at low frequencies, a waveguide coupler remains attractive for very high power (> 100 kW) low-frequency applications, because the rf power density at the waveguide wall is much lower than at the center conductor of a coax.

Programs such as HFSS [290] and MAFIA [73] are very useful in coupler and window design to help achieve a good match over the desired bandwidth, as well as to locate the position of the voltage and the current maxima in the input line.

18.2.2 Coaxial Couplers

The coaxial coupler has been adopted by the storage rings TRISTAN, HERA, and LEP-II to provide 50–100 kW to the beam. Figure 18.1 shows two coaxial coupler designs, one used for HERA [307] and the other for TRISTAN [51]. The LEP-II [310] and HERA input coupler designs are similar, while the TRISTAN design has some significant differences.

The dimensions of the inner and outer conductors of the coaxial line should be chosen to avoid multipacting at the desired power level, as discussed in Chapter 10. The predicted power levels for multipacting are given in Figure 10.16. In the first version of the LEP-II coaxial coupler, the impedance of the coaxial line was 50 ohm. Later, calculations predicted single-point multipacting on the outer conductor, confirming some of the difficulties observed in the high-power performance of this coupler [311, 168]. The coaxial line impedance was subsequently changed to 75 Ω, by reducing the diameter of the inner conductor. This change shifted the multipacting levels up in power by 50%. The remaining multipacting in the modified coupler is dealt with by other techniques, to

COUPLERS

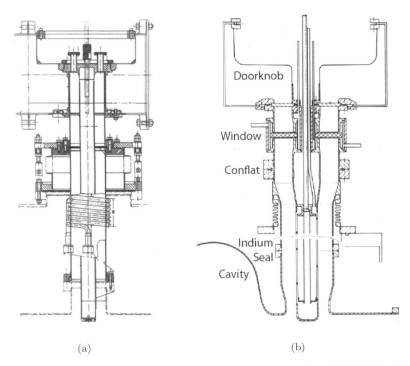

Figure 18.1: Coaxial input couplers for (a) HERA and (b) TRISTAN. The HERA coupler has a room-temperature air-cooled cylindrical ceramic window in the input waveguide. Both center and outer conductor are cooled by 80 K helium gas. The TRISTAN coupler has a room-temperature water-cooled disk window in the coaxial region. The center conductor is also cooled by water. Both versions have fixed coupling. (Courtesy of DESY and KEK.)

be discussed below. The size of the coaxial line must also be chosen to avoid propagation in a non-TEM mode.

The rf losses, conduction heat leak, and radiation heat leak must be estimated for the chosen geometry of the inner and outer conductors. Often counter flow cold He gas is used. The temperature of the cooling gas must be optimized to minimize the heat leak to the liquid helium. The heat load must be kept below tolerable limits by cooling the inner and outer conductors. The heat leak to the cooling bath is also reduced by making sections of the coupler from thin-wall stainless steel coated on the appropriate side with copper of a thickness equal to several penetration depths.

All the couplers start with a waveguide input at the room temperature end. For the LEP-II and HERA versions, the window is a ceramic cylinder integrated

Figure 18.2: A three-stub waveguide tuner used to adjust the input coupling and to compensate for phase errors in the rf distribution system. (Courtesy of DESY.)

with the room temperature waveguide-to-coaxial transition ("door-knob"). The cooling of the ceramic window is increased by forced air flow. For the TRISTAN coaxial coupler, the high power window is a coaxial disk at room temperature. The outer diameter of the ceramic window is enclosed in a water jacket. In all cases, the coaxial line extends between 300 K and 4.2 K, ending in an electric antenna which couples to the cavity electric field. The center conductor probe penetration into the beam pipe is chosen to reach the desired Q_e for beam loading. For HERA, both the inner and outer conductors are cooled by 80 K helium gas, whereas for LEP-II the inner conductor stays at room temperature. For the KEK input coupler, the complete inner conductor is cooled by flowing water. The inner conductor is fabricated from thin-wall copper tubing, while the outer conductor is copper plated thin-wall stainless steel.

All three couplers provide fixed coupling due to the fixed antenna. For HERA, where the intention was to increase the beam current from the initial operation of 30 mA to the final goal of 60 mA, the need to match over the evolving range of beam currents was addressed by a small coupling variation (factor of 10) with a three-stub tuner in the waveguide section between the klystron and the waveguide input. The waveguide coupling tuner has three cylindrical plungers spaced $\lambda/4$ apart (Figure 18.2). This approach is useful not only to adjust the coupling factor for individual cavities, but also to compensate for phase errors in the room temperature waveguide distribution system.

The input coupler design for TESLA [312] has several new features. For matched conditions to the beam, the requirements are to operate in a pulsed mode at 210 kW and 1.3-ms pulse length, at a repetition rate of 10 Hz. For HPP, the peak power desired to reach the TESLA goal of $E_{\text{acc}} = 25$ MV/m is between 500 kW and 1 MW. A unique feature of the TESLA cavity is that the input

COUPLERS

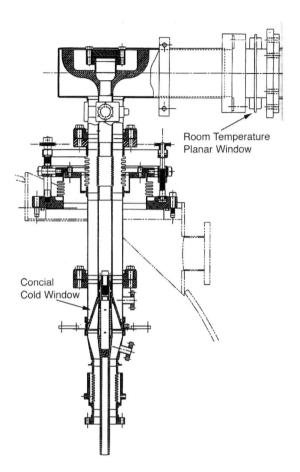

Figure 18.3: One version of the input coupler under development for TESLA. The planar waveguide window is at room temperature. There is conical ceramic window at 70 K. The bellows allow the coupling to be adjusted by factor of 10. (Courtesy of Fermilab.)

coupler port on the beam tube is expected to move about 1.5 cm in the direction of the cavity axis when the cavity is cooled down. Therefore, the coupler design has to incorporate significant mechanical flexibility to accommodate this motion.

In one version of the coupler, being developed at Fermilab (Figure 18.3), there are two windows, a conical one at 70 K and a planar waveguide, warm window at room temperature. Having two windows provides additional protection to the cavity vacuum. The coupler flexibility needed for shrinkage and Q_e adjustability is achieved by hydroformed bellows.

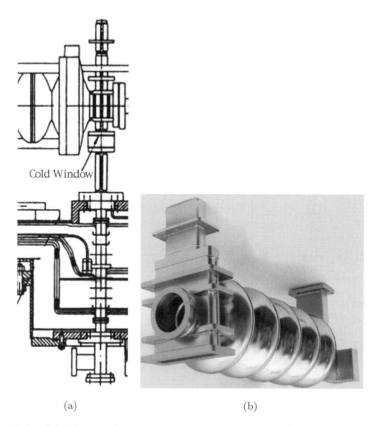

(a) (b)

Figure 18.4: (a) Waveguide input coupler for the Cornell/TJNAF cavity. There is one planar waveguide window at 2 K, and there is a second polyethylene window at room temperature. In the cavity picture (b) the geometry of the shorted waveguide stub is adjusted to reach the desired coupling to the accelerating mode, as well as some of the low-frequency HOMs. (Courtesy of TJNAF.)

18.2.3 Waveguide Couplers

Figure 18.4 shows the waveguide coupler for the 1500-MHz Cornell/TJNAF cavity used in the CESR beam test [26], and in the TJNAF accelerator. For the CESR beam test, the desired power was several tens of kW, while for TJNAF the power demand is 4 kW. The coupler is conceptually similar to the one used for the 1.5-GHz muffin-tin cavity [281]. The Q_e is fixed. The desired coupling to the accelerating mode is achieved by adjusting the length of the matching waveguide stub. The input waveguide also couples to the four other modes of the fundamental passband. Even though the R_a/Q_0 of these modes is much lower than the accelerating mode, they could present a dangerous impedance to the beam if their Q_e remained high. A shorted narrow waveguide stub extends

COUPLERS 409

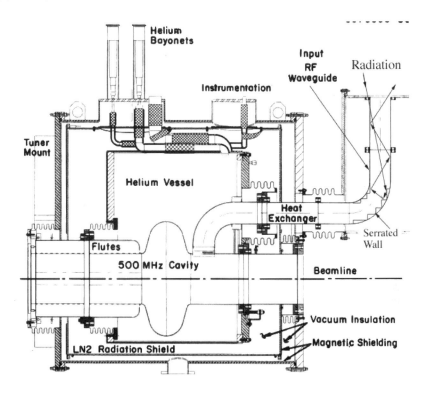

Figure 18.5: Waveguide input coupler for the CESR-III 500-MHz cavity. A reduced-height niobium waveguide couples to the cavity beam tube via a specially shaped aperture. The reduced height helps to cut down the heat leak and make the coupler more compact. Outside the helium vessel, the waveguide heat exchanger is cooled by helium gas flowing through a tracer tube. The elbow is cooled by liquid nitrogen flowing through a tracer tube, attached to the waveguide wall. The inner wall of the elbow is serrated to reflect the room temperature radiation, and thus to prevent it from reaching the helium bath.

from the fundamental input waveguide short to enhance the coupling to the low-frequency HOMs.

Figure 18.5 shows the waveguide input coupler for the luminosity upgrade of CESR (see Chapter 20) where the beam power desired is 325 kW [293]. The coupler consists of a superconducting niobium reduced height waveguide, coupled to the cavity via a specially shaped aperture on the beam tube.

The shape and size of the aperture (Figure 18.6) between the beam tube and the waveguide are chosen to reach the desired coupling of $Q_e = 2 \times 10^5$. A simple hole is insufficient to reach the desired coupling factor; the "tongue" helps to enhance the electric field in the coupler region. The coupling strength increases as the gap between the tongue and the opposing wall is made narrower. The

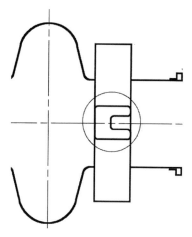

Figure 18.6: Coupling iris for the CESR-III waveguide input coupler. The helium cooled tongue helps to increase the coupling strength. The electric and magnetic fields in the coupling iris regions are significantly lower than the fields in the cell.

peak surface electric and magnetic fields at the coupling iris are well below the fields in the accelerating cell.

18.3 WINDOWS

18.3.1 Design Issues

The primary function of the window is to protect the cavity vacuum. Several important issues must be addressed in the design of the window. The first concerns the location of the window. For high average power application, the preference has been for a warm window, far from the superconducting cavity. With this choice, the rf dissipation from window components does not increase the cryogenic heat load. Electrons and x rays emerging from the cavity cannot reach the window. A disadvantage of the remote window, however, is that the window and the input coupler to which it is attached cannot be put into place when the cavity is first sealed in the clean room. The input coupler/window assembly must be attached later. Great care must be exercised at the later stage to avoid the introduction of dirt into the cavity. Figure 6.25 showed the special dust-free box surrounding the TRISTAN input coupler during final installation into the cavity. It is also desirable to arrange the input coupler to attach to the cavity from below, in order to reduce the risk of dust contamination.

For low average power applications (< 10 kW), when window associated rf losses are less of a concern, a cold window located near the cavity permits final sealing of the cavity in a clean room environment. This minimizes the risk of adding microparticle contaminants that may later cause field emission.

One example of this window choice is the TJNAF planar waveguide window (Figure 18.4), which is located at 2 K and mounted only 8 cm away from the cavity axis [308].

However, there is a negative aspect to the nearby window. Experience at TJNAF has shown that with the window very close to the accelerating structure, field emitted electrons and x rays from the cavity can hit the nearby window ceramic, charge it up and cause arcing [313]. Therefore, it is best to arrange for a location of the window that will avoid bombardment from field emitted electrons and x rays. The options under study at TJNAF to reduce the arcing problems are to increase the distance between the window and the cavity or to introduce an intervening U-bend in the waveguide.

Another question that must be considered in choosing the location of the window is where to place it in the voltage standing wave pattern. There are three aspects: rf heating of the ceramic, voltage breakdown at the ceramic, and multipacting in the coax (or the waveguide) near the window. Ideally, at the design beam current and design cavity voltage, the coupler is matched, so that there are only traveling waves in the input line. Under these conditions, the ceramic heating will be maximum when the power level reaches its highest (i.e., its design, value). However, with no beam current and full cavity voltage, β will become $\gg 1$. In this mismatched condition, the power required to establish the design cavity voltage is one-fourth the design power (see Chapter 17). If a planar window is positioned at the voltage maximum in the standing wave, the power dissipated in the ceramic will be four times larger than the case of the pure traveling wave, i.e., the same as the design value. From a ceramic heating point of view, it is therefore acceptable to place a window at the voltage maximum. Indeed the window may be placed at any position.

However, for processing the cavity and the window, it may be necessary to reach fields higher than the cavity design field. We have seen in Chapter 13 that to process away field emission in the cavity the goal is to reach twice the design cavity field. Since the processing is carried out in the pulsed mode at a duty factor well below unity, the power dissipation and accompanying window heating is not an issue for the higher peak powers involved. However, the high voltage may cause the window to prematurely arc during cavity processing. Therefore it is not recommended to put the window at the voltage maximum. Regarding multipacting, recent studies in a coaxial line (Chapter 10) show that the dangerous region is at the high electric field. Therefore, to avoid multipacting difficulties, the window should be placed at a voltage minimum in the standing wave. In a waveguide, the relationship between multipacting and the best window location is not yet clear, especially in the presence of a ceramic window, so that there is no obvious prescription for window location to avoid multipacting.

Finally, it is important to consider the phase change in the standing wave during the transient condition of cavity filling. If $\beta > 1$, the standing wave pattern in the input line is different by $\lambda/4$ between the time when the rf is first turned on and the time when the stored energy in the cavity reaches its steady-state value. For example, when HPP is carried out, the forward power pulse is

Figure 18.7: A section of the HERA input coupler showing center conductor and cylindrical ceramic window. (Courtesy of DESY.)

the full processing value at the beginning of the pulse. As the cavity fills, the reflected power decreases to zero. After this time, the standing wave pattern in the line will change, and the voltage minimum will become a maximum for the remainder of the rf pulse. If the window is placed at the voltage null for beginning of the rf pulse, the rf should be turned off soon after the reflected power reaches zero. Otherwise the voltage at the window will start to rise and may cause arcing near the ceramic.

18.3.2 Windows for Coaxial Input Couplers

There are several options on the choice of window geometry for the coaxial coupler. Two varieties have been used: a 10 mm thick disk with a central hole for the inner conductor (the annular design, as for TRISTAN), or a ceramic cylinder inside the feed waveguide (as for the HERA and LEP windows). (See Figures 18.1 and 18.7)

One positive feature of the cylindrical window is that it avoids a direct line of sight between the ceramic and the cold cavity surface. "Hiding" the window in this manner may prove to be an advantage because vacuum degradation near the rf window during high power conditioning may be less harmful to the cavity surface. As we discussed in Chapter 12, condensed gases can enhance field emission in superconducting cavities, if the gases reach the cold rf surface.

For the TESLA coupler of Figure 18.3, both windows are positioned so that they will be located at the null of the voltage standing wave pattern that occurs at the beginning of the rf pulse. As we discussed, this is judged to be the best situation for HPP. The conical window was chosen because it allows broad band matching. However, the length of the cold conical window is nearly a quarter

WINDOWS

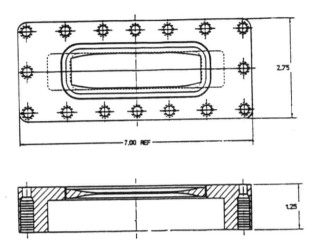

Figure 18.8: (Above) Top view and (below) sectional side view of the TJNAF 2 K planar waveguide window. (Courtesy of TJNAF.)

wave. Thus the connection to the inner conductor is at a voltage minimum but the connection to the outer conductor is at a voltage maximum. It is not clear whether a conical window is a good choice to avoid multipacting.

18.3.3 Windows for Waveguide Couplers

The 2 K window for TJNAF (Figure 18.8) consists of a thin high-purity alumina ceramic mounted in a niobium waveguide flange [308]. The ceramic is shaped to be thin at the center (high electric field region) so as to be almost self-matching.

A planar waveguide BeO (dielectric constant ≈ 7) window was developed as the first prototype for CESR-III (Figure 18.9). The principal idea was to use a high thermal conductivity ceramic for the highest power performance [314]. The thermal conductivity (250 W/m-K) of beryllia at room temperature is much higher than that of alumina (30 W/m-K). Because large-size BeO disks were not available to make the best rf match, three smaller BeO disks were used. These were brazed into a water-cooled copper frame in a reduced-height waveguide on the vacuum side of the window. The mismatch was tuned out by two metal posts placed vertically on either side of the window. Because the match with the small BeO disks alone was far from optimum, there was a significant standing wave component between the ceramic and the posts. The associated high stored energy can lead to frequent discharge during multipactor events. Conditioning to 250 kW took nearly 200 hours.

A planar waveguide alumina window was also developed for CESR-III (Figure 18.10). Unlike the BeO disks, alumina disks were available in the larger sizes needed for a pill-box window design. The matching posts, which were needed only for fine tuning, were much smaller in diameter than in the BeO window

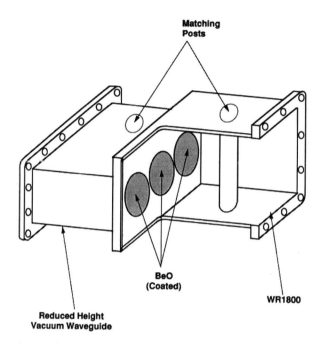

Figure 18.9: A BeO planar waveguide window used in the high-current (220 mA, cw) CESR beam test.

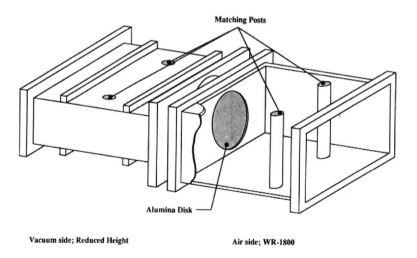

Figure 18.10: An alumina planar waveguide window to be used in CESR-III.

and were placed in a low electric field region. As a result, there was less stored energy between the window and the posts than in the design of Figure 18.9. Conditioning to 300 kW took only 20 hours [315].

18.3.4 Materials Aspects for Windows

Choice of Ceramic In spite of its low thermal conductivity (30 W/m-K), alumina is the prevailing choice for windows because of its wide availability in the needed purity and sizes. It should be mentioned that the thermal conductivity of alumina rises at low temperature reaching a maximum of 100–340 W/m-K, at 70 K, depending on purity. This is an advantage for cold ceramic windows. The loss tangent of a ceramic depends on the purity of the material. At present, both beryllia and alumina are available with low loss, i.e., $\tan\delta \approx 10^{-4}$.

The maximum average power of a window is limited by thermal stress. At high power, the low thermal conductivity of the ceramic causes a large temperature gradient between the interior of the window and the cooled boundary. If the resulting thermal stress exceeds the strength ($\approx 45,000$ psi for alumina) of the material, cracking results. It is standard practice to design a window for a maximum stress value that is only a fraction (e.g., 0.25) of the ultimate strength, in order to allow for variation in ceramic quality. To ensure low losses, it is important to use high-purity alumina; but very high purity makes metallizing of the ceramic (for joining to the waveguide) more difficult. Alumina also has a high dielectric constant (≈ 10). In fabricating the ceramic, voids and impurities in the ceramic must be avoided, as they can become starting points for charging, arcing, and punctures. The thermal conductivity of beryllia is an order of magnitude higher than that of alumina, but the yield strength is lower ($\approx 30,000$ psi). There is still a net gain with respect to thermal stress [316].

Ceramic-to-Metal Joining A coaxial (annular) window requires two brazes, one for the inner conductor and one for the outer conductor. Each braze has different stress conditions, making the brazing operation somewhat more complex than brazing a planar ceramic into a metal frame, or brazing a cylindrical window at each end. A planar waveguide window needs one braze joint at the outer perimeter.

A standard metallization for the ceramic is molybdenum, or molybdenum-manganese, plated with nickel. This is suitable for the subsequent brazing of the ceramic to the metal using the standard braze alloys of copper, copper-silver, copper-gold, or copper-silver-palladium as the filler material in the braze. There are some pitfalls to avoid during brazing. Diffusion of metal into the ceramic during metallization or brazing can produce semiconducting regions with additional rf losses. Sharp edges in the braze regions should be avoided to prevent field emission at low field.

For the TJNAF 2 K window, a silver-copper braze alloy is used to join the ceramic to a thin niobium transition element and the metalllized ceramic. The use of superconducting niobium helps to keep the cold window rf losses low, typically < 0.5 W for 2.4 kW forward power.

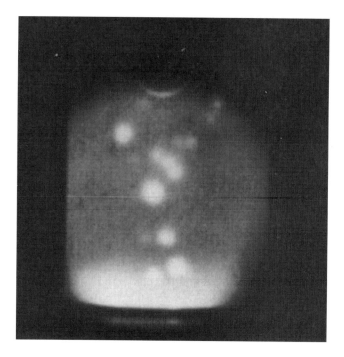

Figure 18.11: Discharges (probably initiated by multipacting) observed on the surface of a cylindrical ceramic window during processing.

18.4 ELECTRON ACTIVITY IN COUPLERS AND WINDOWS

Electronic activity, such as field emission and multipacting, usually limit the performance of couplers and windows on the vacuum side. The maximum peak power is limited by arcing. Multipacting takes place when there are resonant electron trajectories impacting surfaces with a high secondary emission coefficient (see Chapter 10). Figure 18.11 shows the light produced by discharges and multipacting inside a cylindrical ceramic during high power conditioning [317].

If we understood the multipacting trajectories, then it would be possible to design coupler and window geometries to avoid multipacting and improve their performance. Presently, the complex geometry of couplers and windows make it difficult to reliably predict multipacting levels, although much progress has been made in this area.

Vacuum electronic phenomena related to field emission may also play an important role in window behavior. Electrons from the metal–vacuum–ceramic-interface can strike the ceramic and produce a large number of electrons, since the secondary emission coefficient is typically greater than 5 for alumina and beryllia. Excess electron emission leaves the ceramic surface positively charged, increasing the electric field near the triple junction between ceramic, metal and vacuum, and further enhancing the field emission. Electron-induced gas

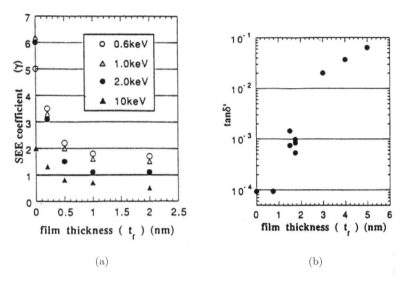

Figure 18.12: (a) Secondary emission coefficient and (b) loss tangent of TiN layer on an alumina substrate, as a function of coating thickness.

desorption from the ceramic and metal surfaces can degrade the vacuum, leading to a glow discharge, which in turn can cause metal deposition on the ceramic by rf sputtering. This can lead to inhomogenous heating, thermal stress, and cracking of the ceramic.

18.4.1 Antimultipactor Measures

It is essential to reduce the secondary emission coefficient of the ceramic by applying a thin coating of special materials. The most used coating is TiN. Others are TiNO, SiN, Cr_2O_3, copper black, and gold black. If they are slightly conductive, these anti-multipacting coatings also help to provide charge drainage, which reduces the risk of voltage breakdown. The thickness of the coating layer must be carefully controlled because the rf loss of the coating increases the surface losses of windows. Figure 18.12 shows the drop in the secondary yield and the increase of loss tangent with coating thickness. Evidently 1 nm is the optimum thickness for TiN [317].

At CERN the conditioning of couplers has been greatly improved by applying a bias voltage of 2.5 kV between the center and outer conductors [311, 168]. The inner conductor is isolated from the outer conductor at the waveguide to coax transition. The multipactor trajectories are altered by the dc field. After operation for some time, the multipactor levels may reappear, presumably due to gas recondensation, and re-conditioning may be required [318].

Figure 18.13: High-power conditioning setup for conditioning coaxial input couplers developed for TRISTAN. (Courtesy of KEK.)

18.4.2 Conditioning and Diagnostics

It is important for the window/coupler to be conditioned with high rf power before attaching the assembly to the clean cavity. If proper conditioning under well-protected conditions is omitted and high power is applied prematurely, the result can be window cracking due to severe arcing. Slight degradations in window performance result in large temperature gradients and window failure due to thermal stress.

The usual setup for conditioning is to place two assemblies back-to-back (as in Figure 18.13) so that the rf power enters the vacuum waveguide region through one coupler and then exits through the second coupler to a room temperature matched load.

The use of fast interlocks during the conditioning stage as well as during normal operation is essential to avoid catastrophic window failure. It is important to have an electron pickup probe and a light detector close to the window to diagnose and protect against arcing and multipacting. The rf should also be turned off fast (within several microseconds) if the vacuum level near the window degrades to above 10^{-7} torr, so as to avoid sputtering and coating the ceramic with metal or removing the anti-multipactor coating from the window.

For best results during window conditioning, the power level should be increased slowly until one exceeds the maximum desired operating level. The vacuum system should provide a high pumping speed for effective conditioning in a reasonable time. There are usually pressure increases induced by vacuum electronic activity, glow discharge, and arcing. The dominant gas evolved during these events is H_2, but H_2O, CO_2, and N_2 are also observed. Pulsed power conditioning is very helpful in limiting the total quantity of gas evolved as the power is raised. The short pulse length also limits the duration of the arc, as well as the amount of energy deposited during a discharge event.

Figure 18.14 shows KEK/Toshiba experience in conditioning cylindrical and coaxial windows [319]. In the uncoated case, a small amount of copper was found to be deposited on the ceramic after the conditioning. The antimultipactor coating improved performance, but even after coating, the cylindrical window was not capable of delivering more than 800 kW. The coaxial (annular) window design was able to handle 1 MW.

After the high-power test in the traveling wave mode, the load should be replaced by a short and the window/couplers retested under full reflection in the standing wave mode to the power level necessary to reach the operating field in the superconducting cavity without beam. Typically this is one-fourth of the beam power because β is usually $\gg 1$.

To limit the severity of vacuum surges during conditioning, a method called "tickle processing" may be applied. While processing in the cw or pulsed mode, 20–50 kW of narrow (e.g., 100 μs) are applied on top of the primary rf pulses. This limits the magnitude of the vacuum bursts and improves the efficiency and speed of processing [315].

After conditioning, the windows should be kept under vacuum until the coupler is ready for attachment to the cavity. The benefits of conditioning have been shown to be preserved for a short (e.g., one hour) exposure to dry nitrogen gas, so that quick, but careful, final assembly should be possible without the need for excessive reconditioning of couplers. During accelerator operation, couplers need to be reconditioned at regular intervals to overcome multipacting barriers that reappear, most likely due to gas migration from warm to cold regions. Conditioning with beam may also be necessary because the voltage standing wave pattern is different than conditioning with full reflected power.

The deconditioning problem in couplers attached to superconducting cavities can be caused by condensation of gases from the ceramic window onto the cold outer conductor of the coupler. CERN experience with LEP-II cavities shows that baking the ceramic window at 200°C after installation of the coupler

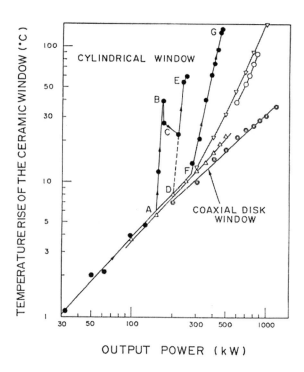

Figure 18.14: Heating of ceramics during high power window conditioning. The experiences for both cylindrical and coaxial disk windows are presented. In the uncoated cylindrical window case, the temperature rises rapidly at 200 kW (point A). Substantial conditioning through the points B-C-D-E-F-G improves the performance. The heating increases abruptly when multipacting starts. After conditioning for some time, heating returns to a normal level, and the power can be raised. Even after coating, the cylindrical window is limited to 800 kW. The coaxial window reached one megawatt.

minimizes the necessary in situ re-conditioning [318]. At HERA, when reconditioning is carried out, the cavity is slowly tuned about its resonance so as to shift the standing wave pattern in the input coupler to permit conditioning of a larger region.

In most applications, there is a back-up window in case of failure of the primary window. The second window provides a vacuum barrier, or a clean gas barrier, to limit the amount of gas exposed to the superconducting cavity. A popular option is a polyimide foil, or kapton foil, squeezed between two waveguide flanges with a metal seal (e.g., indium). Another option is to use a back-up ceramic window of the same design as the main window. If the space between the two windows is filled with clean dry nitrogen gas rather than a vacuum, it reduces the possibility of multipacting in the regions between the windows

Table 18.1: Input couplers and performance

	TRISTAN	HERA	LEP-II	CEBAF	TESLA (2 ms)	CESR-III	KEK-B
RF Frequency (MHz)	508	500	352	1,500	1,300	500	508
Beam Power (kW)	40 – 70	100	60	4	200	325	400
Power Reached on Test Stand (kW)	200	300	300	60	2,000	3,00	800
Power Delivered to Beam (kW)	75	65	40	1.5	—	155	168
Coupler Type	coaxial	coaxial	coaxial	waveguide	coaxial	waveguide	coaxial
Coupling strength	1×10^6	2.4×10^5	2×10^6	6×10^6	3×10^6	2×10^5	5×10^4
Window Temp (K)	300	300	300	2	70/300	300	300

and also provides better cooling for the ceramic. On the other hand, a vacuum between the two windows provides greater protection for the cavity.

18.5 PERFORMANCE OF INPUT COUPLERS AND WINDOWS

Table 18.1 summarizes the performance of the various couplers/windows used in several applications. The performance of the couplers on the high-power test stand as well in the accelerator are given. The highest power ever reached during conditioning was 800 kW for the KEK-B assembly [320]. It took eight hours to condition the couplers to 200 kW and there were many trips. However, most of the trips occurred below 200 kW. Between 200 and 800 kW there were very few trips and the conditioning time was less than two hours. Under full reflection, the KEK-B couplers were tested successfully up to 150 kW. Several other couplers/windows have successfully reached 200 – 300 kW under traveling wave conditions and have reached above 100 kW under standing wave conditions.

The power requirement for the KEK B-factory is 400–500 kW and the CESR-III upgrade is 325 kW. In recent beam tests, superconducting rf systems developed for these high-intensity accelerators have delivered 168 and 155 kW to high-current beams, 220 mA during the CESR beam test [301] and 500 mA during the KEK-B beam test [302]. The maximum power was limited by the available beam current.

18.6 COUPLERS FOR HIGH-PULSED-POWER PROCESSING

Ideally, the main input coupler should be designed and tested to the peak power level and pulse lengths desirable for effective HPP to permit in situ processing of field emission. As we discussed in Chapter 13, HPP is a technique that

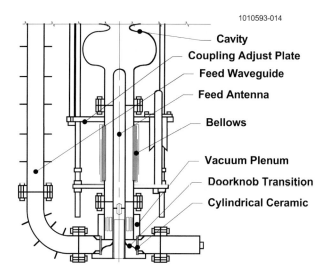

Figure 18.15: High pulsed power test stand for processing field emission in a superconducting cavity.

eliminates field emitters provided that the input coupler can operate at the required high power and pulse length. Experience with HPP on 3-GHz and 1.3-GHz cavities shows that in order to achieve field emission free behavior, it is necessary to reach, during pulsed conditioning, a field level twice as high as the operating field level. During the conditioning, the Q_0 of the cavity often falls to 10^6 to 10^7 due to field emission; therefore the input coupler must be adjustable over a large range.

During investigation of the HPP method, special input couplers were designed for vertical test cryostats. In Chapter 8, we described one high-power test setup used for processing with megawatt 250-μs pulses [249]. We revisit this coupler in Figure 18.15, but in the context of the present chapter, we cover some of the hardware aspects. The high power enters the waveguide on the left through a warm window (not shown) via a reduced height waveguide. Near the bottom of the cryostat is a waveguide to coax doorknob transition with an integrated cylindrical ceramic window to isolate the waveguide vacuum from the cavity vacuum. The ceramic is coated with TiN on both sides. The penetration of the antenna into the cavity is adjustable over 10 cm by flexing a copper-plated hydroformed bellows in the outer conductor. The coupler Q_e can thus be adjusted over the necessary range between 10^5 and 10^{10}. The slotted region of the outer conductor above the doorknob is connected to the vacuum pumping line.

For the best high-power performance of the cold window, it is essential that the length of the antenna be such that there is a standing wave voltage minimum at the side of the ceramic closest to the waveguide elbow. During the development of this coupler, the correct antenna length had to be chosen with the help of HFSS calculations. When the antenna length was wrong by a quarter wavelength, there were frequent breakdown events at 300–400 kW incident power during which the incident power was completely absorbed, but not in the cavity. We found that on the coax side of the waveguide-to-coax transition, the doorknob was coated with silver (from the braze material used for construction) presumably from sputtering in the degraded vacuum during breakdown.

For the warm back-up window, we first used a 1.5 mm thick Teflon sheet with indium wire vacuum joints. There were breakdown events between 300 and 700 kW, accompanied by vacuum bursts in the waveguide. Tracks of metal deposits (presumably from the hold-down flanges) were found on the Teflon sheet near the high electric field region at the center of the waveguide cross section. Placing the Teflon window at the standing wave voltage minimum (for the initial part of the fill) allowed higher power operation. After replacing the warm Teflon window with a warm ceramic (alumina) window we were able to increase the power to 1 MW without breakdown.

CHAPTER 19

Tuners and Frequency Related Issues

19.1 INTRODUCTION

After presenting tuner requirements in the first part of this chapter, we discuss special issues, such as microphonics, Lorentz force detuning, and pondermotive oscillations. Finally, we present several tuner designs and examples realized. Reviews of tuner design and performance can be found in References [321, 322].

19.2 REQUIREMENTS FOR TUNERS

The role of the tuner is to ensure that the cavity resonance frequency will match the desired accelerator operating frequency. There are several considerations in setting the range of the tuner. One can only estimate the final cavity frequency to a certain degree. To make an accurate prediction from the room temperature bench measurement, one must take into account the influence of vacuum, the cavity length change due to cool-down, and the differential contraction of the cavity support system. The predictability of these factors is of the order of one part in 10^4, and the required range of the tuner is correspondingly similar. In the application to storage rings, when the cavity is not being used for acceleration, it may be necessary on occasion to shift the fundamental mode frequency by half a revolution harmonic in order to avoid resonant buildup of beam-induced power in the fundamental mode. For a small circumference (800 m) storage ring such as CESR, the revolution frequency is ≈ 400 kHz. During the accelerator operation the tuner must correct for slow changes in the cavity frequency, as for example, due to changes in the bath pressure, or in the length of the support system (arising from variation in the temperature of the warm sections). Experience shows that the typical pressure sensitivity of the low-frequency storage ring cavities is 8 to 80 Hz/mbar. For the higher frequency linac cavities, the sensitivity is -15 to -60 Hz/mbar. The difference in magnitude and sign comes from the slightly different cavity shapes and the variety of ways in which the cavities are supported inside their cryostats.

In some applications, it is helpful to supplement the main tuner with an additional fine tuner that also has a faster reaction time. The fast feature is used to deal with reactive beam loading or to fast detune a cavity after an rf

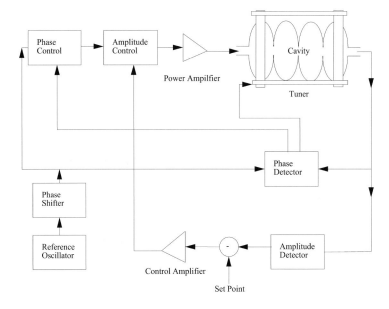

Figure 19.1: Simplified diagram of an rf control system.

trip, so that the beam-induced power into the superconducting cavity does not become excessive. Once the condition that caused the trip is corrected, the fast tuner may prove helpful if it is desired to bring the cavity back into operation without dumping the beam. However, the frequency correction speed is limited by mechanical resonances of the cavity-tuner system.

Tuners are generally made an active part of a complete rf control system which fulfills several functions. It stabilizes the frequency, amplitude, and phase variations induced by sources such as the rf drive, beam current variations, Lorentz force detuning, and microphonics. The level to which the combined frequency, amplitude, and phase control can be achieved determines the energy spread of the beam in a linac. Although our treatment here will be restricted to discussing the cavity tuner aspects, we give one example of a control system in Figure 19.1 to illustrate the overall concept of the complete low-level rf control system [322].

After a signal from a reference oscillator passes through a phase shifter, which determines the reference phase, it is processed by phase and amplitude controllers and is finally amplified to the required power level to drive the cavity. For amplitude control, a signal from the cavity field monitor probe is taken to the amplitude detector, its output compared with a desired set point and the error signal, after amplification and filtering, is used to drive the amplitude control unit. The phase of the probe signal is compared with the reference signal and errors detected are corrected by the phase control unit. The frequency control unit drives the slow and fast tuners which ensure that the cavity frequency agrees with the desired operating frequency. Storage ring rf control systems achieve amplitude stability of 1–2% and phase stability of a few degrees. If each cavity has its individual klystron and rf control system, as in the recirculating linear accelerator at TJNAF, then the performance is remarkably better; fractions of a degree in phase stability and 10^{-4} in amplitude stability have been achieved [323].

19.3 MICROPHONICS

As we discussed in Chapter 18 (Table 18.1), the typical loaded Q of superconducting cavities is chosen to be a few$\times 10^6$, except for high current applications which use a loaded Q of the order of 10^5. In either case the Q_L of a superconducting cavity is higher than for a normal conducting cavity. The resulting narrow bandwidth makes superconducting cavities more sensitive to mechanical vibrations. The effect is enhanced because superconducting cavities are made from thin-walled cells.

Heavy machinery can transmit mechanical vibrations through the beamline, ground, supports and cryostat to the cavity. Mechanical vacuum pumps can interact with the cavity through the beam tubes. The compressors and pumps of the refrigerator will generate vibrations that travel along the pipes of the refrigerator, as well as the helium transfer line, into the cryostat to reach the cavity. Pressure variations generated by machinery can interact with the cavity. In some cases, even boiling helium can cause noise in the phase that is significant for the most demanding applications, such as FELs. However, operating in superfluid reduces bath pressure fluctuations as well as bubble noise.

The noise sources generate a spectrum of mechanical vibrations which is filtered by the transfer medium and finally interacts with the cavity. Together with its tuner and support fixture in the cryostat, the cavity has its own characteristic response to the external perturbations. As a typical illustration [324], measurements of ground vibrations at the SCA (HEPL, Stanford) are shown in Figure 19.2. There is a broad low-level vibrational background and large spikes composed of a harmonic series generated by two large reciprocating compressors that are part of the 1.8 K helium refrigerator.

The cavity amplitude and phase jitter from mechanical vibrations must be reduced to tolerable levels by the rf control system. To avoid instabilities in the frequency control system, one should push the mechanical resonance frequencies as high as possible so that they do not couple to lower-frequency vibrations of the ground motion and the cryogenic system. This can be done in a variety of ways, if properly anticipated, for example by stiffening the cavity or by adding supporting points.

The lowest frequency longitudinal and transverse mechanical modes of a 4-cell 1300-MHz structure are shown in Figure 19.3(a). For FEL application, where amplitude and phase regulation demands are quite severe, the cavity is stiffened as shown in Figure 19.3(b) to raise the frequencies of mechanical resonant modes. For example, the first longitudinal mode was raised from 233 Hz

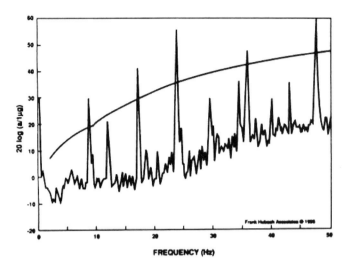

Figure 19.2: Ground vibrations near the Stanford superconducting linac. The spike at the highest frequency corresponds to a ground motion amplitude of 0.1 mm, and the background motion has an amplitude of 2×10^{-3} mm. (Courtesy of Stanford.)

to 744 Hz [324].

Resonators for heavy-ion accelerators, which essentially consist of drift tubes supported by pipes, are more sensitive to microphonics and Lorentz force detuning [325]. Since the beam currents are low, Q_e is usually chosen in the 10^7 range to avoid the need for large klystrons and wasted reflected power. Phase stabilization here is accomplished by using an external voltage-controlled reactance electrically coupled to the resonator. By adjusting the duty cycle between the two states, the average phase can be fixed with respect to an external reference.

19.4 LORENTZ FORCE DETUNING AND PONDERMOTIVE OSCILLATIONS

The rf magnetic field in a cavity interacts with the rf wall current resulting in a Lorentz force which can become important at high accelerating fields [321, 326]. The radiation pressure,

$$P_L \propto \mu_0 H^2 - \epsilon_0 E^2, \tag{19.1}$$

causes a small deformation of the cavity shape resulting in a shift of the cavity resonant frequency:

$$\Delta f \propto (\epsilon_0 E^2 - \mu_0 H^2)\Delta V. \tag{19.2}$$

Here ΔV is the change in the volume of the cavity region that is undergoing deformation. The typical coefficient is a few Hz/(MV/m)^2 and lower for the

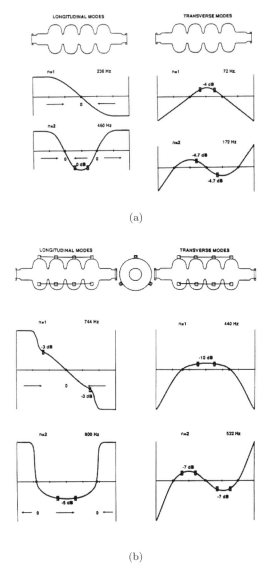

Figure 19.3: Amplitude of mechanical vibrations and their reduction by introducing fixed points. (a) Lowest longitudinal and transverse mechanical modes of a 4-cell structure. (b) Effect on modes shown in (a) after introducing bracing rods as shown in the cavity sketch. (Courtesy of Stanford.)

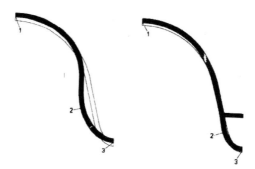

Figure 19.4: The calculated effect of the Lorentz force on the cell shape at $E_{\text{acc}} = 25$ MV/m with a wall thickness of 2.8 mm. The open line is the original cavity shape and the dark line is the deformed shape. On the left, the cavity is slightly deformed without the stiffening ring present on the right. To make the effect visible, the physical deviation is enhanced by 10^5. (Courtesy of DESY.)

pulsed mode.

The iris wall is bent inwards and the equator is bent outwards as shown in Figure 19.4. Both effects lower the resonant frequency. The Lorentz force detuning is particularly important at high fields and for a pulsed accelerator, such as TESLA [277]. Figure 19.4 shows the deformation for the TESLA cavity (wall thickness of 2.8 mm) at $E_{\text{acc}} = 25$ MV/m. Note that the distortion shown is magnified by a factor of 10^5. The predicted frequency change is 1200 Hz. This is signficant compared to the 360 Hz cavity bandwidth.

The regions of the cavity that experience the strongest pressure are the same regions that produce the largest change in frequency. Where the Lorentz force changes sign, the frequency shift also changes sign. This means that Lorentz force detuning cannot be eliminated by optimizing the cavity shape. The only remedy is to increase the mechanical stiffness of the cavity. If the iris region is stabilized by a stiffening ring, the deformation is reduced to 320 Hz. Additional feedforward techniques [327] are required to control the frequency and phase to the required level for TESLA.

Recently, while operating the superconducting cavity in LEP-II, an instability was observed [328] which involved several effects: the mechanical resonances of the cavity, Lorentz force detuning, and the cavity detuning needed to compensate for the reactive part of the beam loading. Any amplitude modulation of the cavity leads to a frequency modulation, which, in turn, results in further amplitude modulation, provided the cavity is detuned. Since the cavities and cryostats are already in place, the only effective solution now is not to detune the cavity in response to beam loading. The price is a modest increase in rf power.

TUNER DESIGNS

Another effect is pondermotive oscillations. When the Lorentz force deforms the cavity its frequency changes, and it loses its coupling to the power source. The field decays, the cavity relaxes, and the field builds up again; the oscillations continue. These pondermotive oscillations must be controlled by feedback.

19.5 TUNER DESIGNS

The standard plunger tuner for copper cavities is generally avoided in superconducting cavities because it introduces moving parts into a cavity vacuum and increases the risk of dust contamination. There is also the risk of multipacting in the vicinity of the tuner piston. A plunger tuner would need an rf filter to reject the fundamental mode, and it would need sliding fingers to make an rf seal required to screen out the HOM and fundamental mode power from the tuner bellows.

The usual tuning method for superconducting cavities is to change the length of the cells by mechanical adjustment of the overall length of the cavity. Since all the cells are mechanically very similar, each cell changes by the same amount and the overall field profile is not affected. A consideration with respect to the extent of the tuning range is the elastic limit of niobium. Staying within the elastic limit gives a tuning range of several ×10 kHz, but a wider range is generally required. At room temperature, this implies exceeding the elastic limit, with some resulting, but tolerable, cavity deformation. At low temperature, the elastic limit is an order of magnitude larger, so that long-term motion of the tuner over a larger range during operation should not present a problem. With a stepping motor, the desired frequency can be reached within a fraction of a bandwidth. However, the drive mechanism has to achieve the required precision in the presence of considerable forces. Because of the high spring constant of a cavity, the vacuum forces are in the range of tons.

The tuner motor is located outside the cryostat in some applications, inside the insulation vacuum in others, and, in some cases, even inside the helium vessel. In the last case, it is important to free the motor of lubricants that will seize in the cold and to shield the cavity from magnetic fields produced by the motor.

An alternative to using a motor as the tuner driver is to use thermal expansion of the bars of a tuner fixture [329]. Details of the design and mechanical aspects are presented in the next section. The thermal method avoids the mechanical problems of motors (e.g., backlash) but increases the heat load to the cryogenic system, since the tuner bars must be maintained above 80 K to get a significant thermal expansion coefficient.

In tuners developed for most applications, length variations produce a frequency shift of 40–80 Hz/μm for the low-frequency storage ring cavities, and 500 Hz/μm for the high-frequency linac cavities. The overall tuning ranges realized are typically 50–1000 kHz.

As we mentioned earlier, the motor-driven or the thermal-driven tuner is augmented in some cases by a faster tuner with a finer resolution. The supple-

Figure 19.5: Tuner for the TJNAF cavity. (Courtesy of TJNAF.)

mental tuner is based on the piezoelectric [51] or magnetostrictive effect [329]. Such tuners can be intrinsically fast, but the response time is limited to tens of milliseconds by mechanical resonances in the accelerating structure. The piezoelectric version is both fast and precise, but the brittle piezo crystals do not stand up well to tensile and shear forces. Also a high voltage (1–2 kV) supply is necessary for the drive. A magnetostrictive tuner is more resilient and uses the more readily available high current source, but the tuning range is smaller. The driver magnetic field may affect the Q_0 of cavity in case of quench. Also, the ohmic losses due to the excitation current may increase the cryogenic heat load. Finally, the response time is limited by eddy currents.

19.6 TUNER EXAMPLES

19.6.1 Mechanical Tuners

TJNAF For the 1500-MHz TJNAF cavities, tuner screws drive a pair of aluminum tuner yokes which grab the end cells of the cavity at the weld step on the equator (Figure 19.5). A stepping motor is located outside the cryostat and the gear trains are located inside. The tuning sensitivity of the cavities is 500 Hz/μm. The frequency range of the tuner is 400 kHz and the precision is 2 Hz [322].

TRISTAN The tuner shown in Figure 19.6 consists of a coarse mechanical tuner and a fine piezotuner with a lever arm to magnify the tuning range by a factor of two [51]. The mechanical tuner consists of a stepping motor, gears, and a jack bolt; it changes the length of the cavity by 0.21 μm/pulse. The tuning

TUNER EXAMPLES 433

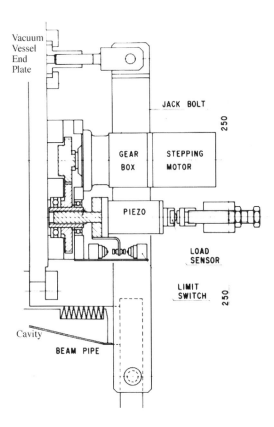

Figure 19.6: Coarse mechanical tuner and fine piezoelectric tuner for TRISTAN. (Courtesy of KEK.)

sensitivity is about 75 Hz/μm. The maximum permissible load is 2 tons, and this allows a wide tuning range of 10 mm. The fine tuner is a stack of 60 layers of piezoelements, each 32 mm in diameter and 1 mm thick. The stroke is about 100 μm with a bias of 1600 V. Fine frequency corrections down to 20 Hz are possible.

CESR-III The waveguide end of the cavity (Figure 18.5) is held fixed by locking down the bellows, and the tuner is attached to the beam pipe at the HOM load [330]. A special feature is the flex hinge, which is machined out of a single piece of stainless steel. As the lever arm rotates, the opposite faces of the hinge stay parallel as they move along the beam axis. The idea was conceived at LANL to eliminate bearings, thereby improving alignment and reducing backlash [331]. The photograph of Figure 19.7 shows the tuner in a test configuration, working against a compressed spring equivalent to the atmospheric load.

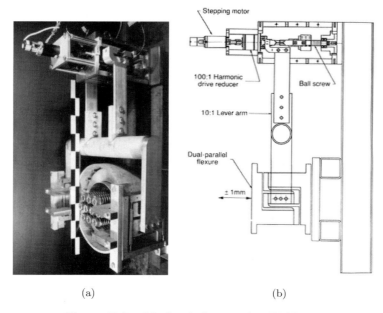

Figure 19.7: Mechanical tuner for CESR-III.

The tuning coefficient is 320 kHz/mm (cavity axis) and the tuning range is 600 kHz. The resolution is 10 Hz, as compared to the 2.5 kHz resonance bandwidth.

19.6.2 Thermal Tuner

LEP-II The CERN tuner [329] uses three Ni tubes as tuner bars located in the insulation vacuum and spaced azimuthally by 120 degrees. Each bar is attached by yokes to both ends of the cavity at the beam tube. The drive mechanism forms a cage that acts on the length of the cavity (Figure 19.8). The cage also increases the resonance frequency of the lowest transverse mechanical vibration mode. An additional support cradle at the middle of the cavity raises the first longitudinal mechanical resonance of the system to 100 Hz. The tuning sensitivity is 40 kHz/mm.

For slow tuning in one direction (constriction), the temperature of the tubes is lowered by flowing cold helium gas drawn from the liquid helium bath. For tuning in the opposite direction (expansion) the temperature is raised by central electric heaters. The typical tuning range is 50 kHz and the speed is 10 Hz/s. Heat losses are minimized by counter-flow helium gas through the tuner tubes. The tuning speed during constriction is limited by the available cooling power to about 10 Hz/s when 0.1 g/s of helium gas is consumed. Expansion presents no response-time problem.

For fast tuning, six coils are wound around the Ni tubes to produce a mag-

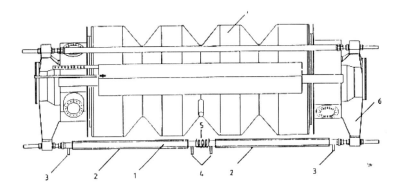

Figure 19.8: Thermal tuner for LEP-II. (1) Nickel tube, (2) coil to activate magnetostrictive effect, (3) cold helium gas inlet, (4) helium gas outlet, (5) heater, (6) supporting frame, (7) cavity welded to helium vessel. (Courtesy of CERN.)

netic field which changes the length of the tubes. A tuning range of ± 1 kHz is adequate for LEP, as the cavity bandwidth is narrow (80 Hz.)

Part IV

Frontier Accelerators

CHAPTER 20

High-Current Accelerators

20.1 THE NEED FOR FRONTIER ACCELERATORS IN HIGH-ENERGY PHYSICS

A recent review of high energy physics topics and high energy physics accelerators can be found in [332]. Our present understanding of the fundamental nature of matter is embodied in the so-called Standard Model. This model envisions all matter as composed of 24 particles, divided into three families (generations) of *quarks* (up/down, strange/charm and top/bottom) and *leptons* (electron/e neutrino, muon/μ neutrino and tau/τ neutrino). Each particle also has its antiparticle. The interactions, or forces, among these particles which bind them into the material world are the electromagnetic force, the weak force and the strong force. Current theory is unable to encompass gravity in this same framework. It is well known that electric and magnetic forces have a common origin, thus the name electromagnetic force. Likewise, electromagnetism and the weak interaction have a common origin, which becomes explicitly manifest at energies above about 1 TeV. It is speculated that the electroweak force may ultimately be seen to have a common origin with the strong force. This unity may become manifest at energies of the order of 10^{12} GeV, the so-called Grand Unification Theory (GUT) scale. Furthermore, even gravity may seen to be part of the same picture above 10^{19} GeV, the Planck scale. It is expected that if these speculations are correct, the signs of this unity will be seen at TeV scale energies.

Along with theory, progress in accelerators has been a vital element in reaching the present level of understanding. Although the Standard Model is one of the triumphs of our century, there remain deep, unanswered questions. Why are there so many "fundamental" particles? (There are a total of 24 quarks, leptons, and their respective antiparticles. In addition, there are 12 force carrying bosons.) Does the apparent complexity of the Standard Model indicate a hidden underlying substructure? What is the origin of the mass of the particles? Why is there such a wide disparity between the masses of the quarks and leptons? Why are there three families? Why are the strengths of the fundamental forces so different? When quarks decay, they can couple to other generations, a phenomenon called "quark mixing". But when leptons decay they do not couple among generations. What is the reason for the fundamental differences between leptons and quarks?

Present accelerators, such as the TEVATRON, HERA, LEP, and the SLC, are continuing to probe for answers to these questions. But to make further progress in the fundamental theory of matter, it will be necessary to use higher energy and higher luminosity accelerators. Superconducting rf technology is already an essential part of two of these machines. We will see how the continued advances described in the previous chapters have made superconducting technology a viable candidate for advancing both the luminosity and the energy frontiers in the next generation of accelerators.

20.2 HIGH-CURRENT STORAGE RINGS

One of the outstanding problems of elementary particle physics is the very small asymmetry between the properties of matter and antimatter. Theory suggests that it is this slight imbalance in nature's otherwise symmetric order that led, during the first moments after the big bang, to the now observed predominance of matter over antimatter in the universe. The asymmetry is related to a phenomenon known as "charge-parity (CP) violation." The behavior is associated with differences in certain elementary processes under reversal of the direction of time. In the absence of CP violation, the behavior of matter and antimatter is completely symmetric. Under such circumstances, there could be no stable universe since antiparticles would annihilate all particles. CP violation is therefore the fundamental reason for a stable, matter-dominated universe.

The origin of CP violation is still a mystery, but it can be accommodated within the Standard Model. Although it is not a necessary consequence of the Standard Model, when CP violation is allowed in the model, the theory predicts a small, but significant, asymmetry in the decay of B mesons and $\bar{B}$ mesons. B mesons are bound pairs of the B quarks with other quarks. CP violation has already been observed in the K meson system, which contains the strange quark. However, the K meson system provides too limited a set of measurements to determine all the mixing parameters between the three quark generations. Data on CP violation in the K meson system also cannot distinguish between the competing theoretical models for the mechanism. The decays of B quarks already provide a wealth of data to test the Standard Model. Future studies of CP violation in the B quark system may help to establish the complete mechanism. But CP violation in B decays has yet to be observed. Indeed it is expected to be a very rare process. Current estimates indicate that 10^7 to 10^8 B decays would have to be produced. Therefore it is necessary to produce a very high intensity of B mesons.

Today CESR is the exclusive and prolific source of B decays providing about 10^6 Bs a year. At a luminosity of 3×10^{32} cm^{-2}s^{-1}, CESR holds the record for the highest luminosity for any collider. The luminosity required in a future B factory will need to be increased by at least another factor of 10 beyond this

HIGH-CURRENT STORAGE RINGS

world record. The needed intensity demand can be reduced by using unequal beam energies so that decay products are moving in the laboratory. The moving frame of reference will allow a more precise measurement of the difference in decay times between B mesons and $\bar{B}$ mesons. This is expected to be the most efficient way to detect asymmetries arising from CP violation.

To test the Standard Model's description of CP violation, and to continue studies of rare B meson decays, several laboratories are planning higher luminosity e^+e^- storage colliders at the B quark threshold — i.e., about 10.6 GeV energy in the CM. To reach the desired luminosity, the beam currents will have to be in the ampere range. For a recent review of the rf issues for high current machines see Reference [302]. At the SLAC and KEK B factories, copious quantities of B mesons will be produced with unequal energy electron and positron beams. KEK plans to use superconducting cavities in the 8 GeV HER (high energy ring) and a special, low impedance normal conducting cavity in the 3.5 GeV LER (low energy ring). SLAC plans to use only normal conducting cavities together with powerful feedback systems in both the high energy and low energy rings.

In parallel with the asymmetric B factories, CESR will continue with symmetric collision energies. It plans to upgrade its record luminosity by raising the beam current in stages, from the present 150 mA/beam to 300 mA/beam to 500 mA/beam. The last stage will incorporate superconducting cavities to avoid multibunch instabilities. Another high current storage ring that is considering the use of superconducting cavities is the Beijing tau charm factory which will look for CP violation in tau lepton and charm decays.

Superconducting cavities are also planned for the next accelerator at the energy frontier, the Large Hadron Collider (LHC) at CERN. According to the Standard Model, the electromagnetic force and the weak force have a common origin. Their connection is only manifest at about 1 TeV. At this energy scale, which LEP is currently exploring, the two forces appear symmetric. But at lower energies the weak force manifests itself quite differently from the electromagnetic force. Several schemes for extending the Standard Model to include yet higher energy phenomena have been put forward. One is the Higgs mechanism, which posits a very heavy force-carrying particle. If it exists, the Higgs would have a mass between 150 GeV and 1 TeV. Other schemes, such as supersymmetry, have also been put forward. These other schemes also make predictions about interesting new particles that may be seen at energies higher than now achievable with today's accelerators. If these predictions are borne out with accelerators that may be built in the next few years, the LHC among them, we may be able to get a glimpse of the physics all the way to the GUT or even the Planck scales.

To be built in the same tunnel as LEP, the LHC will collide 7 TeV high current proton beams. It will be a significant push on the energy frontier beyond the Fermilab Tevatron, which now provides 1 TeV proton collisions at a luminosity of 1×10^{31} cm^{-2}s^{-1}. The luminosity goal for the LHC is 1000 times higher than the Tevatron.

Table 20.1: Accelerator and superconducting cavity parameters for future high current storage rings

Accelerator	CESR-III	KEK-B(HER)	LHC	BTC
RF Frequency(MHz)	500	508	400.8	499.7
Voltage/Cavity(MV)	1.8 – 3	0.9 – 1.6	2	2 – 3
No. of Cavities	4 – 6	10	8	3
Total Beam Current(A)	1	1.1	0.85	0.57
Input Power/coupler(kW)	325	400 – 500	100	60
No. of Bunches	45	5120	2835	30
No. of particles/bunch(10^{11})	1.7	0.14	1	1
Bunch spacing(m)	4.2	0.6	7.2	12
Bunch Length (cm)	1.4	0.4	7.5	1
HOM power/cavity(kW)	10	14	0.1	7.5
Q_{HOM}	< 100	< 100	< 10^4	< 100

20.3 THE BENEFITS OF SUPERCONDUCTING RF FOR HIGH-CURRENT STORAGE RINGS

Future high luminosity, electron–positron and proton–proton colliders stand to gain many advantages from the use of superconducting cavities. To achieve the needed luminosities, ampere size beam currents need to be stored in a very large number of bunches, which are spaced very closely together, as, for example, 0.6 m for KEK-B. The high current and the tight bunch spacing makes the control of multibunch instabilities one of the most important issues for these new machines. Table 20.1 shows the beam and rf parameters for the high luminosity projects underway [302]. Note the large number of bunches, the short bunch length, and the short bunch spacing for the B factories. For the LHC, despite the high beam current, the power that must be delivered to the beam is quite small compared to electron rings because of the negligible synchrotron radiation from the circulating protons.

To avoid multibunch instabilities in the presence of such large numbers of bunches, it is imperative to reduce the beam–cavity interaction. Since superconducting cavities can economically provide higher cw gradients than copper cavities, the necessary voltage can be provided by fewer cells, which means reduced impedance. Table 20.2 compares the superconducting system that is intended for the KEK-B high energy ring with a possible alternative normal conducting system based on the same normal conducting cavities as intended for the LER [274]. The superconducting system reduces the number of cavities by a factor of 3.6 and the total RF power installed by a factor 2.2. Calculations show that the growth time of longitudinal multibunch instabilities for a superconducting system would be 420 ms which is very much greater than the 23 ms radiation damping time. Similarly, the growth time for transverse instabilities is 110 ms, also much longer than the radiation damping time (46 ms).

As discussed before, superconducting cavities have large beam holes com-

BENEFITS OF SRF FOR HIGH CURRENT STORAGE RINGS

Table 20.2: A comparison of normal conducting and superconducting systems to fulfill the need for HER at KEK-B

	NC	SC
No. of Cavities	36	10
Q_0	1.3×10^5	1×10^9
Voltage(MV/cav)	0.45	1.6
Beam Power(kW/cav)	114	400
Wall Loss (kW/cav)	119	0.1
Input power (kW/cav)	233	400
Total RF power (kW)	8388	4000
Highest R_a/Q_0 monopole HOM (Ohm)	34.8	6.4
Highest R/Q_0 HOM dipole (Ohm)	20	15

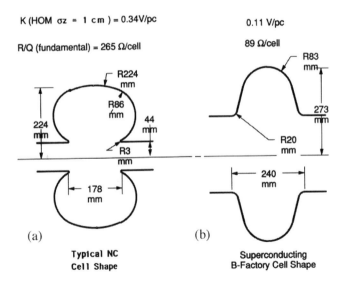

Figure 20.1: A comparison of cell shapes between a typical normal conducting and a large beam hole superconducting cavity, designed for a high current storage ring. The impedances comparison in the next figure is based on these shapes.

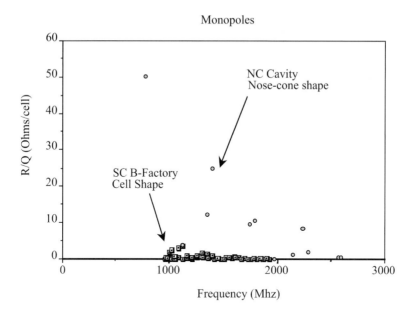

Figure 20.2: A comparison of monopole geometric impedance values for the cell shapes of Figure 20.1.

pared to normal conducting cavities (see Figure 20.1), which means a substantial reduction in the impedance of HOMs. Figure 20.2 compares the R_a/Q_0 for monopole and Figure 20.3 compares the R/Q_0 for dipole modes for typical normal conducting and superconducting 1-cell cavities [293]. Note that the impedance of the dominant monopole HOM is reduced by factor of 5. If high impedance normal conducting cavities are used, as in the case of the SLAC B factory, then the larger number of cells together with their higher R/Q_0 demands not only strong damping for HOMs ($Q < 100$) but also a powerful, wide-band, bunch-to-bunch feedback system to deal with longitudinal and transverse multibunch instabilities.

In a high current storage ring the fundamental mode can also give rise to a multibunch instability. Since the beam is accelerated off the crest of the rf wave for phase stability, the beam-induced voltage is out of phase with the generator voltage. To compensate for reactive loading, and thereby to minimize the required generator power, the cavity is detuned — i.e., operated off resonance. The amount of detuning is given by:

$$\delta f = I_{\text{beam}} \sin \phi_s \frac{R_a}{Q_0} \frac{f_0}{2V_c} = \frac{P_b \tan \phi_s}{4\pi U}. \qquad (20.1)$$

Here I_{beam} is the beam current, ϕ_s is the synchronous phase, f_0 is the accelerating mode frequency, V_c is the cell voltage, U is the cell stored energy and P_b is the beam power.

If δf is a significant fraction of the bunch revolution frequency, f_r, then

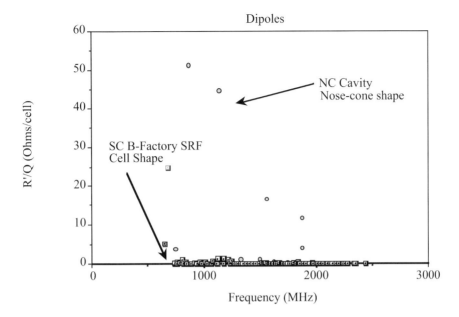

Figure 20.3: A comparison of dipole geometric impedance values for the cell shapes of Figure 20.1.

multibunch instabilities driven by the fundamental mode can become a problem. The coupling impedance at the upper synchrotron side band of the revolution harmonic frequency becomes high. The instability growth time can become 10–100 μs, much faster than the radiation damping time. In the case of the SLAC B factory, which will use normal conducting cavities, the required detuning is large, $0.3 f_r$. The anticipated instability will be addressed with multiple level feedback loops around the klystron-cavity-beam system [333]. In the case of CESR-III with superconducting cavities, the lower impedance in the fundamental mode, coupled with the higher voltage per cell (and accompanying higher stored energy), reduces the extent of detuning required to $0.03 f_r$. In the KEK-B HER the detuning is also reduced to $\approx 0.03 f_{\text{rev}}$, but in the LER it is still quite high. Even with a superconducting cell, the voltage per cell is low because the LER has less synchrotron radiation and needs less total voltage. When combined with the high beam current, the resulting detuning would be large. Therefore special normal conducting cavities have been being developed with fundamental mode R_a/Q_0 reduced to 15 Ohm. This drastic reduction is achieved by coupling the accelerating cavity to a storage cell, which increases the stored energy of the accelerating mode.

For the LHC, the serious issue connected with the use of high shunt impedance normal conducting cavities is transient beam loading. The beam induced voltage, which is nearly all reactive, is compensated by detuning the cavity according to Equation 20.1. When the equilibrium of the beam is disturbed, as when a new beam pulse is added during injection, the transmitter must compensate rapidly until the relatively slow tuner settles to its new position. A similar effect is caused by gaps in the bunch train, such as abort or injection gaps, which cause excessive phase modulation of the beam given by

$$\delta f \propto I_{\text{beam}} \frac{R_a}{Q_0} \Delta t \frac{2\pi f_0}{2V_c}. \tag{20.2}$$

Here Δt is the gap in the beam bunch train. The phase modulation causes capture losses during injection. In the B factory, the phase modulation by the gaps in the train shift the longitudinal position of the collision point and decrease the luminosity. Superconducting cavities reduce these transient effects by virtue of their low R/Q_0 and high cell voltage V_c.

Of course, when superconducting cavities are used in high current storage rings, capital and operating cost savings are also realized from the reduced rf installation, which does not need to provide for rf power dissipation in the cavity walls. But this cost saving is only a small fraction of the total installation required to provide the large beam power. *The main incentive to use superconducting cavities in the high current machines is to lower impedance.* A superconducting rf system may also prove more suitable for future luminosity upgrades, which call for more bunches or higher bunch current.

Another high current storage ring that will use superconducting cavities is the proposed light source, SOLEIL, under study at Orsay in France. Here the use of a superconducting cavity will reduce emittance and energy spread from multibunch effects, yielding a higher brilliance [334].

20.4 SYSTEMS UNDER DEVELOPMENT

Superconducting cavities for the present generation of storage rings (TRISTAN, HERA and LEP) have been upgraded in several key aspects to make them suitable for higher current accelerators. Besides lowering the impedance of the HOMs, another benefit of enlarging the beam pipe is that the HOMs propagate out of the cell, down the beam pipe and into HOM loads located outside the cryostat. Figures 20.4 and 20.5 show the beam pipe of the KEK-B [54] and CESR-III cavities [335]. By lining a short section of the warm beam pipe with an excellent microwave absorber, such as ferrite, the HOM Q_L values are reduced to below 100. Many kilowatts of beam-induced HOM power can be safely dissipated in these absorbers. As we discussed in Chapter 16 on HOM couplers, ferrite-lined beam pipe couplers have been tested to 15 kW. These beam pipe HOM couplers are significant advances over the loop- and antenna-type couplers used in the present generation storage ring superconducting cavities, which lower HOM Q_L values of 10^3 to 10^4 and handle a few hundred watts of beam-induced HOM power.

Figures 20.6(a) and 20.6(b) show the KEK-B and CESR-III layouts for the superconducting cavity, cryostat, input coupler, window, ferrite HOM coupler, and other components. As we discussed in the section on input couplers, there

SYSTEMS UNDER DEVELOPMENT 447

Figure 20.4: To damp the nonpropagating, low frequency dipole modes, one of the two beam pipes is made larger in diameter. In the case of the KEK-B superconducting cavity, the 32 cm diameter beam pipe must also be made longer to provide sufficient cutoff for the accelerating mode. The coaxial input power coupler port is on the small diameter beam tube. (Courtesy of KEK.)

Figure 20.5: In the case of the CESR-III cavity, four large diameter "flutes" are introduced into the beam pipe on the right to propagate the dipole modes. This approach avoids increasing the length of the beam pipe, at the cost of increased mechanical complexity. The waveguide input power coupler is attached to the opposite beam pipe.

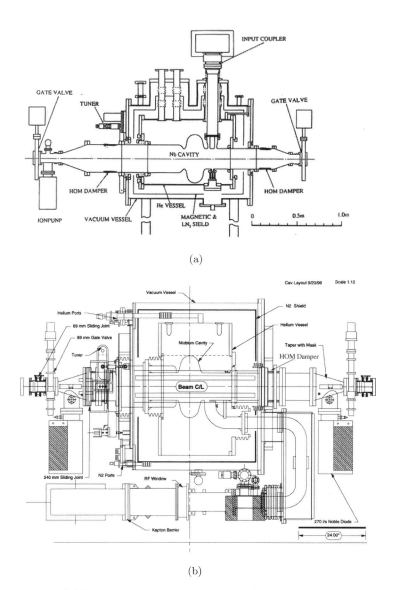

Figure 20.6: (a) Layout of KEK-B superconducting cavity inside cryostat, showing input coupler, ferrite beam pipe HOM dampers, tuner, and beam line components. (Courtesy of KEK.) (b) Layout of CESR-III superconducting cavity inside cryostat, showing input coupler, ferrite beam pipe HOM dampers, and other beam line components.

SYSTEMS UNDER DEVELOPMENT

has been much progress toward increasing the capability of waveguide and coaxial input power couplers, and their associated windows, toward delivering 300–500 kW to the beam through the cryogenic environment.

Prototype superconducting cavities have been successfully tested at high current in CESR (August 1994) [301, 300] and in the TRISTAN Accumulation Ring (March - November 1996). In CESR, the superconducting cavity accelerated a 220 mA beam in 27 bunches without causing any beam instabilities and without any adverse effects on the cavity performance and machine operation. This was the maximum current of CESR at the time. The cavity ran with beams at gradients between 4.5 MV/m (220 mA) and 6 MV/m (100 mA), coupled 155 kW of power (limited by the window) to a 118 mA beam, and removed an expected 2 kW of beam-induced power (limited by the maximum beam current). The high circulating beam current did not increase the heat load or present any danger to the cavity. Beam stability studies were conducted for a variety of bunch configurations.

The maximum single bunch current of 44 mA showed that the additional broad band impedance (short range wake) from the ferrite HOM load is not a problem. The HOM power deposited by the beam was carefully monitored. No beam instabilities or resonant excitation of HOMs were observed. A different method of searching for cavity-engendered beam instabilities was to change the frequencies of HOMs in the presence of a large beam current. While supporting a 100 mA beam, the cavity was axially deformed with the tuner to sweep the HOM frequencies over a few revolution harmonics. There were no beam instabilities. During this experiment there was also no change in the temperature of the HOM loads, confirming the absence of resonant excitation of HOMs. These tests confirmed that all dangerous HOMs were damped sufficiently to be rendered harmless.

In the beam test at KEK [302] the maximum gradient reached in the superconducting cavity was 12 MV/m and the maximum beam current stored was 500 mA (E_{acc} = 9.6 MV/m). The maximum power coupled to the beam was 164 kW and the maximum HOM power extracted was 2.5 kW. These tests are continuing to higher currents.

Figure 20.7 shows the layout envisioned for a pair of LHC superconducting cavities [336]. Because of the long bunch length in the LHC, the HOM damping and beam induced power are not as severe as for the high current electron machines. The designs for LEP-type HOM couplers are expected to be adequate. The demand of the input coupler is also in the range of LEP technology (100 kW). A unique feature of the LHC superconducting cavity is the tuning system which must provide fast tuning over at least 25 kHz (i.e., 150 mm of tuner stroke) as required for LHC operation and for the fast compensation of beam loading. The tuner demands a substantial increase in performance over the LEP cavity tuning system which provides a maximum of 5.5 kHz using magnetostriction on 2 m long nickel bars.

A prototype superconducting cavity has already been fabricated, tested, and installed in the SPS. It was also based on the LEP technology of sputtered

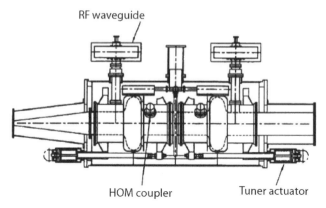

Figure 20.7: Layout of LHC superconducting twin cavities inside cryostat showing location of input coupler, HOM coupler, and tuner. (Courtesy of CERN.)

niobium on copper. The cavity operated at a gradient of 3 MV/m with the usual current for the SPS, which is 100 mA peak and $\approx$ 50 mA average.

20.5 CRAB CAVITIES FOR BUNCH ROTATION

Since crabbing is an unusual, nonaccelerating application for a superconducting cavity, we will discuss it at some length.

To obtain the highest possible luminosity in a B factory, the most desirable method to separate the e^+e^- beams is a non-zero crossing angle, typically 10–12 mradians. This scheme avoids parasitic collisions, permits the smallest spacing between bunches, and therefore allows the largest number of bunches. The absence of separation magnets near the interaction point minimizes synchrotron radiation incident on the detector region. However, prior experience shows that an angle crossing may cause harmful coupling between synchrotron and betatron motion. Another consequence is luminosity reduction due to the geometrical effect. By rotating the bunches before collision so they collide head-on, and then rotating them back so they pass normally through the rest of the machine, the harmful effects can be eliminated. The scheme is illustrated in Figure 20.8. The tilted motion of the bunches after deflection resembles the crawling motion of crabs; hence the name "crab crossing" [337, 338].

In the KEK-B factory ring, the bunches are very closely spaced ($\approx$ 0.6 m), eliminating room for any conventional separators. Instead, the small ($\approx$ 11 mr) crossing angle will be explored. The crab cavity will be located at an odd multiple of quarter betatron wavelengths from the interaction point. For deflection, the bunch is synchronized to pass through the center of the cavity at the zero crossing of the magnetic field. Thus the centroid of the bunch is not

CRAB CAVITIES FOR BUNCH ROTATION

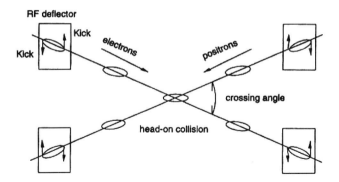

Figure 20.8: The idea of crabbing to provide head-on collisions with a crossing angle. Two deflecting cavities are needed for each beam to supply and remove the tilt. The crossing angle shown is magnified for clarity.

affected, while the head and tail of the bunch are kicked in opposite directions. They then oscillate under the influence of the magnetic field of the focussing quadrupoles and reach their peak amplitude at an odd multiple of quarter betatron wavelengths, which matches the distance from the cavity to the interaction point.

The required transverse kick is about ≈ 1.5 MV. The TM_{110} deflecting mode is the most suitable because it has the highest transverse geometric shunt impedance [278]. For the same reasons of impedance and power economy as for accelerating cavities, it is desirable to minimize the number of crabbing cells. A single-cell superconducting cavity, operating in the TM_{110} mode at ≈ 500 MHz, can provide the needed kick and still have a surface electric field below 15 MV/m to avoid field emission degradation of Q_0. Splitting the degeneracy of the two polarizations of the TM_{110} mode is accomplished by breaking the axial symmetry of the cell as shown in Figure 16.1. The deformation is built into the cell-forming die. The mode separation is 14 MHz for the TM_{110} mode, and 1 MHz for the TE_{111} mode. The cell design allows all modes higher in frequency than the crab mode to propagate out the beam pipe and to be damped outside the cryostat with ferrite beam pipe absorbers.

Four unwanted modes remain trapped in the cell region even with the large beam pipe: the fundamental (TM_{010}), two polarizations of the TE_{111} mode, and the unwanted polarization of the TM_{110} mode. To damp these modes [339], a coaxial beam pipe with a notch filter is adopted as shown in the sketch of Figure 20.9 and the photograph of Figure 20.10. Thus the lower frequency parasitic modes can be extracted from the cavity and damped in a section of the coax, lined with ferrite on the outer conductor. In a coaxial transmission line there is no cutoff for TEM mode mode waves, but there is a cutoff for dipole modes. Every monopole resonant mode in the cavity couples to the coaxial beam pipe as a TEM wave and propagates out. In addition, every dipole

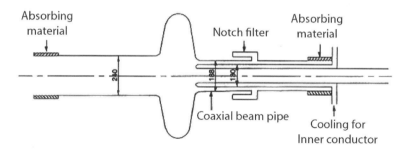

Figure 20.9: Schematic for a crab cavity. On the left is a beam pipe ferrite HOM damper as in a superconducting accelerating cavity. On the right is a beam line coaxial HOM damper to damp the unwanted TM_{010} and TE_{111} modes. The coaxial filter cavity ensures rejection of the TM_{110} operating mode.

Figure 20.10: Niobium crab cavity corresponding to schematic of Figure 20.9.

mode couples to the coaxial beam pipe as a dipole mode and will propagate, as long as the mode frequency is higher than the cutoff frequency of the coaxial pipe. The cell shape and the coaxial pipe dimensions are chosen so that $f(TM_{110}) < f(\text{cutoff}) < f(TE_{111})$. If the cutoff frequency is 600 MHz, the attenuation for the crabbing mode is 60 dB/m. In a real cavity, some asymmetry due to manufacturing errors or a misalignment of the coaxial beam pipe can couple the crab mode to the TEM wave of the coaxial beam pipe. To avoid this, a notch filter cavity is added to reject the TEM coupled crabbing mode. If the resonant frequency of the filter is tuned to ±0.5 MHz, a rejection rate higher than 50 db is obtained.

Table 20.3 gives the measured Q_L values of the parasitic modes along with the calculated dipole R/Q_0. The only mode that does not propagate is the orthogonal polarization of the crab mode. Since this mode has a high shunt impedance, it must be tuned to a safe frequency. Another possibility is to increase the polarization of the cell to increase the frequency of the unwanted TM_{110} polarization to be above the cutoff of the dipole wave in the coaxial

Table 20.3: R/Q_0 and Q_L for the crab cavity parasitic modes.

Mode	Frequency (MHz)	R/Q_0 (Ohm/cell)	Q_L
Monopole Modes			
TM$_{010}$	342	135	108
TM$_{020}$	731	26	40
TM$_{011}$	914	21	38
TM$_{030}$	1107	1	76
TM$_{021}$	1200	4	< 100
Dipole Modes			
TE$_{111}$	720	6	18
TM$_{120}$	898	6	30
TE$_{112}$	1048	1	161

beam pipe. Such a cavity with very eccentric cell shape, is under development at KEK.

One-third-scale (1.5 GHz) niobium cavities have been tested to study the multipacting behavior of the TM$_{110}$ mode [339]. At a surface field of 1 MV/m, multipacting was encountered and processed away in about one hour. It is believed that this multipacting took place in the coaxial beam pipe. A Q_0 value $> 10^9$ was reached at the desired surface field of 15 MV/m. The maximum surface field was 22 MV/m, above which value field emission dominated the Q_0.

In the crab mode of operation, there is no power transferred from the cavity to the beam, provided that the beam goes through the cavity exactly on axis. Only the power to sustain the fields in the crabbing mode will be required. However, there will always be some orbit errors. For a particle passing through off-axis at a phase corresponding to the zero crossing of the deflecting magnetic field, as is necessary for crabbing, the electric field E_z off axis is at the maximum phase. The cavity will behave as an accelerating cavity, albeit with a small accelerating field. For a 1 mm displacement and a transverse voltage maximum of 2 MV, the effective accelerating voltage is 2×10^4 V. A 1 A beam current extracts or gives back about $20 \times b$ kW for b mm displacement. The required coupling of $Q_e = 6.4 \times 10^6/b$ (mm) can be accomplished by a coaxial coupler similar to input couplers presently used for storage ring superconducting cavities.

20.6 INTENSE PROTON ACCELERATORS

High intensity proton linear accelerators are presently under study as spallation neutron sources for a variety of applications: materials research, nuclear transmutation, and energy production [340, 341, 342]. The second application mentioned is the treatment of nuclear waste and the production of nuclear materials, such as tritium. Superconducting cavities offer improved economy and performance for such high intensity proton linacs.

20.7 PULSED NEUTRON SOURCES FOR MATERIALS RESEARCH

Neutron scattering is an important tool for material science, chemistry and life science. Past studies have led to major breakthroughs in the understanding of the structure and dynamics of condensed matter. For example, neutron studies have played a major role in elucidating the structure of high temperature superconductors. With neutrons it is possible to sensitively detect the presence of the light oxygen atoms in the presence of heavy neighboring atoms of Cu and Y. The oxygen atoms play a dominant role in the origin of superconductivity. In polymers, hydrogen atoms can be located precisely with neutrons, but not with x rays. Neutrons are also used to probe structure, dynamics, and properties of magnetic materials. Since the penetration of neutrons is relatively deep, it is possible to study strain distribution in bulk materials, such as welded sections.

To increase the capability for such studies, higher neutron flux is desired since the available intensity limits the signal-to-noise ratio. For many years, nuclear reactors, such as the ones operating at BNL, ORNL and ILL, have been the main source of neutrons, producing a continuous flow of particles. Neutrons emerging from reactors must be slowed down in a hydrogen-rich moderator to energies of 0.1–1000 mV so that they may be useful for the study of condensed matter. The need to provide a monochromatic beam of thermal neutrons means that a large fraction is lost in the moderator. The development of high flux reactors has now reached a level where further enhancement of flux is difficult. The reactor-based ANS (advanced neutron source) proposed a factor-of-5 increase in neutron flux, but it was very costly and technologically challenging. The more serious obstacle, however, is that a reactor-based neutron source is an environmentally unpopular option. It would add to the large inventory of fissile material that has already built up to a dangerous level. Therefore there is a very strong incentive to develop an accelerator-based neutron source [343].

Although lower in flux, accelerator-based pulsed neutron sources have begun to compete with reactors. High energy (1-GeV) protons produce neutrons by hitting a heavy metal target and exciting nucleii to energies where neutrons are "evaporated" in a process referred to as spallation. Typical production rates are 20 neutrons per 1-GeV proton. Accelerator based neutron sources provide high peak intensity and very short (μs) pulses at rep rates of the order of 50 Hz. The advantage of short pulses is it that becomes possible to use time-of-flight measurements to determine the incident neutron energy, eliminating the need for monochromatization and the accompanying waste of neutrons.

The highest intensity accelerator-based neutron sources in operation today are LANSCE at Los Alamos and ISIS at Rutherford Lab, providing $2\text{--}5 \times 10^{13}$ protons per pulse at target. The flux of these pulsed sources is still a factor of 30 below what reactors can provide. Therefore, several laboratories are exploring designs for an advanced pulsed spallation source with 5 MW of beam power at the target. This would correspond to the average flux of the highest flux reactor at Grenoble. In the United States, the laboratories involved are ORNL, LANL,

Argonne, and BNL; in Europe, KFK (Julich), and the University of Wuppertal are conducting studies for the European Spallation Source.

20.8 TRANSMUTATION APPLICATIONS

20.8.1 Reduction of Nuclear Waste

Over the last 50 years, nuclear power production and nuclear materials production have produced a large quantity of radioactive wastes, a stockpile which continues to grow while 17% of the world electricity supply continues to be based on nuclear energy. The safe disposal of this waste is a technical problem. To store the highly concentrated wastes in a geological repository is fraught with dangers because of the 10,000-year life time of some of the radioactive waste products. This danger may be reduced if the long-lived species can be transmuted to isotopes with short life.

In Accelerator Based Transmutation of Waste (ATW), spallation neutrons transmute long-lived actinide isotopes and fission products to stable isotopes, or to isotopes that decay to stable products over 100 years instead of 10,000 years. No additional transuranic waste is produced. This approach can lessen the technical problems of storing long-lived high level radioactive waste. In an optimistic design, a single accelerator can burn the waste from ten 1 GW reactors, while providing enough power to run itself. ATW is also under consideration for converting weapons plutonium, and excess plutonium from dismantling nuclear weapons, to a form that cannot be subverted for use in other nuclear weapons. With beam powers of 100–200 MW, the accelerator approach is ultimately capable of providing a flux of thermal neutrons which is 100 times higher than reactors can provide.

20.8.2 Tritium Production

Tritium production is related to U.S. national security and fusion applications. Because tritium has a half-life is 12.4 years, it needs to be continuously replenished. Tritium is produced by placing lithium in a neutron flux from a reactor, or an accelerator based spallation source. The advantages of accelerator-based production of tritium over reactor-based production are that the accelerator uses no fissile materials, has no chance of a criticality accident, produces no high level radioactive waste, and can be easily scaled up or down depending upon tritium needs.

20.9 ACCELERATOR BASED FISSION REACTORS

Today's nuclear energy is based on fission reactors which use ^{235}U as fuel. Since this crucial isotope is only 1% of ordinary uranium, a costly and complicated isotopic enrichment is required. One of the alternative fuels frequently discussed is ^{232}Th, which is more abundant than uranium and converts to readily fissionable ^{233}U with the absorption of one neutron. Thorium is available in nature as a pure isotope so that no expensive enrichment would be required. Compared to conventional reactors, which produce large quantities of long-lived, highly toxic actinides, the products of thorium decay in a few hundred years to levels well below the toxicity levels of natural uranium ores. This makes the safe depository idea sensible, environmentally acceptable and at a lower long-range cost. In a thorium-based reactor, plutonium is produced in very small quantities, reducing the risk of nuclear proliferation. As mentioned before, a neutron-based power reactor is noncritical because the accelerator that supplies the needed neutrons can be turned off.

To summarize the energy production possibilities, the overall potential of a thorium-based power reactor is to provide environmentally clean energy from abundantly available materials, operate without danger of criticality, and yield minimal radioactive waste without risk of nuclear proliferation. However, a thorium-based thermal reactor cannot operate in a satisfactory way because of insufficient neutron breeding. Neutrons are captured by other materials and by fission fragments. An external supply of neutrons removes the problem, opening the way to efficient utilization of this promising fuel.

20.10 ADVANTAGES OF THE SUPERCONDUCTING APPROACH TO HIGH-INTENSITY PROTON LINACS

The most expensive part of a 1 to 2 GeV proton linac is the high energy section—i.e., the part of the linac above 100 MeV. Since superconducting cavities provide 5–10 MV/m in cw operation vs. the 1 MV/m typical for copper cavities, the superconducting approach can substantially reduce the capital cost. By eliminating the power dissipated in the cavity wall, superconducting cavities also offer significant savings on rf power installation and operating power. For pulsed neutron sources, a superconducting machine offers the possibility of designs with a larger duty factor, although there is an upper limit to the duty factor based on the growth rate of instabilities in the proton accumulation ring that follows the linac. The instability threshold of a ring is also higher for higher proton energy. If the target beam power is reached with higher energy, rather than with higher current, the savings are even greater.

In a high current proton accelerator, the beam loss must be kept down to ensure hands-on maintenance after waiting a reasonable length of time for the radiation to fall to safe levels. Remote handling devices drive up the operating cost and reduce the availability of the accelerator. Therefore the projected use for remote manipulators must be restricted as much as possible. High current proton accelerators demand stringent control of beam halos that activate accelerator components. The higher the current, the larger the risk of residual radioactivity from spilled beam. Consequently there is some advantage to reaching the target beam power with high energy, rather than with high current. One must also bear in mind that the performance of a high intensity linac is dominated by the low energy section. Keeping the current low will also yield benefits here.

To keep the activation levels acceptable, it is anticipated that the allowed beam loss must be as low as 10^{-8}/m. At LAMPF, where the average beam current is 1 mA, beam loss is kept below a workable 1 nA/m. If the average current in a future high intensity machine is increased to 100 mA, beam loss will need to be lowered by another factor of 100 over LAMPF. At present, the understanding and computation of beam halo formation is not advanced enough to make a reliable extrapolation to this extreme level. Therefore it is safest to design the high intensity proton linac by using low peak current and large aperture cavities, so that the beam is minimally disturbed, and the least possible amount is scraped. In this respect, the large opening of superconducting cavities and the higher gradient offer the greatest flexibility in the design and its optimization. Finally, the modularity of the superconducting design makes it possible to adopt a machine configuration that is more fault-tolerant as needed for a routine production-type operation. A single klystron failure, or the trip of a single cavity, or even the voltage loss of an entire cryomodule will not interrupt operation, thereby increasing the overall availability of the accelerator — a key concern of users of high intensity neutron beams.

20.11 PROGRESS IN SUPERCONDUCTING CAVITIES FOR HIGH-CURRENT PROTON ACCELERATORS

In recent years there has been considerable progress in addressing some of the basic issues facing the use of superconducting cavities in high intensity proton linacs. The first concerns the ability of superconducting cavities to withstand the effects of stray beam impingement from the beam halo. The next is the design of multipactor-free structures that can accelerate protons with velocity less than c. Finally, there is the issue of coupling high beam power which is closely related to the steady progress forthcoming in the area of high current electron-positron storage rings for high energy physics.

At LANL [344], 3-GHz cavities have been shown tolerant to proton radiation of a dose of 6×10^{16} protons, beam energy 800-MeV without any observable degradation in Q_0 or any drop in the achievable accelerating field. Q_0 values up to 10^{10} and $E_{pk} = 60$ MV/m were sustained after the intense exposure.

Single and multi-cell cavities with $\beta = 0.4$–0.8 cavities and suitable for high intensity proton accelerators are under development at LANL, JAERI and CERN. Single cell-cavities, such as the one in Figure 20.11 have been shown in laboratory tests to be free of multipactor up to surface magnetic field of 800 Oe. These results are very encouraging for the application of superconducting cavities to proton linacs.

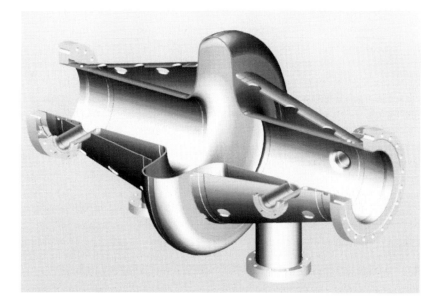

Figure 20.11: Cut-away of $\beta = 0.64$ single-cell cavity developed at LANL for a future high intensity proton linac. In laboratory tests, a niobium cavity of this design reached $E_{\text{acc}} = 10.7$ MV/m with corresponding peak surface fields of $E_{\text{pk}} = 38$ MV/m and $H_{\text{pk}} = 900$ Oe. The cavity is reinforced by conical stiffeners which reduce the risk of collapse under atmospheric load and raise the mechanical resonant frequency. LANL is now exploring multicell designs as well as trying to eliminate the stiffeners. (Courtesy of LANL.)

CHAPTER 21

High-Energy Accelerators

21.1 INTRODUCTION

Over the last two decades the quest to understand the fundamental nature of matter has proceeded on two parallel paths with accelerators that provide proton–proton and electron–positron collisions (see Reference [332] for an overview of high energy physics and accelerator topics). The CM energy of the two generations of accelerators has been growing exponentially; roughly a factor of 7 every decade. With 14 TeV in the CM, the LHC at CERN is planned to be the next push on the energy frontier. It will keep pace with the historical rate of energy growth. The highest energy electron–positron collider is the 100-GeV storage ring LEP presently undergoing an upgrade program to 200 GeV CM by using the largest superconducting rf system ever built. As of this writing, the CM energy has already reached 165 GeV.

As a rough rule, the energy required to do complementary physics with an electron-positron collider is about one-tenth that of a proton machine. Since the proton consists of multiple quarks and gluons, the beam energy is shared among the constituent particles and only a fraction is available as particle-to-particle collision CM energy. The interactions provided by lepton collisions are also cleaner than proton–proton collisions. When the elementary electrons and positrons annihilate, they give up all their energy to the production of new particles. As a result, interesting events stand out more clearly above the competing background. However, the total event rate in a lepton collider at a given luminosity is much lower than for a proton collider; the nonresonant cross sections for interesting physics fall as $1/E^2$. Therefore, to carry out thorough investigations of new phenomena and to extend the discovery potential, electron–positron colliders will need to increase luminosity as the square of the energy and aim for an overall luminosity 10^{34} cm^{-2}s^{-1} at 1 TeV.

To complement the LHC, the CM energy for the next lepton collider will have to be in the range of 1–1.5 TeV, a factor of five to seven over LEP-II. In the past, however, the energy steps for lepton colliders have been smaller, because the needed technologies are only mastered in incremental stages. According to the above arguments, the widespread consensus which has emerged is that the next lepton collider should have an *initial* center-of-mass energy of 500 GeV and luminosity in excess of 10^{33} cm^{-2}s^{-1}. It should eventually be capable of reaching 1–1.5 TeV with luminosity in excess of 10^{34}. The initial collider would have the potential for discovery of the Higgs. It will also provide well-controlled experimental conditions for precise measurements to elucidate why the electromagnetic and weak forces are so different in nature and strength.

At the TeV energy scale it is expected that storage rings will become unaffordable as lepton colliders because energy losses from synchrotron radiation increase with the fourth power of the beam energy

$$U_{\text{sr}} = \frac{4\pi r_e}{3} mc^2 \frac{\gamma^4}{R_{\text{sr}}}. \tag{21.1}$$

Here γ is the beam energy divided by the rest energy, mc^2, r_e is the classical electron radius, and R_{sr} is the storage ring bending radius.

Because of the steep energy dependence, the circumference, the rf power, and the cost of a circular machine will increase with the square of the energy. At a CM energy of 500 GeV, a circular machine would have a circumference greater than 200 km and an operating ac power of many hundreds of megawatts. On the other hand, for a linear e^+e^- collider, the length and rf power scale linearly with energy for a fixed gradient. At some energy above 200 GeV, but well below 500 GeV, a crossover point is expected between the capital cost of a linear and a circular collider. The ac power of a linear collider can also be kept below 200 MW.

21.2 ISSUES IN OPTIMIZING THE DESIGN PARAMETERS OF LINEAR COLLIDERS

Linear colliders were first proposed by M. Tigner [345] who called them "clashing linacs." Even in this early paper he pointed out the advantages of a superconducting version to avoid the high rf power needed to establish the accelerating field. The pioneer linear collider is the SLC at SLAC, which delivers 100 GeV in the CM with a three kilometer, 2.86-GHz, copper linac operating at 17 MV/m [346]. For the next linear collider, the beam energy will need to be increased by at least a factor of five over the SLC, a significant challenge. But an even greater challenge is that the luminosity will need to be increased by a factor of 10^4 over the SLC. The luminosity demand dominates the design of linear colliders. A discussion of the design issues for linear colliders can be found in reference [347]. A review of the various linear collider designs can be found in [347].

To explore the approaches available to reach high luminosity ($\mathcal{L}$), we start with the definition

$$\mathcal{L} = \frac{N^2 f_c}{4\pi \sigma_x \sigma_y}, \tag{21.2}$$

where N is the number of particles per bunch, f_c is the bunch collision frequency, and σ_x and σ_y are, respectively, the horizontal and vertical spot size at collision point. Equation 21.2 shows that there are three strategies to achieve high luminosity : Increase the collision frequency, increase the number of particles per bunch, or squeeze down the beam spot size at the collision point. The first approach increases the ac wall plug power needed to operate the machine. The limit to this avenue is set by the desire to keep the operating power well below 200 MW.

The second approach (increase N) increases the beam power:

$$P_{\text{beam}} = NeE_{\text{beam}}f_c. \qquad (21.3)$$

The limit is set by the efficiency of conversion of wall plug ac power to beam power. More importantly, increasing N increases the unwanted background generated from beam–beam collisions. These cause a number of deleterious effects. When the electron bunch collides with the oncoming positron bunch, the collective fields of one bunch act like a lens to focus the particles of the opposing bunch toward the axis. As the particles are deflected, they emit synchrotron radiation, called "beamstrahlung," which causes an undesired energy spread, as well as production of spurious e^+e^- pairs and hadrons. Some of the unwanted particles have a transverse momentum large enough to hit the detector. Background particles reduce the inherent advantage that electron–positron colliders hold over their proton counterparts. For example, the number of e^+e^- pairs produced by one photon is given by:

$$n_\gamma = \frac{2\alpha r_e N}{\sigma_x}. \qquad (21.4)$$

Here $\alpha = 1/137$ is the fine structure constant. Note that n_γ only serves as a gauge for the overall background. A common technique adopted by all the approaches to reduce background is to use flat, ribbon-like beams, with horizontal size σ_x much larger than the vertical size σ_y. This reduces the magnetic fields within the colliding beams, and hence the beamstrahlung radiation. In choosing the design parameters of a linear collider, the bunch charge must be adjusted so that the noise background of these spurious particles does not drown the signal particles generated by the interesting reactions.

The third approach to high luminosity is to squeeze the collision spot size. This requires starting with very low emittance beams (i.e., very low ϵ_y) and providing very strong final focusing systems (i.e., very low β_y). The final spot size is

$$\sigma_{x,y} = \sqrt{\frac{\epsilon_{x,y}\beta_{x,y}}{\gamma}}. \qquad (21.5)$$

A small spot size also increases the background by intensifying the magnetic field within the bunch. To produce and collide beams with spot size σ_y of 1 to 100 nm will require not only very low emittance sources and high demagnification final focus systems but also very tight alignment of structure and linac, high precision in beam position measurement and orbit control, and tight control of rf amplitude and phase. The choice of N and σ_y must therefore be balanced to reach the desired luminosity while keeping both the background and the ac

Table 21.1: Main parameters for three distinct approaches for the next linear collider (500 GeV CM)

Parameter	TESLA	NLC	CLIC
RF Freq (GHz)	1.3	11.4	30
Gradient (MV/m)	25	37	78
Total Active Length (km)	20	14.2	6.3
$N(10^{10})$	5.15	0.65	0.8
Beam Power (MW)	16.5	4.2	3.9
AC Power to make rf (MW)	164	103	100
Wall Plug Power to Beam Power Efficiency (%)	20	8.2	7.8
Final Spot Size σ_x, σ_y (nm)	1000, 64	320, 3.2	247, 7.4
Emittance $\gamma\epsilon_x, \gamma\epsilon_y$ (m-rad $\times 10^{-8}$)	2000, 100	500, 5	300, 15
Final Focus β_x, β_y (mm)	25, 2	10, 0.1	10, 0.18
Background n_γ	2.7	0.8	1.35

power within acceptable range. These are but a few of the challenges facing the next linear collider.

Today there is a worldwide research activity toward TeV e^+e^- colliders [348]. The large number of designs under development can be classified into three major groups:

1. The short-wavelength, normal conducting linac, pursued as SBLC at DESY, JLC at KEK, NLC at SLAC, and VLEPP in Novosibirsk. The SBLC distinguishes itself from the rest as the long-wavelength normal conducting approach.

2. The long-wavelength superconducting linac, TESLA, pursued by an international collaboration led by DESY.

3. The two-beam accelerator, pursued as CLIC at CERN, and as the RK-TBA (LBL & LLNL).

Table 21.1 lists one program, along with its key parameters, from each of the three distinct approaches. One must keep in mind that the design parameters for all the approaches are still being optimized and there is considerable cross-fertilization of ideas among groups.

All approaches have ongoing programs to build prototype accelerators to help optimize their designs and to determine the component costs. At the present stage of the development, one of the main challenges for all the approaches is to show that costs can be lowered substantially from the present state of the art. Toward the end of the 1990s a choice is likely to be made from among these proposals. The choice will be influenced by the experience gained from building and running the prototypical linacs. It is expected that the overall cost of a TeV collider facility (expected to be in the range of a few billion dollars) will be high enough that an international collaboration will be necessary to fund and build the final design chosen. To improve the likelihood for funding, the aim is to keep the cost *substantially* below ten billion dollars.

21.3 THE SUPERCONDUCTING LINEAR COLLIDER (TESLA)

We will focus our discussion on the approach called TESLA which stands for TeV Energy Superconducting Linear Accelerator[1]. Over the course of the evolution of the design parameters for this accelerator, there have been several international workshops on TESLA [349, 350, 351] during which the strategies and trade-offs have emerged. The resulting "baseline" parameter set [348, 192, 352] is based on 20 km active length superconducting cavities to operate at an initial accelerating gradient of 25 MV/m at a Q_0 of 5×10^9. Since DESY adopted the TESLA Test Facility (TTF) project, the design has continued to evolve under its auspices. For the latest parameter set see reference [353].

The baseline rf parameters for TESLA are given in Table 21.2, and are compared to those of NLC (also from reference [348]). In contrast to cw operation of rf superconductivity discussed for most applications in other chapters, the rf in a superconducting linear collider must be pulsed [354] so as to keep the refrigerator cost affordable. The accelerating gradient of 25 MV/m chosen to make a superconducting linear collider economically viable demands a Q_0 of 5×10^{11} to bring the rf dissipation at liquid helium temperature down to an acceptable level (see Problem 50). This is an inordinately high Q_0 value, higher than the BCS Q_0 at 2 K. Therefore a superconducting linear collider needs to be pulsed. Based on the capital cost of TJNAF or the LEP-II refrigeration systems, an 80 kW refrigerator required for TESLA is expected to cost less than 0.3 billion dollars out of a total expected collider cost of a few billion dollars. Thus we emphasize that the capital cost of the refrigerator is *not* the main component of the cost, a common misapprehension. Even at 1%, the duty factor of TESLA is much larger than all normal conducting versions ($< 0.01\%$). The large duty factor provides many benefits towards the goal of high luminosity, as we will discuss.

In the generic layout (Figure 21.1) we show the major components of a TeV linear collider [355]. Electron and positron sources produce the beam. For TESLA, the large beam power demands a more copious source of positrons than any of the other approaches. Damping rings reduce the starting beam emittances by synchrotron radiation. The emittances needed can be achieved in damping rings with design parameters in use today. Bunch compressors reduce the length of the bunch. The main linacs must accelerate electrons and positrons to the desired energy without significantly degrading the low values of emittances from the damping rings. The final focus system demagnifies the beam size for final collision.

The two linacs for TESLA will house at total of 2500 cyromodules. The

---
[1]The name TESLA was coined by W. Hartung

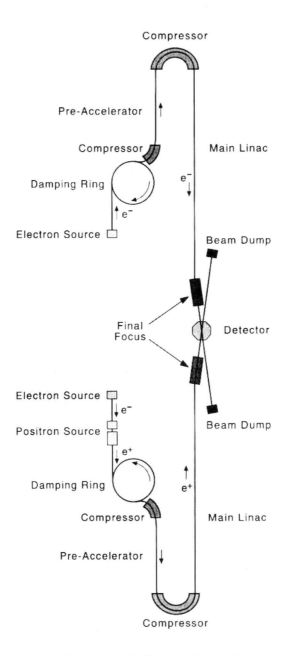

Figure 21.1: Layout for a generic linear collider showing sources, positron damping ring, bunch compressor, electron linac, beam handling, and collision points. (Courtesy of SLAC.)

THE SUPERCONDUCTING LINEAR COLLIDER (TESLA)

Table 21.2: A parameter comparison between TESLA and NLC

	TESLA	NLC
Frequency (GHz)	1.3	11.4
Peak power/m (MW/m)	0.2	70
Klystron power (MW)	10	50
No. of klystrons	600	4000
Klystron pulse length (ms)	1315	1.2
RF pulse length (ms)	1315	0.24
Q_0	5×10^9	7000
Structure length (m)	1	1.8
Iris aperture (cm)	3.5	0.2 – 0.29
No. of cells/m	9	114
Repetition rate (Hz)	10	180
Number of bunches per rf pulse	800	100
Bunch spacing (m)	300	0.42
Duty factor (%)	0.8	0.0025
Peak beam current (mA)	8.3	740
Typical HOM damping Q	10^4	10^3

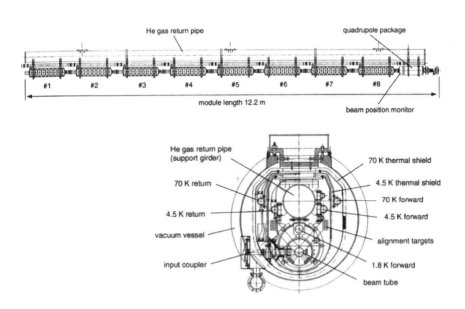

Figure 21.2: Schematic for the TESLA cryomodule showing eight superconducting cavities supported by the cold gas return pipe and a superconducting quadrupole magnet. The cross section of the cryostat is shown at the location of the input coupler. (Courtesy of DESY.)

Figure 21.3: TESLA 9-cell niobium cavity with stiffeners to reduce Lorentz force detuning. (Courtesy of DESY.)

cryomodule, shown in Figure 21.2, will carry eight 9-cell cavity units supported by the cold gas return pipe in the cryostat [356]. A superconducting quadrupole focusing magnet is housed inside the same cryostat. Figure 21.3 shows the TESLA 9-cell niobium cavity [277]. The cavities are stiffened near the iris to counteract the Lorentz force detuning effect at 25 MV/m.

21.4 ATTRACTIVE FEATURES OF TESLA

One key difference between TESLA and the other approaches is its route to high luminosity. TESLA proposes to reach high luminosity with a large charge per bunch, N, but with a relatively large final spot size of 64 nm. This spot size has already been achieved by the Final Focus Test Beam, a collaborative effort led by SLAC. By contrast, the NLC intends to reach a 3 nm final spot. To produce, measure, and keep in collision such incredibly small spot sizes is a major technological challenge for the next linear collider. To set the scale, one has to keep in mind that the SLC vertical spot size is 1 μm.

In TESLA, the large number of particles per bunch means that the beam power is higher than for the competing avenues. The basic reason that TESLA can afford the high beam power stems from the high Q_0 of the superconducting cavities. With the very low losses, the rf pulse length can be made long enough to accelerate a 1000 bunches, rather than 10–100 bunches per rf pulse, typical of the normal conducting designs. The large number of bunches makes it possible to efficiently convert rf energy stored in the structure to beam energy. The stored energy is used more effectively when many bunches are accelerated in a single rf pulse. The typical overall efficiency for wall plug ac power to beam power conversion for TESLA is 25%, as compared to 3–10% of the various normal conducting designs.

The second attractive feature of the superconducting approach is that the peak rf power/m is quite low as compared to all other options. Since the Q_0 of a superconducting cavity is very high (10^9), it is possible to fill the cavity slowly (ms), which means that modest peak powers (200 kW/m) can be used.

Note that the peak power per unit length is

$$P_g^{\text{peak}} = \frac{u}{\tau_L}. \qquad (21.6)$$

Here u is the stored energy/length and τ_L is the filling time. The peak rf power is essentially the same as the peak beam power; little additional power is required to establish cavity fields.

Existing klystron technology is adequate to fulfill the peak rf power needs for TESLA. Nevertheless, efforts are underway to develop higher efficiency klystrons and modulators to further reduce the overall operating power.

Copper cavities, on the other hand, must be filled very fast (< 100 ns) to avoid wasting energy. To drive copper cavities to gradients of 50 MV/m, the peak power requirements are of the order of 100 MW/m. A suitable technology is under development to meet the demand of high power, high efficiency klystrons at 11.4 GHz. In this development, the usual klystron focusing solenoids are replaced with periodic permanent magnet focusing. To ease the burden on klystron peak power, pulse compression systems with compression ratios of 3–5 are envisioned. The klystron, its HV power supply and modulator are major items of equipment that need to operate with high reliability. Given the large number of klystrons required and the typical failure rate of klystrons, it is expected that about one klystron will have to be replaced per day. This is equivalent to increasing the operating cost by several tens of MW ac power. In the proposed TESLA rf distribution layout, one 10 MW, 1.3-GHz klystron feeds 32 cavities, so that the total number of klystrons required is 600. This is drastically lower than the 2000–4000 klystrons required for the competing normal conducting approaches. It is expected that the substantially smaller number of rf power units of the TESLA approach will be helpful toward commissioning and operation.

With respect to the need for high luminosity, a crucial feature that puts TESLA at an advantage over the other approaches is its low rf frequency (1.3 GHz). The resultant large aperture ($0.27\times$ wavelength) yields the pleasant consequence of substantially reduced short- and long-range wakefields, fighting the main enemies of high luminosity. When the cavity walls are close to the beam, the wakefields increase, making tolerances tighter and emittance dilution effects stronger. Once again, it is superconductivity that permits the use of a long wavelength. For the same length of accelerating structure, and the same accelerating voltage, the rf energy stored in a structure increases as the square of the wavelength, but the large amount of structure stored energy becomes affordable with superconducting cavities because the amount of energy dissipated is drastically lower.

Consider the adverse effect of short range wakefields on luminosity. Transverse wakefields cause emittance growth along the linac, spot size degradation at the final focus, and corresponding loss of luminosity. Since transverse wakes scale with aperture (a) as $1/a^3$ and since the aperture of a 1.3-GHz superconducting cavity is ten times larger than for the high frequency designs, this source of luminosity degradation is negligible. Longitudinal wakefields produce energy

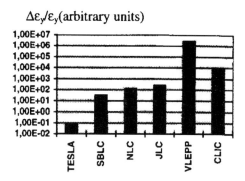

Figure 21.4: Relative vertical emittance growth from transverse wakes if cavities are misaligned by the same degree. The comparison shown for the different approaches is without application of any emittance damping schemes, such as BNS damping. (Courtesy of DESY.)

spread within a single bunch and also produce bunch-to-bunch energy variations in a train of bunches. Such spreads cause acceptance problems in the final focus system which typically tolerate spreads of only a few parts per thousand. Again, since the longitudinal wake scales with aperture as $1/a^2$, this problem is virtually absent in TESLA.

Consider next the effect of wakefields on alignment and vibration tolerances necessary to build and operate the linacs. For the NLC, the requirement to steer the beam to the middle of the position monitors imposes extremely tight (5 μm) tolerances on the alignment of the cells, the structures in the accelerator, quadrupoles, and beam position monitors. For TESLA, because of the low frequency and concomitant lower wakefields, the tolerances can be relaxed on the mechanical properties of the cavities, accelerator alignment as well as vibration, and beam alignment, resulting in cost savings.

The vertical emittance growth, $\Delta\epsilon_y/\epsilon_y$, from random structure misalignments increases with the number of particles per bunch, rf frequency, and bunch length, but decreases with the gradient. However, the dependence on rf frequency and aperture is so strong, that it dominates. Figure 21.4 compares the emittance growth among several designs [355] without the application of emittance damping schemes such as BNS damping [357]. Even though TESLA has the highest bunch charge and longest bunch length, its low rf frequency makes it three orders of magnitude less sensitive to random structure misalignments. To maintain the tight tolerances, high frequency designs demand strong focusing along the linac and very tight tolerances. With the TESLA baseline parameter set (Table 21.1), cavity alignment errors of 1 mm produce an emittance growth of only 4%. Typical expected random uncorrelated 0.1 mm quad motion yields an emittance growth of 6% [271]. These numbers are substantially superior to emittance growths near 100% projected for the NLC after allowing for tighter alignment tolerances and after application of emittance damping techniques [358].

The issue of long-range wakefields from cavity HOMs is greatly ameliorated in TESLA by virtue of the long rf pulse length. Even with a thousand bunches per rf pulse, the time between bunches is very long (microseconds) compared to nanoseconds for normal conducting versions. For such large bunch separation, HOMs can be fully damped between bunch passages. Techniques to damp HOMs in superconducting cavities to the required Q_L values of 10^4 are already well developed, as discussed in Chapter 16.

The long time between bunch passages for TESLA also offers the possibility to measure individual bunch position variations, and to make corrections to subsequent bunches. Such corrections could be made at the end of the linac, or a few times along the linac. This scheme could eliminate any problems arising from pulse-to-pulse orbit variation. The TESLA approach is the only one that allows this form of correction.

Compared to the short-wavelength, normal conducting routes, there are many additional attractive features of the long-wavelength, superconducting approach to TeV energy and high luminosities. As mentioned before, the relaxed spot size of TESLA eases the burden on the source and final focus systems. The relatively large source emittance opens the option to generate the required beam quality for the electron linac directly from a laser-driven rf gun and thereby eliminate the cost of the electron damping ring. (A damping ring for the positron beam would still be required.) Because of the larger emittance, more dilution can be tolerated in the linac, in the optics after the linac, and in the final focus. The focusing strength, optical quality, and alignment needed for the final focus are also not as stringent.

The TESLA configuration also offers advantages for the detectors. With the considerable time (1 μs) between bunch interactions, there can be more longitudinal space after the last focusing element. The free space available on either side of the interaction point is 3 meters, a valuable feature for detector designers. Individual bunch crossings are easier to resolve. The large bunch separation allows more time between collisions for detectors to resolve particle tracks and to sort out the desired physics from the background noise.

Another potential source of background is from the dark current in the linac. At high accelerating gradient, stray electrons from field emission in the accelerating structure can be captured by the accelerating field. The unwanted electrons move along the linac toward the collision point. To some extent this effect is ameliorated by the intervening focusing quadrupole magnets. The dark current electrons are lower in energy than the main linac beam, and thus they tend to be over-deflected by the fields in the focusing magnets along the linac; they get swept out of the beam aperture. But a superconducting cavity is prepared clean and must have low dark current if a Q_0 of several 10^9 is desired. At the lower gradients envisioned for TESLA, the field emission currents will also be lower due to the exponential dependence of emission current on field.

Table 21.3: Input parameters and resulting performance expectations for a 0.5 TeV CM superconducting linear collider design exercise that *combines* high beam power with small spot size to maximize luminosity

Parameter	Unit	TESLA	Combined[a]	
Chosen input				
Q_0		5×10^9	5×10^9	
E_{acc}	MV/m	25	25	
N	10^{10}	5.15	5	
Emittance ϵ_x	m-rad 10^{-8}	2000	500	
Emittance ϵ_y	m-rad 10^{-8}	100	5	
Final Focus β_x	mm	25	140	
Final Focus β_y	mm	2	1	
Bunch Length σ_z	mm	1	0.2	
Bunch Spacing	ms	1	1	
n_b		800	800/3200	
Rep rate	Hz	10	10/2.5	
Performance		TESLA	Combined	Range[b]
Luminosity	10^{33} cgs	6.1	25.6	4.3 – 7.1
Beam Power	MW	16.5	16	1.3 – 16.5
σ_y	nm	64	12	3.0 – 64
R		15	100	15 – 100
D_y		8.7	8	7.3 – 25
Y		0.03	0.095	0.03 – 0.22
δ	%	3.3	5.5	2.4 – 12.7
n_γ		2.7	1.7	0.8 – 2.7
Peak Power	MW/m	0.2	200	12 – 200
AC Power	MW	164	152/119	100 – 164
Efficiency	%	20	21/27	3.0 – 20

[a] High beam power and small spot size.
[b] For most linear colliders.

By using HPP, dark current from accidental contamination can also be reduced, as we discussed in Chapter 13. A normal conducting linear collider designed with gradients of 50 MV/m, and later upgraded to 100 MV/m, will have to deal with the exponential growth of dark current.

21.5 DESIGN FLEXIBILITY AND ENERGY UPGRADES

The designs associated with the various approaches to linear colliders are still in the process of refinement. The superconducting option offers important design flexibilities. Starting from the baseline TESLA parameter set, reducing the source emittance and reducing the final spot size opens benefits in several areas: ac power reduction, luminosity increase, or lower collision-induced energy spread for better resolution of interesting events. Because of the low wakefields, TESLA is ideally suited to deliver a small vertical emittance required for nanometer size beams, leading to attractive scenarios for future luminosity upgrades. By incorporating the many advances in producing low emittance beams, and high demagnification final focus systems, future luminosity upgrades will be possible by squeezing the final spot size down from 64 nm toward 10 nm or lower.

For example, the TESLA approach considers reduction of spot size to 20 nm with ac wall power reduction to 88 MW [353]. Another example *combines* the high beam power of TESLA with the small emittance of the NLC design. The superconducting linac is the only approach that allows one to use both high beam power *and* small spot size because it offers both high efficiency (ac power to beam power) and low wakefields. With this strategy, it is possible to design a 0.5 TeV collider with a luminosity of 2.6×10^{34} at an ac operating power of 120–150 MW, depending on the repetition rate chosen [359]. Table 21.3 gives a short parameter list. The final vertical spot size turns out to be 12 nm, still larger than NLC. The luminosity is a factor of four higher than the baseline design for all machines. The starting emittance has to be a factor of 20 smaller than TESLA baseline parameters and the bunch length a factor of 5 smaller. As a result, the transverse emittance growth will be a factor of 4 larger if no other measures are taken. If needed, the growth may be corrected with BNS damping.

It is desired that the next linear collider be upgradable in energy to 1–1.5 TeV. The options for the superconducting approach are to exercise a combination of length and gradient increases. The gradient of 25 MV/m is a factor of two below the theoretical limit set by the rf critical magnetic field. The prospects of reaching 40 MV/m are improving steadily. As proof of principle, the maximum surface magnetic field of 1880 Oe has been reached in a single-cell niobium cavity at 1.3 GHz [141]. This corresponds to $E_{acc} = 45$ MV/m. The maximum surface electric field of 100 MV/m has been reached [360] in a two-cell niobium cavity at 3 GHz, corresponding to $E_{acc} = 50$ MV/m. When techniques to reliably upgrade cavities become available, they can be applied to the existing niobium structures. For example, niobium purification research is already showing ways to increase the RRR to several thousand, which may turn out to the proper avenue to avoid quench and to allow field emission processing. Another option is to adiabatically convert niobium cavities into Nb_3Sn cavities by vapor diffusion of tin. As discussed in Chapter 14, the theoretical maximum rf surface magnetic field for Nb_3Sn is 4000 Oe, corresponding to $E_{acc} = 95$ MV/m. The theoretical prediction is not yet verified, however. The technology of producing good Nb_3Sn films still needs to be advanced.

The competitive normal conducting approaches to TeV energy also propose strategies based on combining length and gradient increases. It is not economical to simply fix the length and double the gradient because the needed rf power increases as the square of the gradient. Since the detail cost coefficients of the

Table 21.4: Combining high beam power and small spot size to maximize luminosity at 1 TeV CM energy

Parameter	Unit	TESLA	Combined	
Input				
E_{acc}	MV/m	25		
Q_0		5×10^9	1×10^{10}	
N	10^{10}	1.8	5	
Emittance ϵ_x	m-rad 10^{-8}	1400	500	
Emittance ϵ_y	m-rad 10^{-8}	6	5	
Final Focus β_x	mm	25	250	
Final Focus β_y	mm	0.7	1	
Bunch Length σ_z	mm	0.5	0.2	
Bunch Spacing	ns	354	1000	
n_b		2260	3200	
Rep rate	Hz	5	2.5	
Resulting Performance		TESLA	Combined	Range (For Most Linear Colliders)
Luminosity	10^{33} cgs	13	42	6.3 – 14.5
Beam Power	MW	16.5	32	4.2 – 16.5
σ_y	nm	6.5	7.1	2.3 – 6.5
R		92	158	55 – 179
D_y		14	7	7.6 – 16.2
Y		0.053	0.2	0.053 – 0.33
δ	%	2.5	9.4	2.5 – 9.6
n_γ		1.2	1.8	1.1 – 1.52
Peak rf Power	MW/m	0.2	200	$\approx$ 200
Total AC Power	MW	184	208	152 – 284
Efficiency	%	17.9	30.8	4.1 – 20

linear (structure) and quadratic (rf power) terms are unknown at this time, the exact trade-offs and the optimum gradient are as yet uncertain for either normal conducting or superconducting approaches.

It is important to remember that future energy upgrades demand that the luminosity must be increased simultaneously as the square of the energy. With the superconducting approach there is margin to achieve higher luminosity by squeezing the starting spot size. The spot size of 64 nm is relatively large for the baseline 0.5 TeV TESLA design. By combining the high beam power with the small spot size, a parameter list [359] has been worked out at 1 TeV CM for a luminosity of 4.2×10^{34} with a final spot size of 6.5 nm. This luminosity is a factor of three higher than competing 1 TeV designs. Table 21.4 gives a brief parameter list.

To conclude this section, we briefly summarize the challenges for the next linear collider. With their high frequency and high gradient, the normal conducting approaches *simultaneously* push many parameters into new and difficult regions: peak rf power, final spot size, beam spot aspect ratio, starting emittance, emittance preservation through the linac, bunch spacing, bunch length, peak current, final focus apertures, alignment tolerances, vibration tolerances, beam position monitor precision, pulse-to-pulse jitter, cavity damping, cavity gradient, and dark current at high gradient.

The long-wavelength, superconducting approach relaxes all of these demands. The challenges for TESLA are to reliably achieve high gradients at high Q_0, produce high bunch charge, a large number of bunches and intense positron beams. It is clear that there is a fundamental limit of $E_{acc} = 60$ MV/m to niobium cavities which pushes the superconducting approach to a long machine. Producing high gradients and high Q_0 in superconducting cavities demands excellent control of material properties and surface cleanliness. The spread in gradients that arises from the random occurence of defects and emitters must be reduced. It will be important to improve installation procedures so as to preserve the excellent gradients now obtained in laboratory tests in vertical cryostats. Refrigerators and cryogenic distribution systems will add complexity to the facility. However the operating experience at TJNAF, LEP-II, HERA, RHIC and the LHC should go a long way to providing an excellent experience base.

There have been many advances in superconducting cavity technology toward meeting the gradient challenge, as we discussed in Chapter 13. An international collaborative effort is launched to build a TESLA Test Facility (TTF) at DESY [361]. Its aims are to establish a technological base needed to build TESLA, to demonstrate progress toward the needed gradient and cost goals, and to build and beam test a 500 MeV linac with high gradient superconducting cavities.

21.6 THE TWO-BEAM ACCELERATOR WITH SUPERCONDUCTING LINAC

In the CLIC two-beam approach to linear colliders purused by CERN (Figure 21.5), there are two parallel linacs: a high gradient, normal conducting linac, and a parallel, high-current drive beam that generates rf power to supply the main linac [362]. The high energy beam is accelerated with 30-GHz room temperature structures that will operate at 80 MV/m. Instead of separate high frequency, high peak power sources which would need to feed ≈ 150 MW/m to the 30-GHz structures, the rf power is supplied by transfer structures which extract energy from a 3-GeV high intensity electron beam running parallel to the main linac in the same tunnel. The 3-GeV drive beam is generated in a 350-MHz superconducting linac complex which could use LEP-type supercon-

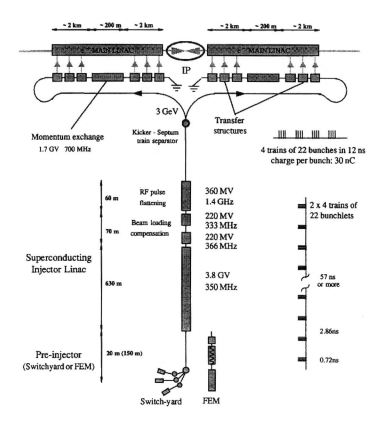

Figure 21.5: Accelerator layout for CLIC showing superconducting drive beam generation. (Courtesy of CERN.)

ducting cavities. No re-acceleration of the drive beam is necessary, but there is a momentum rotation section using 700-MHz superconducting cavities to maximize the energy extraction efficiency. In addition, 360 MV of 1400-MHz superconducting cavities are envisioned for pulse flattening, and 220 MV sections operating at 333 MHz and 366 MHz are to be used to compensate the effects of beam loading. All these sections are laid out in Figure 21.5.

21.7 MUON COLLIDERS

Looking beyond the generation of high energy accelerators now under development (i.e., the 14-TeV LHC and the 1-TeV linear collider) we are faced with the realization that there may be both physical and economic constraints to proton–proton colliders beyond 20 TeV, as well as to electron–positron colliders well beyond 1 TeV. Apart from the inherent performance drawbacks which

MUON COLLIDERS

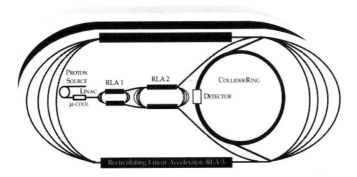

Figure 21.6: The muon collider layout showing cascade of recirculating superconducting linacs to raise the beam energy to 2 TeV. (Courtesy of LBL.)

stem from their complicated many-particle collisions, hadron colliders may already be reaching their size and cost limits with the LHC. Similarly, extension of e^+e^- linear colliders to multi-TeV energy may also be constrained by size and cost, because they are based on full energy linacs. Perhaps new and efficient technologies based on laser-driven plasma waves will arrive someday with GeV/m acceleration of a respectable number of particles/bunch [363]. Since the luminosity demand increases with the square of the energy, collision spot sizes in e^+e^- colliders will have to approach the enormously difficult regime of angstrom-size spots. But eventually the unwanted background from beamstrahlung effects, which increase as $(E/m_e)^4$, are likely to limit the performance electron–positron colliders.

For these reasons, the interest in muon colliders has been increasing over the last few years [364]. Muons, which have 200 times the mass of electrons, can be accelerated without synchrotron radiation in recirculating devices, stored in rings with negligible beamstrahlung, and collided in sufficient numbers to allow using micron-size spots. However, muons have a short lifetime. At 2 TeV, the lifetime of 4.4 ms is already long enough for respectable luminosity and is longer at higher energy. The background issue from decaying muons is a serious uncertainty and is under study. However, it does not increase as strongly with energy as electron–positron beamstrahlung.

A muon collider concept has been put forward for 4 TeV CM at a luminosity of 3×10^{34} cm^{-2}s^{-1}. There are many aspects of the system in need of substantial exploration, such as schemes for cooling the muons. After cooling and preacceleration, the beam will be accelerated to full energy using a cascade of superconducting recirculators (Figure 21.6) that would accelerate the beam in stages, as shown in Table 21.5. The rf frequency would increase as the bunch length decreases. An early stage could be based on 350-MHz LEP-type superconducting cavities operating at 10 MV/m. The final stage would be based

Table 21.5: Cascaded recirculating linacs for the muon collider [365]

	RLA1	RLA2	RLA3	RLA4
RF Frequency (MHz)	100	350	800	1300
Gradient (MV/m)	5	10	15	20
Linac Length (m)	100	300	533	2800
N turns	9	11	12	16
Energy out GeV	9.6	70	250	2000

on 2.8 km of 1.3-GHz TESLA-type superconducting cavities operating in the pulsed mode with 1% duty factor and a gradient of 20 MV/m [365].

21.8 CONCLUDING REMARKS ON FUTURE PROSPECTS

The last two chapters have discussed a variety of prospects for the applications of superconducting cavities to future accelerators. In the "low gradient" class, there are the high current, high luminosity colliding beam storage rings for high energy physics. The high intensity proton linacs would generate neutrons for a variety of purposes. The challenges for superconducting cavities to serve this class of accelerator applications are to provide high power for the beam and to demonstrate good performance in the new structure geometries. These have to be tailored to meet the new demands, for example "single mode" cavities in which all higher order modes are strongly damped, cavities suitable for "crabbing" a beam, short-gap structures for accelerating protons with $\beta = v/c > 0.3$. In the "high gradient" class there are the linear collider and the muon collider and possibly a high energy linac driver for a high intensity x-ray free electron laser. These applications demand substantial progress in the state of the art for gradient and Q.

We are confident that the rf superconductivity community has the both the creativity and the determination to face these challenges and successfully realize these exciting prospects.

Problems

1. Sketch the electric field lines of the TM_{110} and TM_{210} modes in the transverse cross section of a pill-box cavity. Why is the TM_{110} mode called a deflecting or dipole mode? Why is the TM_{210} mode called a focusing or quadrupole mode?

2. Consider using a TM_{012} pill-box mode for acceleration.

    (a) Determine the condition placed on the ratio d/R (d = length of the cavity, R = radius of the cavity) if a bunch is to always see an accelerating field as it passes through the cavity. Is it physically possible to build such a cavity?

    (b) What should the phase between a bunch and the cavity fields be if the bunch is to always see an accelerating field during transit?

    (c) From now on use $d/R = 0.5$. What is the length of the cavity if the mode frequency is 1.5 GHz?

    (d) Calculate the accelerating voltage.

    (e) Derive an expression for the dissipated power.

    (f) Derive an expression for the stored energy.

    (g) What is the cavity quality if the surface resistance is 20 nΩ?

    (h) What is the shunt impedance?

    (i) What is the R_a/Q_0? Show that this is independent of frequency.

3. Calculate the stored energy in a 4-cell LEP cavity with $R_a/Q_0 = 464 \ \Omega$ at 5 MV/m.

4. Using the London equation, along with Maxwell's equations for static fields, derive that

$$\nabla^2 \mathbf{H} = \frac{1}{\lambda_L^2} \mathbf{H} \tag{P.7}$$

and

$$\nabla^2 \mathbf{j}_s = \frac{1}{\lambda_L^2} \mathbf{j}_s, \tag{P.8}$$

where

$$\lambda_L^2 = \frac{m}{n_s e^2 \mu_0} \tag{P.9}$$

is called the London penetration depth.

5. In the static case, show that London equations lead to $\mathbf{E} = 0$ inside the superconductor.

6. Determine the maximum value of the improvement factor that can be expected for the Q_0 of a 500-MHz copper cavity if it is cooled down from room temperature to liquid helium temperature. The $\rho\ell$ product of copper is 6.8×10^{-16} Ωm^2. The resistivity of copper at room temperature is 1.76×10^{-8} Ωm and the RRR of good-quality copper is 100.

7. Calculate the surface resistance of niobium of RRR = 30 and 300 at 500 MHz in the normal conducting state at 10 K. What is the typical Q_0 of a niobium cavity at room temperature? What is the improvement factor for a niobium cavity on cooling from room temperature to 10 K? The resistivity of niobium at room temperature is 15×10^{-8} Ωm, and the $\rho\ell$ product of niobium is 6×10^{-16} Ωm^2.

8. Assuming a value $n_s = 5.6 \times 10^{28}$ m^{-3} for the density of superelectrons, make an estimate of the following.

    (a) the coefficient of the normal state electronic specific heat γ.
    (b) Fermi velocity of niobium
    (c) the thermal conductivity of reactor grade niobium at 10 K.
    (d) the thermal conductivity of RRR = 300 grade niobium at 10 K.
    (e) the coherence distance ξ_0, the London penetration depth λ_L, and the Ginsburg–Landau parameter κ for niobium.
    (f) the thermodynamic critical field H_c of niobium.
    (g) the upper and lower critical fields H_{c1} and H_{c2}.
    (h) the rf frequency corresponding to the gap energy.

9. Using the London equations, the total current $\mathbf{j} = \mathbf{j}_s + \mathbf{j}_n$ and Maxwell's equations, show that

    (a)
    $$\nabla \times \mathbf{j} = -\frac{1}{\lambda_L^2}\mathbf{H} - \sigma\mu_0 \frac{\partial \mathbf{H}}{\partial t} \quad \text{(P.10)}$$

    and
    $$\frac{\partial \mathbf{j}}{\partial t} = \frac{1}{m_0 \lambda_L^2}\mathbf{E} + \sigma \frac{\partial \mathbf{E}}{\partial t}. \quad \text{(P.11)}$$

    (b) If E and H vary with time as
    $$E = E_0 e^{i\omega t} \quad \text{and} \quad H = H_0 e^{i\omega t}, \quad \text{(P.12)}$$
    show that
    $$\nabla^2 \mathbf{H} = k^2 \mathbf{H} \quad \text{and} \quad \nabla^2 \mathbf{E} = k^2 \mathbf{E}, \quad \text{(P.13)}$$

PROBLEMS

where
$$k^2 = \frac{1}{\lambda_L^2}(1 + i\sigma\mu_0\omega\lambda_L^2 - \epsilon_0\mu_0\omega^2\lambda_L^2). \tag{P.14}$$

Estimate the order of magnitude for the second and third term of k^2.

10. In the London two-fluid model, $\mathbf{j} = \mathbf{j}_s + \mathbf{j}_n$, where $\mathbf{j}_n = \sigma_1 \mathbf{E}$ and $\mathbf{j}_s = -i\sigma_2 \mathbf{E}$ for sinusoidally time-varying fields. In analogy with normal conductors, show that
$$\nabla^2 \mathbf{E} = \tau^2 \mathbf{E}, \tag{P.15}$$

where
$$\tau = \sqrt{\mu_0 \omega i(\sigma_1 - i\sigma_2)}. \tag{P.16}$$

Then show that the surface impedance for a superconductor is given by
$$Z_s = \sqrt{\frac{i\omega\mu_0}{\sigma_1 - i\sigma_2}}. \tag{P.17}$$

11. With the approximations
$$\sigma_1 = \frac{n_u e^2 \tau}{m} \ll \sigma_2 = \frac{n_s e^2}{m} \frac{\ell}{\omega}, \tag{P.18}$$

where n_u stands for the number of unpaired electrons, show that if
$$Z_s = R_s + iX_s \tag{P.19}$$

then
$$R_s = \frac{1}{2}\sigma_1 \omega^2 \mu_0^2 \lambda_L^3 \tag{P.20}$$

and
$$X_s = \omega\mu_0 \lambda_L. \tag{P.21}$$

12. Show that the expression for the London penetration depth remains unchanged if single charge carriers are replaced by Cooper pairs.

13. Show that at 1 GHz the reactance of a superconductor is large compared to the rf surface resistance.

14. You have constructed a 3-cell cavity and want it to have a flat field profile in its π mode. The resonant frequencies of the 3 modes in the TM_{010} passband of the cavity are measured to be:
$$f^{\pi/3} = 1\,009.698 \text{ MHz}$$
$$f^{2\pi/3} = 1\,029.802 \text{ MHz}$$
$$f^{\pi} = 1\,039.239 \text{ MHz}$$

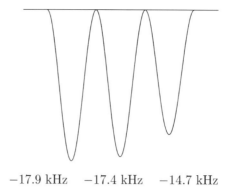

-17.9 kHz -17.4 kHz -14.7 kHz

Figure P.1: Frequency versus perturbing bead position for a 3-cell cavity.

A bead is pulled through the cavity while the π mode is excited. Figure P.1 shows a plot of the frequency versus bead position.

By how much should each cell be stretched or squeezed to make the fields flat at $f^\pi = 1.039.230$ MHz? Give your answer in terms of the frequency changes of the π mode while each cell is altered in the tuning stage. To remove some of the tedius math, take as given that $H^{(3)}_{21} = 3.17736$ and $H^{(3)}_{22} = -6.35472$.

15. (a) Derive an expression for f_0, the frequency of an unperturbed cell in isolation, in terms of the measured frequencies of two of the modes.

    (b) What is f_0 for this cavity?

16. When a multicell cavity is out of tune, how does this affect its Q_0 relative to a perfectly tuned cavity with the same stored energy?

    (a) With only Ohmic losses?
    (b) With field emission or other anomalous losses?

17. In the development of Chapter 7 we verified that our eigenmode solution $\mathbf{v}^{(m)}$ satisfied the equation for the middle cells of the cavity in the process of finding the eigenvalue. To be sure that there is nothing fishy about the way that the end cells were detuned to make a flat π mode, verify that $\mathbf{v}^{(m)}$ satisfies the end-cell equation too.

18. In Chapter 7, the resistances in the circuit model for the multicell cavity were taken to be zero because the cavity is low loss. Put in resistance and compute its effect on a single cell cavity.

19. In the detailed analysis of the cavity behavior in Chapter 8 we added a transmitted power probe to the cavity but it was very weakly coupled. What if it were not weakly coupled? This problem involves working out the effect of having $\beta_t \approx 1$.

PROBLEMS 481

(a) Rederive $d\sqrt{U}/dt$ to include $\beta_t = Q_0/Q_t$.

Hint: Define
$$\frac{1}{Q_L} \equiv \frac{1}{Q_0} + \frac{1}{Q_e} + \frac{1}{Q_t} \qquad \text{(P.22)}$$

(b) In analogy with Chapter 8, for a constant Q_0, give expressions for $U(t)$ and $P_r(t)$ for the cases of turning on and turning off the rf.

(c) What input coupling, β, will yield the maximum stored energy in steady state?

(d) What can you say about the case of having N such additional non-negligible transmitted power couplers?

20. In Chapter 8 the cavity was modeled as a mismatch in admittance to the transmission line driving the sytem. It can be shown from (8.42) that for a high Q_0 cavity excited $\Delta\omega$ away from its resonance when $\Delta\omega \ll \omega$, the general expression for the voltage reflection coefficient in our transmission line and cavity system is
$$\Gamma = \frac{1 - \beta + 2i\tau_0\Delta\omega}{1 + \beta + 2i\tau_0\Delta\omega}. \qquad \text{(P.23)}$$

(a) How would you measure the cavity Q_0 for a normal conducting cavity while measuring only reflected power.

(b) The locking circuit presented in Chapter 8 relied upon the transmitted power signal for feedback. Determine a method of frequency locking for a cavity that has only an input coupler.

(c) Outline the approach one would use to measure the Q_0 versus E of this cavity.

(d) What is the disadvantage of this approach?

21. High peak power processing (HPP) refers to driving the cavity with short pulses of high peak power rf in order to process away field emission sites. Pulses are ~ 100 μs and ~ 1 MW peak. The cavity fields are raised to very high levels during this short time and field emitters are forced to burn themselves out.

(a) Considering that the Q_0 of superconducting cavities is about 10^{10}, what can you say about the coupling that is required to make HPP effective?

(b) Given that the rectangular input pulse is of length T_p, what coupling is required to maximize the pulse efficiency? Pulse efficiency is defined as
$$\eta = \frac{\text{peak energy in cavity}}{\text{total energy in pulse}}. \qquad \text{(P.24)}$$

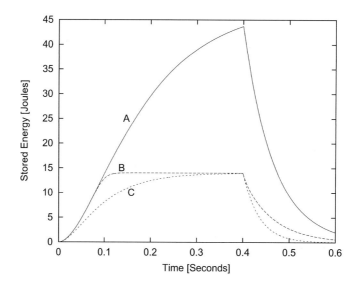

Figure P.2: Stored energy as a function of time for a cavity in three different scenarios. $Q_e = 5.74 \times 10^8$ for all three cases. (A) Overcoupled cavity with constant $Q_0 = 7 \times 10^9$. (B) Field emission loaded cavity that changes its Q_0 from 7×10^9 at low field to 5.74×10^8 at high field. (C) Unity-coupled cavity with constant $Q_0 = 5.74 \times 10^8$.

Assume (unrealistically) that Q_0 does not drop during the pulse. What is the peak efficiency? What is the coupling required for a 1.3-GHz cavity with $Q_0 = 7 \times 10^9$ and $T_p = 100\,\mu s$? What is Q_e in that case?

(c) Calculate the time it takes the reflected power to go to zero during HPP for a given Q_e.

(d) What is the stored energy at this time (relative to the steady-state stored energy, U_0)? Why is this reasonable?

(e) Draw a qualitative picture of the reflected power marking the appropriate times.

22. (a) Sketch a graph that shows the reflected power traces for the three curves in Figure P.2. Note that $P_f(t)$ is the same rectangular pulse for all three cases.

(b) Convince yourself that the presence of field emission still allows one to measure the steady-state β in both of the two ways presented in Chapter 8.

23. In this problem you will explore another method of measuring Q_0.

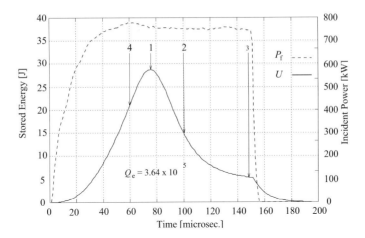

Figure P.3: Cavity response to a high peak power pulse.

(a) Assume that you can make measurements of $U(t)$, $P_f(t)$, and Q_e and that your measurements are good enough to let you compute $d\sqrt{U(t)}/dt$. Use the differential equation for $d\sqrt{U}/dt$ to construct an expression for $Q_0(t)$.

(b) Figure P.3 shows the stored energy of a cavity as a function of time in response to a high-power pulse of rf. These measurements from a single-cell 1.3-GHz cavity were collected using a digital oscilloscope. During the pulse a defect in the cavity caused a thermal breakdown. Calculate Q_0 of the cavity at points 1, 2, and 3 on the graph in Figure P.3. Explain why it is very difficult to get a good value for point 4 and earlier on the curve.

(c) If the normal conducting Q_0 of this cavity is 8.5×10^4, what fraction of the cavity surface is normal conducting at the peak of $U(t)$. Assume that the surface magnetic field in the cavity is uniform.

(d) If Q_e had not been measured previously, some of the data in the plot could be used to determine it. How would that be done?

(e) If all of the measured data were known to very high precision, one could obtain some idea of the field emission behavior characterized by $Q_0(E)$. How?

24. Calculate the value of H in Oersted and $E_{\rm acc}$ in MV/m at which first-order multipacting may take place for a pill-box cavity at 1500 MHz.

25. Calculate the Q_0 value of a 1-cell, 1.3-GHz niobium cavity at 1.8 K if there is defect of 1 mm² of copper on the surface. Repeat the calculation to estimate the Q_0 change due to a typical normal conducting crater obtained after HPP.

26. Explain the shape of the temperature dependence of the thermal conductivity of copper shown in Figure 11.11.

27. Show that for one-point multipacting the kinetic energy of the electrons should follow:
$$K \propto \frac{e^2 E_\perp^2}{m\omega^2}. \tag{P.25}$$

28. Estimate the dc magnetic field allowed if a Q_0 value higher than 10^{10} is desired for 1.3-GHz niobium cavity at 2 K. Remember to correct for the increase of surface resistance with RRR.

29. Determine the length of the cutoff tube to achieve a field attenuation factor of 10 for a cavity with beam hole diameter of 24 cm, operating in the TM_{010} mode at 500 MHz.

30. The equilibrium concentration of hydrogen in niobium is given in wt. ppm by
$$C_H = 0.153\sqrt{p_{H_2}} \exp\left(\frac{4766.4}{T}\right). \tag{P.26}$$

    If niobium is heated in a furnace with partial pressure of $H_2 = 1 \times 10^{-5}$ torr, calculate the temperature at which the hydrogen removal should be carried out to achieve a hydrogen concentration < 10 ppm atomic.

31. Show that a storage time of 3 hours at 150 K will have the same ill effect for hydrogen-associated Q_0 degradation as 11.2 hours at 120 K. Diffusion constant $= D_0 = 9 \times 10^{-5}$ cm^2/sec, $D(T) = D_0 \exp(-E_a/kT)$, $E_a/k = 790$ K.

32. Using the surface resistance of niobium in the normal state at 1.3 GHz, along with the thermal conductivity of pure niobium as 10 W/(m-K), calculate the limiting thermal breakdown field (in oersteds) for a 0.1-mm radius defect.

33. Calculate the RRR of niobium with the following impurities (typical reactor grade niobium).

    (a) Ta: 500 wt. ppm
    (b) O: 100 wt. ppm
    (c) N: 20 wt. ppm
    (d) C: 5 wt. ppm
    (e) H: 1 wt. ppm

34. The degassing of oxygen takes place by the evaporation of NbO and NbO$_2$ from the metal surface, and the concentration $c(t)$ is given by

$$\log\left(\frac{c(t)}{c(0)}\right) = -\frac{F}{m} t \times 5.87 \times 10^{10} \exp\left(\frac{-6.53 \times 10^4}{T}\right). \quad (P.27)$$

Here t is the time in minutes, F is the surface area of the molten pool in cm^2, m is the mass of the ingot in grams, and T in temperature in K.

Calculate the melt rate necessary to drop the oxygen content by a factor of 10 at the melting temperature if the ingot has a diameter 25 cm. Melting point of niobium = 2470 °C.

35. Assuming a constant partial pressure of 1×10^{-5} torr for O$_2$ and N$_2$ in the electron beam melting furnace, calculate the equilibrium concentrations of oxygen and nitrogen in niobium at the melting temperature. Calculate the highest RRR of niobium purified in such a furnace, if it is not limited by other impurities.

36. If the diffusion coefficient of oxygen in niobium is

$$D = 0.02 \exp\left(\frac{-1.354 \times 10^4}{T}\right) \text{ cm}^2/\text{sec}, \quad (P.28)$$

calculate the temperature necessary for the diffusion length to reach 2 mm in 2 hours.

37. Show that the barrier lowering $\Delta\phi$ due to the image charge effect is given by

$$\Delta\phi = e\sqrt{\frac{eE}{4\pi\epsilon_0}}. \quad (P.29)$$

38. Calculate the value of E (in MV/m) required to completely lower the barrier for niobium ($\Delta f = 4.3$ eV) so that electrons can escape from the metal (without invoking quantum mechanical tunelling).

39. Calculate the field emission current at $E = 50$ MV/m for a typical emitter with $\beta_{\text{FN}} = 100$ and $A_{\text{FN}} = 10^{-12}$ m^2.

40. Calculate the surface electric field reached with 1 MW of pulsed power in 1 ms for a 1-meter-long cavity at 1.3 GHz, assuming that the input coupler is adjusted to $Q_e = 10^7$. Suppose that for the duration of pulse, (a) Q_0 remains at 10^9, and (b) Q_0 falls to 10^7. Take $R_a/Q_0 = 1080$ Ω/m and $E_{\text{pk}}/E_{\text{acc}} = 2$.

41. Consider a Gaussian bunch of length 2 cm traveling at the speed of light. It is stored in CESR, which has a revolution frequency of 400 kHz. The beam current is 40 mA. Table P.1 gives the data on the ten lowest monopole modes of a *copper model* (with stainless steel beam pipes) of the Cornell

Table P.1: Lowest monopole modes of a copper model of the Cornell 500-MHz cavity.

Mode	Frequency [MHz]	R_a/Q_0 [Ω]	Q_0
1	500	89.7	41000
2	968	0.834	310
3	968	0.18	3000
4	998	0.01	4500
5	1000	1.776	2200
6	1044	0.884	5900
7	1049	2.44	3600
8	1119	0.238	5700
9	1120	6.846	5900
10	1211	4.05	4200

superconducting 500-MHz cavity. Assume that we want to use this in the CESR storage ring.

(Note that these values are for the cavity without ferrite HOM dampers.)

(a) Calculate the loss factor for each mode for the Gaussian bunch.

(b) What is the total energy lost to HOMs in a single pass, if no fields are present (i.e., the bunch passes through the cavity for the first time)?

(c) What is the effective decelerating voltage the first bunch sees if the cavity is *not* being driven by an rf generator?

(d) Calculate the decay times of the fields for each mode.

(e) Which modes in the list are potentially the most dangerous regarding the excitation of bunch-to-bunch instabilities?

(f) By what factor would the R_a/Q_0 scale if the cavity were made of superconducting niobium?

(g) What would the Q_0 of the fundamental mode be?

(h) Which modes are now potentially dangerous?

42. A superconducting cavity is designed to accelerate a current $I_0 = 100$ mA through 1 MV. Assume the synchronous phase angle is zero. The coupling is fixed and is set to yield reflectionless operation when the full current is being accelerated. The shunt impedance of the accelerating mode is $89\ \Omega \times Q_0$ and the unloaded Q is 10^9.

(a) What is the power transferred to the beam during operation?

(b) What is the power dissipated in the cavity walls?

(c) What is the optimum coupling?

(d) Calculate the ratio of the required generator power when $I_0 = 0$ to the power obtained in (a).

(e) What value do you get for the ratio in (d) if the system is designed to accelerate very large beam currents ($I_b \to \infty$) and/or the cavity quality is very large ($Q_0 \to \infty$)?

43. (a) Consider a superconducting cavity being operated on resonance and in a reflectionless manner. In the liquid helium space of the cryostat, fluctuations in the helium boil-off rate result in pressure variations of 2 torr. How much does the demand on the generator power fluctuate, if an accelerating voltage of 1 MV is to be maintained and the beam current is 180 mA? Use the following data:

$$\frac{R_a}{Q_0} = 89 \,\Omega$$
$$Q_0 = 10^9$$
$$\frac{d\omega_0}{dP} = 2\pi \times 0.333 \text{ kHz/torr},$$

where P is the pressure in the helium bath.

(b) Show that Equation 17.46 applies even if the cavity is detuned and $\phi_s \neq 0$, provided that the cavity–beam system represents a real load and β is adjusted to yield zero reflection when accelerating the design current. [In this case, $\delta\omega$ respresents the detuning from the optimum tune given by (17.68).]

44. Consider a window located in the input line to a superconducting cavity. The cavity coupling is fixed to yield zero power reflection when the beam power (P_b) equals the design beam power ($P_{b,\text{design}}$) and the beam current (i_b) equals the design beam current ($i_{b,\text{design}}$). For any beam current, the generator power is adjusted to keep the cavity voltage constant. For currents $i_b \neq i_{b,\text{design}}$ power reflections are incurred and a standing wave exists in the input line. Possible window locations include the voltage maximum (current minimum) and the voltage minimum (current maximum). Show that:

(a) If the window is at the voltage minimum (when $i_b = 0$), the window voltage increases linearly from zero with the beam current reaching some value V_{design} when $i_b = i_{b,\text{design}}$,

(b) If the window is located at the voltage maximum (when $i_b = 0$), the voltage is always V_{design}, regardless of the beam current.

45. In the TRISTAN storage ring, the typical beam current was 10 mA and the typical bunch charge was 20 nC. Calculate the HOM power induced in an HOM of frequency ≈ 1 GHz, if the R_a/Q_0 is 100 ohms/cavity, typical for the most dangerous HOM. What is the contribution to the total loss factor from this mode? If the Q_e of the HOM coupler for this mode is 13000, calculate the cryogenic power dissipated. If the bunch spacing is

10 ms, calculate the minimum damping required for this mode to avoid resonant power buildup.

46. In a typical recirculating linac, such as CEBAF, the design beam current is 200 μA × 5 recirculated beams. The bunch charge is 0.1 pC and the total loss factor of the 5-cell, 1500-MHz cavity is $k = 10^{13}$ V/C for a bunch length of 0.25 mm. Calculate the beam-induced HOM power.

47. In the low-energy ring of a possible asymmetric-energy B-factory, the beam energy is 3.5 GeV, the beam current is 2 A, the bunch charge $q = 20$ nC, and the bunch spacing = 14 ns. If one uses a large aperture superconducting cavity, the typical dangerous mode will have $R_a/Q_0 = 5$ ohms/cell. Calculate the beam-induced power in that mode. If the total loss factor of the cell is 10^{11} V/C at 1 cm bunch length, calculate the total beam-induced power.

48. The high-energy ring (HER) of KEK-B (B-factory at KEK) has a design beam current of 1.1 A and a bunch spacing of 2 ns. Show that the longitudinal and transverse instability rise times for the highest R/Q monople and dipole modes mode are ≈0.5 sec.

49. Present thinking toward a superconducting linear collider calls for two linacs, each having an active length of 10 km. To keep the refrigerator capital and operating cost affordable, the cavities are operated in the pulsed mode, with a pulse length of 1 ms, bunch spacing of 1 μs, and a duty factor of about 1%. The accelerator parameters call for a bunch charge of 8 nC and a peak beam current of 80 mA. The loss factor for a 1-m 1.3-GHz cavity is $k = 9.3$ V/C at 1-mm bunch length. Calculate the average beam-induced HOM power. Compare this power to the dissipated power in the fundamental mode. Show that the energy spread due to longitudinal wakefields is about 0.5%.

50. Calculate the ac power required to operate the refrigerator for TESLA if the accelerator is to be operated cw.

51. Compare the ratio of beam energy/linac stored energy for TESLA and NLC.

52. Calculate the peak power/m required for TESLA from the peak beam power. Assume that the input power coupler is matched for beam-loaded conditions. Calculate the external Q of the coupler and the rise time of the rf pulse.

53. Calculate the typical Q_0 of the fundamental mode of a normal conducting cavity at 11.4 GHz, assuming a pill-box geometry. Compare this to the rise time of the rf pulse for a normal-conducting cavity without beam.

54. Compare the structure stored energy per meter of a 1.3-GHz, superconducting cavity, gradient 25 MV/m, with a normal conducting cavity at 11.4 GHz, gradient 50 MV/m.

PROBLEMS

Table P.2: Accelerator parameters to be determined for a proton linac used as a neutron spallation source

Option (gradient MV/m)	NC (2)	NC (1)	SC (5)	SC (10)
Peak rf power installed (MW)				
Accelerator real estate length (m)				
AC power (MW)				

55. Make an estimate of the total ac power required to operate TESLA (0.5 TeV).

56. Carry out an analysis for a 100-MW beam power neutron spallation source assuming a 1-GeV proton linac operating cw with a beam current of 100 mA. Assume a gradient of 5 MV/m and 10 MV/m for the superconducting (sc) solutions and 1 MV/m and 2 MV/m for the normal conducting (nc) solutions. Note that besides the larger aperture, an important advantage of the superconducting solution for a cw machine is in reducing the real estate length when using a high gradient. Fill in the parameters in Table P.2.

References

[1] S. Ramo et al., *Fields and Waves in Communication Electronics*, John Wiley and Sons, New York, New York (1965).

[2] R. Reif, *Fundamentals of Statistical and Thermal Physics*, McGrawHill (1965).

[3] N. W. Ashcroft and N. D. Mermin, *Solid State Physics*, W. B. Saunders Company (1976).

[4] W. Weingarten, in *Proceedings of the Joint US-CERN-Japan International School, Frontiers of Accelerator Technology*, edited by S. Kurokawa et al. World Scientific, p. 311 (1996).

[5] H. Padamsee and J. Knobloch, in *Proceedings of the Joint US-CERN-Japan International School, Frontiers of Accelerator Technology*, edited by S. Kurokawa et al. World Scientific, p. 101 (1996).

[6] H. Padamsee et al., *Annu. Rev. Nucl. Part. Sci.*, **43**:B635 (1993).

[7] H. Padamsee, in *AIP Conference Proceedings 249, US Particle Accelerator School*, edited by M. Month, p. 1402 (1992).

[8] H. Piel, in *CERN Accelerator School Proceedings*, p. 149 (1989). CERN 89-04.

[9] M. Kuntze, ed. *Proceedings of the 1st Workshop on RF Superconductivity*, KFK, Karlsruhe, Germany, 1980. KfK-3019.

[10] H. Lengeler, ed. *Proceedings of the 2nd Workshop on RF Superconductivity*, CERN, Geneva, Switzerland, 1984. CERN.

[11] K. W. Shepard, ed. *Proceedings of the 3rd Workshop on RF Superconductivity*, Argonne Natl. Lab., Argonne, Il, 1988. ANL-PHY-88-1.

[12] Y. Kojima, ed. *Proceedings of the 4th Workshop on RF Superconductivity*, KEK, Tsukuba, Japan, 1990. Rep. 89-21.

[13] D. Proch, ed. *Proceedings of the 5th Workshop on RF Superconductivity*, DESY, Hamburg, Germany, 1991. DESY-M-92-01.

[14] R. M. Sundelin, ed. *Proceedings of the 6th Workshop on RF Superconductivity*, CEBAF, Newport News, Va., 1994.

[15] B. Bonin, ed. *Proceedings of the 7th Workshop on RF Superconductivity*, Gif-sur-Yvette, France, 1995. CEA/Saclay 96 080/1.

[16] G. Arnolds-Mayer et al., in *Proceedings of the 3rd Workshop on RF Superconductivity*, edited by K. W. Shepard, Argonne Natl. Lab., Argonne, Il, p. 55 (1988). ANL-PHY-88-1.

[17] S. Takeuchi, in *Proceedings of the 3rd Workshop on RF Superconductivity*, edited by K. W. Shepard, Argonne Natl. Lab., Argonne, Il, p. 429 (1988). ANL-PHY-88-1.

[18] L. M. Bollinger, *Annu. Rev. Nucl. Part. Sci.*, **36**:475 (1987).

[19] D. W. Storm, in *Proceedings of the 6th Workshop on RF Superconductivity*, edited by R. M. Sundelin, CEBAF, Newport News, Va., p. 216 (1994).

[20] J. Delayen and K. W. Shepard, *Appl. Phys. Lett.*, **57(5)**:514 (1990).

[21] J. M. Pierce et al., in *Proceedings of the 9th International Conference on Low Temperature Physics*, Vol. A. Plenum, New York, p. 36 (1965).

[22] C. Lyneis et al., *IEEE Trans. Nucl. Sci.*, **28**:3445 (1981).

[23] P. Axel et al., *IEEE Trans. Nucl. Sci.*, **26**:3143 (1979).

[24] R. Sundelin et al., *IEEE Trans. Nucl. Sci.*, **20**:98 (1973).

[25] J. Kirchgessner et al., *IEEE Trans. Nucl. Sci.*, **20**:1141 (1973).

[26] R. Sundelin et al., *IEEE Trans. Nucl. Sci.*, **30**:3336 (1983).

[27] W. Bauer et al., *Nucl. Instrum. Methods Phys. Res.*, **214**:189 (1983).

[28] P. Bernard et al., *IEEE Trans. Nucl. Sci.*, **30**:3342 (1983).

[29] T. Furuya et al., in *Proceedings of the 5th Symposium on Accelerator Science and Technology*, KEK, p. 122 (1984).

[30] S. Noguchi et al., in *Proceedings of the 5th Symposium on Accelerator Science and Technology*, KEK, p. 124 (1984).

[31] R. Sundelin et al., *IEEE Trans. Nucl. Sci.*, **32**:3570 (1985).

[32] B. Dwersteg et al., *IEEE Trans. Nucl. Sci.*, **32**:3596 (1985).

[33] T. Furuya et al., in *Proceedings of the 13th International Conference on High Energy Accelerators*, Novosibirsk, p. 7 (1986).

[34] A. Brandelik et al., *Part. Accel.*, **4**:3 (1972).

[35] R. Benaroya et al., *IEEE Trans. Magn.*, **11**:413 (1975).

REFERENCES

[36] G. J. Dick et al., *Nucl. Instrum. Methods*, **138**:203 (1976).

[37] J. Noé et al., *IEEE Trans. Nucl. Sci.*, **24**:1144 (1977).

[38] K. W. Shepard et al., *IEEE Trans. Nucl. Sci.*, **24**:1147 (1977).

[39] J. Aron et al., in *Proceedings of the 1979 Linear Accelerator Conference*, edited by W. RL. Montauk, New York, p. 105 (1979). Brookhaven Natl. Lab. Rep. BNL 511-34.

[40] K. W. Zieher and H. Hornung, *IEEE Trans. Nucl. Sci.*, **28**:3312 (1981).

[41] B. Cauvin et al., in *Proceedings of the 3rd Workshop on RF Superconductivity*, edited by K. W. Shepard, Argonne Natl. Lab., Argonne, Il, p. 379 (1988). ANL-PHY-88-1.

[42] I. Ben-Zvi and J. Brennan, *Nucl. Instrum. Methods Phys. Res.*, **212**:73 (1983).

[43] D. W. Storm et al., *IEEE Trans. Nucl. Sci.*, **32**:3607 (1985).

[44] S. Takeuchi et al., in *Proceedings of the 4th Workshop on RF Superconductivity*, edited by Y. Kojima, KEK, Tsukuba, Japan, p. 469 (1990). Rep. 89-21.

[45] G. Fortuna et al., in *Proceedings of the 3rd Workshop on RF Superconductivity*, edited by K. W. Shepard, Argonne Natl. Lab., Argonne, Il, p. 399 (1988). ANL-PHY-88-1.

[46] W. Bauer et al., *IEEE Trans. Nucl. Sci.*, **18**:181 (1971).

[47] H. Hahn and H. J. Halama, *IEEE Trans. Nucl. Sci.*, **16**:1013 (1969).

[48] A. Citron et al., *Nucl. Instrum. Methods*, **164**:31 (1979).

[49] K. W. Shepard et al., *IEEE Trans. Nucl. Sci.*, **32**:357 (1985).

[50] J. R. Delayen et al., in *Proceedings of the 6th Workshop on RF Superconductivity*, edited by R. M. Sundelin, CEBAF, Newport News, Va., p. 11 (1994).

[51] S. Noguchi et al., in *Proceedings of the 3rd Workshop on RF Superconductivity*, edited by K. W. Shepard, Argonne Natl. Lab., Argonne, Il, p. 605 (1988). ANL-PHY-88-1.

[52] T. Tajima et al., in *Proceedings of the 4th Workshop on RF Superconductivity*, edited by Y. Kojima, KEK, Tsukuba, Japan, p. 821 (1990). Rep. 89-21.

[53] S. Noguchi, in *Proceedings of the 7th Workshop on RF Superconductivity*, edited by B. Bonin, Gif-sur-Yvette, France, p. 163 (1995). CEA/Saclay 96 080/1.

[54] T. Furuya et al., in *Proceedings of the 7th Workshop on RF Superconductivity*, edited by B. Bonin, Gif-sur-Yvette, France, p. 729 (1995). CEA/Saclay 96 080/1.

[55] B. Dwersteg et al., in *Proceedings of the 7th Workshop on RF Superconductivity*, edited by B. Bonin, Gif-sur-Yvette, France, p. 151 (1995). CEA/Saclay 96 080/1.

[56] G. Geschonke, in *Proceedings of the 7th Workshop on RF Superconductivity*, edited by B. Bonin, Gif-sur-Yvette, France, p. 143 (1995). CEA/Saclay 96 080/1.

[57] G. Cavallari et al., in *Proceedings of the 6th Workshop on RF Superconductivity*, edited by R. M. Sundelin, CEBAF, Newport News, Va., p. 49 (1994).

[58] S. Doebert et al., in *Proceedings of the 7th Workshop on RF Superconductivity*, edited by B. Bonin, Gif-sur-Yvette, France, p. 57 (1995). CEA/Saclay 96 080/1.

[59] H. Heinrichs et al., in *Proceedings of the 2nd Workshop on RF Superconductivity*, edited by H. Lengeler, CERN, Geneva, Switzerland, CERN, p. 141 (1984).

[60] J. Preble, in *Proceedings of the 7th Workshop on RF Superconductivity*, edited by B. Bonin, Gif-sur-Yvette, France, p. 173 (1995). CEA/Saclay 96 080/1.

[61] C. A. Brau, *Science*, **239**:1115 (1988).

[62] D. A. Deacon et al., *Phys. Rev. Lett.*, **38**:892 (1977).

[63] S. Rohatgi et al., in *Proceedings of the 1987 IEEE Particle Accelerator Conference*, p. 230 (1987). IEEE Cat. No. 87CH-2387-9.

[64] M. Shibata, in *Proceedings of the 6th Workshop on RF Superconductivity*, edited by R. M. Sundelin, CEBAF, Newport News, Va., p. 124 (1994).

[65] D. V. Neuffer et al., in *Proceedings of the 1995 Particle Accelerator Conference and International Conference on High Energy Accelerators*, Dallas, TX, p. 243 (1996). Cat. No. 95CH35843.

[66] K. Akai, in *Proceedings of the 4th Workshop on RF Superconductivity*, edited by Y. Kojima, KEK, Tsukuba, Japan, p. 189 (1990). Rep. 89-21.

[67] J. Mammosser, in *Proceedings of the 6th Workshop on RF Superconductivity*, edited by R. M. Sundelin, CEBAF, Newport News, Va., p. 33 (1994).

[68] J. D. Jackson, *Classical Electrodynamics*, Wiley & Sons, New York, New York, 2nd edition (1975).

REFERENCES

[69] S. A. Heifets and S. A. Kheifets, *Rev. Mod. Phys.*, **63**:631–673 (1991).

[70] M. T. Menzel and H. K. Stokes, *User's Guide for the POISSON/SUPERFISH Group of Codes*. Accelerator Theory and Simulation Group, AT-6, Los Alamos National Laboratory, Los Alamos, NM 87545, January 1987. LA-UR-87-115.

[71] U. Lauströer et al., *URMEL and URMEL-T USER GUIDE (Modal Analysis of Cylindrically Symmetric Cavities; Evaluation of RF-Fields in Waveguides)*, Jan 1987.

[72] D. G. Myakishev and V. P. Yakovlev, in *Conference Record of the 1991 IEEE Particle Accelerator Conference*, pp. 3002–3004 (1991).

[73] M. Bartsch et al., in *Proc. of the 1990 Linear Acc. Conf.*, Albuquerque, New Mexico, p. 372 (1990). LA-12004-C, Los Alamos National Lab.

[74] http://www-laacg.atdiv.lanl.gov/componl.html.

[75] D. G. Myakishev and V. P. Yakovlev, in *Proceedings of the 1995 Particle Accelerator Conference and International Conference on High Energy Accelerators*, Dallas, TX, pp. 2348–2350 (1996). Cat. No. 95CH35843.

[76] S. Belomestnykh. Spherical Cavity: Analytical Formulas. Comparison of Computer Codes. SRF SRF941208-13, Laboratory of Nuclear Studies, Cornell University, 1994.

[77] K. Schulze, in *Niobium, Proceedings of the International Symposium*, edited by H. Stuart, San Francsico, The Metallurgical Society of AIME, p. 163 (1981).

[78] H. Kamerlingh Onnes, *Konink. Akad. Wetensch.*, **14**:113–115 (1911). Also Communication Number 122b from the Physics Lab of the University of Leyden.

[79] H. Kamerlingh Onnes, *Electrician*, **67**:657–658 (1911).

[80] W. Meissner and R. Ochsenfeld, *Naturwissenshaften*, **21**:787 (1983).

[81] J. Bardeen et al., *Phys. Rev.*, **108**:1175 (1957).

[82] E. Maxwell, *Phys. Rev.*, **78**:477 (1950).

[83] C. A. Reynolds et al., *Phys. Rev.*, **78**:487 (1950).

[84] H. Frolich, *Phys. Rev.*, **79**:845 (1950).

[85] H. Padamsee et al., *J. Low Temp. Phys.*, **12**:387 (1973).

[86] B. Mühlschlegel, *Z. Physik*, **155**:313–327 (1959).

[87] T. P. Sheehan, *Phys. Rev.*, **149**:368 (1966).

[88] L. P. Kadanoff and P. Martin, *Phys. Rev.*, **124**:670 (1961).

[89] A. Oota and Y. Masuda, *J. Phys. Soc. Jpn.*, **41**:434 (1976).

[90] J. Bardeen et al., *Phys. Rev.*, **113**:982 (1959).

[91] K. R. Krafft, PhD thesis, Cornell University, 1983.

[92] G. E. H. Reuter and E. H. Sondheimer, *Proc. R. Soc. Lond. A, Math. Phys. Sci.*, **195**:336 (1948).

[93] Y. Bruynseraede et al., *Physica*, **54**:137 (1971).

[94] J. Halbritter, *Z. Physik*, **238**:466–476 (1970).

[95] D. C. Mattis and J. Bardeen, *Phys. Rev.*, **111**:412 (1958).

[96] J. P. Turneaure, PhD thesis, Stanford University, 1967.

[97] F. Palmer, in *Proceedings of the 3rd Workshop on RF Superconductivity*, edited by K. W. Shepard, Argonne Natl. Lab., Argonne, Il, p. 309 (1988). ANL-PHY-88-1.

[98] C. Benvenuti, in *Proceedings of the 5th Workshop on RF Superconductivity*, edited by D. Proch, DESY, Hamburg, Germany, p. 189 (1991). DESY-M-92-01.

[99] U. Klein, PhD thesis, University of Wuppertal, 1981. WUB-DI, 83-2.

[100] R. Blashcke et al., *Verh. Dtsch. Phys. Ges.*, **V1**:988 (1982).

[101] J. Halbritter, *J. Appl. Phys.*, **46**:1403 (1975).

[102] M. Tinkham, *Introduction to Superconductivity*, McGraw-Hill (1975).

[103] V. L. Ginsburg and L. Landau, *Zh. Eksp. Teor. Fiz.*, **20**:1064 (1950).

[104] G. Müller, in *Proceedings of the 3rd Workshop on RF Superconductivity*, edited by K. W. Shepard, Argonne Natl. Lab., Argonne, Il, p. 331 (1988). ANL-PHY-88-1.

[105] C. Durand et al., *IEEE Trans. Appl. Supercond.*, **5(2)**:1107 (1995).

[106] J. Matricon and D. Saint-James, *Phys. Lett. A*, **24**:241 (1967).

[107] R. B. Flippen, *Phys. Lett. A*, **17**:193 (1965).

[108] T. Yogi et al., *Phys. Rev. Lett.*, **39**:826 (1977).

[109] T. Hays et al., in *Proceedings of the 7th Workshop on RF Superconductivity*, edited by B. Bonin, Gif-sur-Yvette, France, p. 437 (1995). CEA/Saclay 96 080/1.

REFERENCES

[110] D. Moffat, in *Proceedings of the 4th Workshop on RF Superconductivity*, edited by Y. Kojima, KEK, Tsukuba, Japan, p. 445 (1990). Rep. 89-21.

[111] C. Hauviller, in *Proceedings of the 1989 IEEE Particle Accelerator Conference*, p. 485 (1989). IEEE Cat. 89CH2669-0.

[112] D. Proch, in *Proceedings of the 7th Workshop on RF Superconductivity*, edited by B. Bonin, Gif-sur-Yvette, France, p. 259 (1995). CEA/Saclay 96 080/1.

[113] V. Palmieri et al., in *Proceedings of the 7th Workshop on RF Superconductivity*, edited by B. Bonin, Gif-sur-Yvette, France, p. 595 (1995). CEA/Saclay 96 080/1.

[114] H. Stuart, ed. *Niobium, Proceedings of the International Symposium*, San Francsico, 1981. The Metallurgical Society of AIME.

[115] Cabot Performance Materials, Boyertown, PA 19512-1607.

[116] Teledyne Wah Chang Albany, Albany, Oregon 97321-0136.

[117] W. C. Heraeus, D-63450 Hanau, Germany.

[118] K. K. Schulze, *J. Met.*, **33**:33 (1981).

[119] A. Gladun et al., *J. Low Temp. Phys.*, **27**:873 (1977).

[120] H. Padamsee, *IEEE Trans. Magn.*, **21**:1007 (1985).

[121] H. Padamsee, in *Proceedings of the 6th Workshop on RF Superconductivity*, edited by R. M. Sundelin, CEBAF, Newport News, Va., p. 515 (1994).

[122] A. V. Elyutin et al., in *Proceedings of the 5th Workshop on RF Superconductivity*, edited by D. Proch, DESY, Hamburg, Germany, pp. 354,426 (1991). DESY-M-92-01.

[123] J. Kirchgessner, *IEEE Trans. Nucl. Sci.*, **30**:2901 (1983).

[124] V. Palmieri et al., in *Proceedings of the 4th European Particle Accelerator Conference*, edited by V. Suller et al. World Scientific, p. 2212 (1994).

[125] J. Kirchgessner, in *Proceedings of the 3rd Workshop on RF Superconductivity*, edited by K. W. Shepard, Argonne Natl. Lab., Argonne, Il, p. 533 (1988). ANL-PHY-88-1.

[126] M. G. Rao et al., in *Proceedings of the 6th Workshop on RF Superconductivity*, edited by R. M. Sundelin, CEBAF, Newport News, Va., p. 643 (1994).

[127] M. G. Rao and P. Kneisel. Mechanical Properties of High RRR Nb. DESY Print, TESLA 95-09, March 1995.

[128] R. Sundelin. Private communication.

[129] G. Müller, in *Proceedings of the 2nd Workshop on RF Superconductivity*, edited by H. Lengeler, CERN, Geneva, Switzerland, CERN, p. 377 (1984).

[130] H. Padamsee et al., *IEEE Trans. Magn.*, **19**:1322 (1983).

[131] H. Padamsee, in *Proceedings of the 2nd Workshop on RF Superconductivity*, edited by H. Lengeler, CERN, Geneva, Switzerland, CERN, p. 339 (1984).

[132] P. Kneisel, *J. Less-Common Met.*, **139**:179 (1988).

[133] H. Safa et al., in *Proceedings of the 7th Workshop on RF Superconductivity*, edited by B. Bonin, Gif-sur-Yvette, France, p. 649 (1995). CEA/Saclay 96 080/1.

[134] Q. S. Shu et al., *IEEE Trans. Magn.*, **27**:1935 (1990). Proceedings of the 1990 Applied Superconductivity Conference.

[135] U.S. Federal Standard 209E, September 1992.

[136] B. Bonin and R. W. Röth, in *Proceedings of the 5th Workshop on RF Superconductivity*, edited by D. Proch, DESY, Hamburg, Germany, p. 210 (1991). DESY-M-92-01.

[137] D. Proch, in *Proceedings of the 6th Workshop on RF Superconductivity*, edited by R. M. Sundelin, CEBAF, Newport News, Va., p. 77 (1994).

[138] K. Saito et al., in *Proceedings of the 4th Workshop on RF Superconductivity*, edited by Y. Kojima, KEK, Tsukuba, Japan, p. 635 (1990). Rep. 89-21.

[139] T. Furuya, in *Proceedings of the 4th Workshop on RF Superconductivity*, edited by Y. Kojima, KEK, Tsukuba, Japan, p. 95 (1990). Rep. 89-21.

[140] P. Bernard et al., in *Proceedings of the 1992 European Particle Accelerator Conference*, edited by E. H. Henke et al. Editions Frontieres, p. 1269 (1992).

[141] P. Kneisel and B. Lewis, in *Proceedings of the 7th Workshop on RF Superconductivity*, edited by B. Bonin, Gif-sur-Yvette, France, p. 311 (1995). CEA/Saclay 96 080/1.

[142] K. Yojima et al., in *Proceedings of the 4th Workshop on RF Superconductivity*, edited by Y. Kojima, KEK, Tsukuba, Japan, p. 85 (1990). Rep. 89-21.

[143] T. I. Smith. Technical Report, Stanford University, 1966. Internal Report HEP 437.

REFERENCES

[144] J. Sekutowicz, in *Proceedings of the 4th Workshop on RF Superconductivity*, edited by Y. Kojima, KEK, Tsukuba, Japan, p. 849 (1990). Rep. 89-21.

[145] P. Schmüser. Tuning of Multi-Cell Cavitites using Bead Pull Measurements. SRF SRF920925-10, Laboratory of Nuclear Studies, Cornell University, 1992.

[146] J. C. Slater, *Microwave Electronics*, D. Van Nostrand Company, Inc. (1950).

[147] H. Piel, in *Proceedings of the 1st Workshop on RF Superconductivity*, edited by M. Kuntze, KFK, Karlsruhe, Germany, p. 85 (1980). KfK-3019.

[148] C. Lyneis et al., in *Proceedings of the 1972 Proton Linear Accelerator Conference*, LANL, Los Alamos, p. 98 (1972).

[149] W. Weingarten, in *Proceedings of the 2nd Workshop on RF Superconductivity*, edited by H. Lengeler, CERN, Geneva, Switzerland, CERN, p. 551 (1984).

[150] J. Knobloch et al., *Rev. Sci. Instrum.*, **65**:3521 (1995).

[151] J. P. Turneaure, in *Proceedings of the 1972 Applied Superconductivity Conference*, Annapolis, IEEE, p. 621 (1972). IEEE Pub. No. 72CH0682-5-TABSC.

[152] R. Röth, in *Proceedings of the 5th Workshop on RF Superconductivity*, edited by D. Proch, DESY, Hamburg, Germany, p. 599 (1991). DESY-M-92-01.

[153] J. Halbritter, *J. Appl. Phys.*, **42**:82 (1971).

[154] C. Valet et al., in *Proceedings of the 1992 European Particle Accelerator Conference*, edited by E. H. Henke et al. Editions Frontieres, p. 1295 (1992).

[155] W. Weingarten, in *Proceedings of the 7th Workshop on RF Superconductivity*, edited by B. Bonin, Gif-sur-Yvette, France, p. 129 (1995). CEA/Saclay 96 080/1.

[156] P. Kneisel et al., in *Proceedings of the 1972 Applied Superconductivity Conference*, Annapolis, IEEE, p. 657 (1972). IEEE Pub. No. 72CH0682-5-TABSC.

[157] R. Calder et al., *Nucl. Instrum. Methods Phys. Res. B, Beam Interact. Matter. At.*, **13**:631 (1986).

[158] A. Woode and J. Petit, *ESA J.*, **14**:467 (1990).

[159] D. Proch et al., in *Proceedings of the 1995 Particle Accelerator Conference and International Conference on High Energy Accelerators*, Dallas, TX, pp. 1776–1778 (1996). Cat. No. 95CH35843.

[160] H. Padamsee and A. Joshi, *J. Appl. Phys.*, **50**(2):1112 (1979).

[161] C. M. Lyneis, in *Proceedings of the 1st Workshop on RF Superconductivity*, edited by M. Kuntze, KFK, Karlsruhe, Germany, p. 119 (1980). KfK-3019.

[162] C. M. Lyneis et al., *Appl. Phys. Lett.*, **31**(8):541 (1977).

[163] P. Kneisel and H. Padamsee. Technical Report, Cornell University, 1979. Laboratory of Nuclear Studies report CLNS-79/433.

[164] H. Padamsee et al., *IEEE Trans. Magn.*, **17**:947 (1981).

[165] U. Kelin and D. Proch, in *Proceedings of the Conference of Future Possibilities for Electron Accelerators*, Charlottesville, pp. N1–17 (1979).

[166] P. Kneisel et al., *Nucl. Instrum. Methods Phys. Res.*, **188**:669 (1981).

[167] E. Somersalo et al., in *Proceedings of the 1995 Particle Accelerator Conference and International Conference on High Energy Accelerators*, Dallas, TX, pp. 1500–1502 (1996). Cat. No. 95CH35843.

[168] J. Tückmantel. Multipacting Calculations for a Power DC-Biased 75 Ω Coupler. Technical Report 94-26, CERN, 1994. CERN LEP-2 Notes.

[169] J. Halbritter, in *Proceedings of the 1972 Applied Superconductivity Conference*, Annapolis, IEEE, p. 662 (1972). IEEE Pub. No. 72CH0682-5-TABSC.

[170] J. Knobloch et al., in *Proceedings of the 1995 Particle Accelerator Conference and International Conference on High Energy Accelerators*, Dallas, TX, p. 1623 (1996). Cat. No. 95CH35843.

[171] H. Padamsee et al., *IEEE Trans. Magn.*, **19**:1308 (1983).

[172] W.-D. Möller and M. Pekeler, in *Proceedings of the 5th European Particle Accelerator Conference*, edited by S. Myers et al., Barcelona, Spain, IOPP Publishing, Bristol, p. 2013 (1996).

[173] W. Weingarten, *CERN Accelerator School, RF Engineering for Particle Accelerators*, **CERN 92-03**:318 (1991).

[174] D. Bloess et al., *CERN/EF/RF*, **85-2** (1985).

[175] R. P. Reed and A. F. Clark, eds., *Materials at Low Temperatures*, ASME (1983).

[176] H. Padamsee, in *Proceedings of the 1st Workshop on RF Superconductivity*, edited by M. Kuntze, KFK, Karlsruhe, Germany, p. 145 (1980). KfK-3019.

[177] P. L. Kapitza, *J. Phys.*, **4**(3):181–210 (1941).

[178] A. Boucheffa et al., in *Proceedings of the 7th Workshop on RF Superconductivity*, edited by B. Bonin, Gif-sur-Yvette, France, p. 659 (1995). CEA/Saclay 96 080/1.

[179] J. M. Khalatnikov, *Zh. Eksp. Teor. Fiz.*, **22**:687 (1952).

[180] A. C. Anderson, in *Proceedings Phonon Scattering in Solids*, edited by L. I. Challis et al. Plenum, NY, p. 1 (1976).

[181] H. Tsuruga and K. Endoh, in *Proceedings of the 5th International Cryogenic Engineering Conference, ICEC5*, Kyoto, p. 62 (1974).

[182] C. Johannes, in *Proceedings of the 3rd International Cryogenic Engineering Conference, ICEC3*, p. 97 (1970).

[183] R. Röth et al., in *Proceedings of the 1992 European Particle Accelerator Conference*, edited by E. H. Henke et al. Editions Frontieres, p. 1325 (1992).

[184] H. Padamsee, *IEEE Trans. Magn.*, **21**:149 (1985).

[185] H. Padamsee, *IEEE Trans. Magn.*, **23**:1607 (1987).

[186] H. L. Schick, ed., *Thermodynamics of Certain Refractory Compounds*, Academic Press (1966).

[187] K. A. Gschneider, Jr and N. Kippenhan. Thermochemistry of the Rare Earth Carbides, Nitrides and Sulfides for Steelmaking. Technical Report 1S-RIC-5, Rare Earth Information Center, Iowa State University, Ames, Iowa, 1971.

[188] K. A. Gschneider, Jr et al. Thermochemistry of the Rare Earths. Technical Report 1S-RIC-6, Rare Earth Information Center, Iowa State University, Ames, Iowa, 1973.

[189] C. Z. Antoine et al., in *Proceedings of the 7th Workshop on RF Superconductivity*, edited by B. Bonin, Gif-sur-Yvette, France, p. 647 (1995). CEA/Saclay 96 080/1.

[190] C. Reece et al., in *Proceedings of the 6th Workshop on RF Superconductivity*, edited by R. M. Sundelin, CEBAF, Newport News, Va., p. 650 (1994).

[191] E. Kako et al., in *Proceedings of the 6th Workshop on RF Superconductivity*, edited by R. M. Sundelin, CEBAF, Newport News, Va., p. 425 (1994).

[192] H. Padamsee, in *Proceedings of the 5th Workshop on RF Superconductivity*, edited by D. Proch, DESY, Hamburg, Germany, p. 852 (1991). DESY-M-92-01.

[193] J. Graber, PhD thesis, Cornell University, 1993.

[194] R. J. Noer, *Appl. Phys. A, Solids Surf.*, **28**:1 (1982).

[195] G. Loew and J. W. Wang, in *Proceedings of the XIII International Symposium on Discharges and Electrical Insulation in Vacuum*, Paris, France, p. 480 (1988). SLAC-PUB-4647.

[196] H. Padamsee, in *Proceedings of the 4th Workshop on RF Superconductivity*, edited by Y. Kojima, KEK, Tsukuba, Japan, p. 207 (1990). Rep. 89-21.

[197] D. Moffat et al., *Part. Accel.*, **40**:85 (1992).

[198] R. V. Latham, ed., *High Voltage Vacuum Insulation*, Academic Press (1995).

[199] G. A. Mesyats and D. I. Proskurovsky, *Pulsed Electrical Discharge in Vacuum*, Springer-Verlag (1988).

[200] S. Noguchi et al., *Nucl. Instrum. Methods*, **179**:205 (1981).

[201] R. H. Fowler and L. Nordheim, *Proc. R. Soc. Lond. A, Math. Phys. Sci.*, **119**:173 (1928).

[202] W. P. Dyke and J. K. Trolan, *Phys. Rev.*, **89**:799 (1953).

[203] H. H. Race, *J. Sci. Technol.*, **43**:365 (1940). Now journal is called *GEC Review*.

[204] J. Graber, *Nucl. Instrum. Methods Phys. Res. A, Accel. Spectrom. Detect. Assoc. Equip.*, **350**:582 (1994).

[205] J. Tan, in *Proceedings of the 7th Workshop on RF Superconductivity*, edited by B. Bonin, Gif-sur-Yvette, France, p. 105 (1995). CEA/Saclay 96 080/1.

[206] J. Knobloch, PhD thesis, Cornell University, 1997.

[207] B. Rusnak et al., in *Proceedings of the 1992 Linac Conference*, edited by C. R. Hoffman, p. 728 (1992). AECL-10728.

[208] P. Niedermann, PhD thesis, U. of Geneva, 1986. No. 2197.

[209] C. Chianelli et al., in *Proceedings of the 6th Workshop on RF Superconductivity*, edited by R. M. Sundelin, CEBAF, Newport News, Va., p. 700 (1994).

REFERENCES

[210] B. Bonin, in *Proceedings of the 6th Workshop on RF Superconductivity*, edited by R. M. Sundelin, CEBAF, Newport News, Va., p. 1033 (1994).

[211] E. Mahner, in *Proceedings of the 6th Workshop on RF Superconductivity*, edited by R. M. Sundelin, CEBAF, Newport News, Va., p. 252 (1994).

[212] N. Pupeter, in *Proceedings of the 7th Workshop on RF Superconductivity*, edited by B. Bonin, Gif-sur-Yvette, France, p. 67 (1995). CEA/Saclay 96 080/1.

[213] Q. S. Shu, in *Proceedings of the 4th Workshop on RF Superconductivity*, edited by Y. Kojima, KEK, Tsukuba, Japan, p. 539 (1990). Rep. 89-21.

[214] K. Saito et al., in *Proceedings of the 6th Workshop on RF Superconductivity*, edited by R. M. Sundelin, CEBAF, Newport News, Va., p. 1151 (1994).

[215] M. Jimenez et al., *J. Phys. D, Appl. Phys.*, **27**:1038 (1994).

[216] R. Noer, in *Proceedings of the 6th Workshop on RF Superconductivity*, edited by R. M. Sundelin, CEBAF, Newport News, Va., p. 236 (1994).

[217] R. V. Latham, in *High Voltage Vacuum Insulation*, edited by R. V. Latham, pp. 61–113. Academic Press, 1995.

[218] N. S. Xu, in *High Voltage Vacuum Insulation*, edited by R. V. Latham, pp. 115–164. Academic Press, 1995.

[219] Q. S. Shu et al., *IEEE Trans. Magn.*, **25**:1868 (1989).

[220] W. Weingarten, in *Proceedings of the XIII International Symposium on Discharges and Electrical Insulation in Vacuum*, Paris, France, p. 480 (1988).

[221] C. B. Duke and M. E. Alferieff, *J. Chem. Phys.*, **46**:923 (1967).

[222] T. Hays et al., in *Proceedings of the 6th Workshop on RF Superconductivity*, edited by R. M. Sundelin, CEBAF, Newport News, Va., p. 750 (1994).

[223] J. W. Wang and G. A. Loew, in *Proceedings of the 1989 IEEE Particle Accelerator Conference*, p. 1137 (1989). IEEE Cat. 89CH2669-0.

[224] B. Jüttner, *Physica*, **114C**:255 (1982).

[225] E. A. Litvinov et al., *Sov. Phys.-Usp.*, **26(2)**:138 (1983).

[226] G. A. Mesyats, *IEEE Trans. Electr. Insul.*, **18(3)**:218 (1983).

[227] B. Jüttner, *IEEE Trans. Plasma Sci.*, **15(5)**:474 (1987).

[228] MASK, SAIC Corp., McLean, Va. 22102.

[229] R. Ferraro et al. Guide to Multipacting/Field Emission Simulation Software. SRF SRFD-960703/8, LNS, Cornell University, 1996.

[230] H. A. Schwettman et al., *J. Appl. Phys.*, **45**:914 (1974).

[231] E. L. Murphy and R. H. Good, Jr., *Phys. Rev.*, **102**(6):1464 (1956).

[232] E. Chiaveri, in *Proceedings of the 7th Workshop on RF Superconductivity*, edited by B. Bonin, Gif-sur-Yvette, France, p. 181 (1995). CEA/Saclay 96 080/1.

[233] J. F. Benesch, in *Proceedings of the 6th Workshop on RF Superconductivity*, edited by R. M. Sundelin, CEBAF, Newport News, Va., p. 581 (1994).

[234] H. Padamsee et al., in *Proceedings of the 1993 Particle Accelerator Conference*, Washington, DC, p. 998 (1993). IEEE Cat. No. 3279-7.

[235] Giredmet, 109017, Moscow, Russia.

[236] T. Hays et al., in *Proceedings of the 7th Workshop on RF Superconductivity*, edited by B. Bonin, Gif-sur-Yvette, France, p. 441 (1995). CEA/Saclay 96 080/1.

[237] D. Reschke et al., in *Proceedings of the 6th Workshop on RF Superconductivity*, edited by R. M. Sundelin, CEBAF, Newport News, Va., p. 1095 (1994).

[238] J. Graber, in *Proceedings of the 1995 Particle Accelerator Conference and International Conference on High Energy Accelerators*, Dallas, TX, p. 1478 (1996). Cat. No. 95CH35843.

[239] H. A. Schwettman et al., *J. Appl. Phys.*, **45**:914 (1974).

[240] C. Athwal and W. Weingarten, *CERN/EF/RF*, **84-7** (1984).

[241] J. Knobloch and H. Padamsee, *Part. Accel.*, **53**:53 (1996).

[242] S. Bajic et al., in *Proceedings of the XIII International Symposium on Discharges and Electrical Insulation in Vacuum*, Paris, France, p. 8 (1988).

[243] P. Niedermann et al., *J. Appl. Phys.*, **59**:3851 (1986).

[244] H. Padamsee et al., in *Conference Record of the 1991 IEEE Particle Accelerator Conference*, p. 2420 (1991).

[245] D. Reschke et al., in *Proceedings of the 4th European Particle Accelerator Conference*, edited by V. Suller et al. World Scientific, p. 2063 (1994).

[246] J. Kirchgessner et al., in *Proceedings of the 1993 Particle Accelerator Conference*, Washington, DC, p. 918 (1993). IEEE Cat. No. 3279-7.

REFERENCES

[247] P. Kneisel et al., in *Proceedings of the 6th Workshop on RF Superconductivity*, edited by R. M. Sundelin, CEBAF, Newport News, Va., p. 628 (1994).

[248] J. Graber, *Nucl. Instrum. Methods Phys. Res. A, Accel. Spectrom. Detect. Assoc. Equip.*, **350**:572 (1994).

[249] C. Crawford et al., *Part. Accel.*, **49**:1 (1995).

[250] K. Shepard. Private communication.

[251] I. E. Campisi, *IEEE Trans. Magn.*, **25**:134 (1985).

[252] M. Peiniger, in *Proceedings of the 3rd Workshop on RF Superconductivity*, edited by K. W. Shepard, Argonne Natl. Lab., Argonne, Il, p. 503 (1988). ANL-PHY-88-1.

[253] G. Müller, in *Proceedings of the 4th Workshop on RF Superconductivity*, edited by Y. Kojima, KEK, Tsukuba, Japan, p. 267 (1990). Rep. 89-21.

[254] N. Klein, in *Proceedings of the 5th Workshop on RF Superconductivity*, edited by D. Proch, DESY, Hamburg, Germany, p. 285 (1991). DESY-M-92-01.

[255] C. Liang, in *Proceedings of the 6th Workshop on RF Superconductivity*, edited by R. M. Sundelin, CEBAF, Newport News, Va., p. 307 (1994).

[256] M. A. Hein, in *Proceedings of the 7th Workshop on RF Superconductivity*, edited by B. Bonin, Gif-sur-Yvette, France, p. 267 (1995). CEA/Saclay 96 080/1.

[257] G. Orlandi et al., in *Proceedings of the 6th Workshop on RF Superconductivity*, edited by R. M. Sundelin, CEBAF, Newport News, Va., pp. 718–729 (1994).

[258] E. Chiaveri, in *Proceedings of the 5th European Particle Accelerator Conference*, edited by S. Myers et al., Barcelona, Spain, IOPP Publishing, Bristol, p. 200 (1996).

[259] B. Hillenbrand, in *Proceedings of the 1st Workshop on RF Superconductivity*, edited by M. Kuntze, KFK, Karlsruhe, Germany, p. 41 (1980). KfK-3019.

[260] G. Müller, in *Proceedings of the 5th European Particle Accelerator Conference*, edited by S. Myers et al., Barcelona, Spain, IOPP Publishing, Bristol, p. 2085 (1996).

[261] C. W. Chu, *IEEE Trans. Appl. Supercond.*, **7**(2):80–89 (1997). Proceedings of the 1996 Applied Superconductivity Conference.

[262] P. P. Nguyen et al., *Phys. Rev. B, Condens. Matter*, **48**:6400 (1993).

[263] B. A. Baumert, *J. Supercond.*, **8**(1):6400 (1995).

[264] P. B. Wilson, in *Physics of High Energy Particle Accelerators (Fermilab Summer School, 1981)*, edited by R. A. Carrigan et al., number 87 in AIP Conference Proceedings. American Institute of Physics (1982).

[265] B. Zotter and K. Bane. Transverse resonances of periodically widened cylindrical tubes with circular cross sections. Technical Report, Stanford Linear Accelerator, September 1979. PEP-Note 308.

[266] W. K. H. Panofsky and W. A. Wenzel, *Rev. Sci. Instrum.*, **27**:967 (1956).

[267] A. Chao, *Physics of Collective Beam Instabilities in High Energy Accelerators*, John Wiley and Sons, NY (1993).

[268] J. Bisognano, in *Proceedings of the 3rd Workshop on RF Superconductivity*, edited by K. W. Shepard, Argonne Natl. Lab., Argonne, Il, p. 237 (1988). ANL-PHY-88-1.

[269] T. Weiland, *Nucl. Instrum. Methods Phys. Res.*, **212**:13 (1983).

[270] Y. Chin, in *Proceedings of the 1993 Particle Accelerator Conference*, Washington, DC, p. 3414 (1993). IEEE Cat. No. 3279-7.

[271] A. Mosnier, in *Proceedings of the 1992 European Particle Accelerator Conference*, edited by E. H. Henke et al. Editions Frontieres, p. 777 (1992).

[272] R. Siemann, *IEEE Trans. Nucl. Sci.*, **28**:2437 (1981).

[273] F. J. Sacherer, *IEEE Trans. Nucl. Sci.*, **20**:825 (1973).

[274] National Laboratory for High Energy Physics. KEKB B-Factory Design Report. Technical Report 95-7, KEK, Tsukuba, Japan, 1995.

[275] W. Hartung and E. Haebel, in *Proceedings of the 1993 Particle Accelerator Conference*, Washington, DC, p. 898 (1993). IEEE Cat. No. 3279-7.

[276] M. S. de Jong et al., *J. Microw. Power Electromagn. Energy*, **27**:136–142 (1992).

[277] D. Proch, in *Proceedings of the 6th Workshop on RF Superconductivity*, edited by R. M. Sundelin, CEBAF, Newport News, Va., p. 382 (1994).

[278] H. Padamsee et al., in *Conference Record of the 1991 IEEE Particle Accelerator Conference*, p. 2423 (1991).

[279] J. C. Amato, *IEEE Trans. Nucl. Sci.*, **32**:3593 (1985).

[280] J. Kirchgessner et al., in *Proceedings of the 1989 IEEE Particle Accelerator Conference*, p. 479 (1989). IEEE Cat. 89CH2669-0.

REFERENCES

[281] J. Kirchgessner et al., in *Proceedings of the 11th International Conference on High Energy Accelerators*, Basel, Switzerland, Birkhauser Verlag, p. 886 (1980).

[282] J. C. Amato, in *Proceedings of the 3rd Workshop on RF Superconductivity*, edited by K. W. Shepard, Argonne Natl. Lab., Argonne, Il, p. 589 (1988). ANL-PHY-88-1.

[283] N. Kroll and D. Yu. SLAC-PUB-5171. Technical Report, SLAC, 1990.

[284] I. Campisi, in *Proceedings of the 6th Workshop on RF Superconductivity*, edited by R. M. Sundelin, CEBAF, Newport News, Va., p. 587 (1994).

[285] B. Dwersteg et al., in *Proceedings of the 2nd Workshop on RF Superconductivity*, edited by H. Lengeler, CERN, Geneva, Switzerland, CERN, p. 235 (1984).

[286] D. Proch, in *Proceedings of the 3rd Workshop on RF Superconductivity*, edited by K. W. Shepard, Argonne Natl. Lab., Argonne, Il, p. 29 (1988). ANL-PHY-88-1.

[287] E. Haebel, in *Proceedings of the 5th Workshop on RF Superconductivity*, edited by D. Proch, DESY, Hamburg, Germany, p. 334 (1991). DESY-M-92-01.

[288] A. Mosnier, in *Proceedings of the 4th Workshop on RF Superconductivity*, edited by Y. Kojima, KEK, Tsukuba, Japan, p. 377 (1990). Rep. 89-21.

[289] J. Sekutowicz, in *Proceedings of the 3rd Workshop on RF Superconductivity*, edited by K. W. Shepard, Argonne Natl. Lab., Argonne, Il, p. 597 (1988). ANL-PHY-88-1.

[290] H. P. Company. HFSS: the High Frequency Structure Simulator. Technical Report HP85180A.

[291] P. Bernard et al., in *Proceedings of the 5th Workshop on RF Superconductivity*, edited by D. Proch, DESY, Hamburg, Germany, p. 956 (1991). DESY-M-92-01.

[292] K. Akai, in *Proceedings of the 5th Workshop on RF Superconductivity*, edited by D. Proch, DESY, Hamburg, Germany, p. 126 (1991). DESY-M-92-01.

[293] H. Padamsee et al., *Part. Accel.*, **40**:17 (1992).

[294] D. Moffat et al., in *Proceedings of the 1993 Particle Accelerator Conference*, Washington, DC, p. 977 (1993). IEEE Cat. No. 3279-7.

[295] S. Belomestnykh et al., in *Proceedings of the 5th European Particle Accelerator Conference*, edited by S. Myers et al., Barcelona, Spain, IOPP Publishing, Bristol, p. 2100 (1996).

[296] W. Hartung et al., in *Proceedings of the 1993 Particle Accelerator Conference*, Washington, DC, p. 3450 (1993). IEEE Cat. No. 3279-7.

[297] J. F. DeFord et al., in *Proceedings of the 1989 IEEE Particle Accelerator Conference*, p. 1181 (1989). IEEE Cat. 89CH2669-0.

[298] W. Hartung, PhD thesis, Cornell University, 1996.

[299] T. Tajima et al., in *Proceedings of the 6th Workshop on RF Superconductivity*, edited by R. M. Sundelin, CEBAF, Newport News, Va., p. 962 (1994).

[300] S. Belomestnykh et al., in *Proceedings of the 1995 Particle Accelerator Conference and International Conference on High Energy Accelerators*, Dallas, TX, p. 3391 (1996). Cat. No. 95CH35843.

[301] H. Padamsee et al., in *Proceedings of the 1995 Particle Accelerator Conference and International Conference on High Energy Accelerators*, Dallas, TX, p. 1515 (1996). Cat. No. 95CH35843.

[302] K. Akai, in *Proceedings of the 5th European Particle Accelerator Conference*, edited by S. Myers et al., Barcelona, Spain, IOPP Publishing, Bristol, p. 205 (1996).

[303] W. Hartung et al., in *Proceedings of the 1995 Particle Accelerator Conference and International Conference on High Energy Accelerators*, Dallas, TX, p. 3294 (1996). Cat. No. 95CH35843.

[304] D. A. Edwards and M. J. Syphers, *An Introduction to the Physics of High Energy Accelerators*, John Wiley and Sons, New York (1993).

[305] H. Wiedemann, *Particle Accelerator Physics II*, Springer-Verlag, Berlin (1995).

[306] D. Boussard, *IEEE Trans. Nucl. Sci.*, **32**(5):1852 (1985). Proceedings of the 1985 Particle Accelerator Conference.

[307] B. Dwersteg, in *Proceedings of the 4th Workshop on RF Superconductivity*, edited by Y. Kojima, KEK, Tsukuba, Japan, p. 351 (1990). Rep. 89-21.

[308] L. Phillips, in *Proceedings of the 6th Workshop on RF Superconductivity*, edited by R. M. Sundelin, CEBAF, Newport News, Va., p. 267 (1994).

[309] M. Champion, in *Proceedings of the 7th Workshop on RF Superconductivity*, edited by B. Bonin, Gif-sur-Yvette, France, p. 195 (1995). CEA/Saclay 96 080/1.

[310] H. P. Kindermann et al., in *Proceedings of the 5th European Particle Accelerator Conference*, edited by S. Myers et al., Barcelona, Spain, IOPP Publishing, Bristol, pp. 2091, 2013 (1996).

REFERENCES

[311] J. Tückmantel. Technical Report 94-25, CERN, 1994. CERN LEP-2 Notes.

[312] M. Champion, in *Proceedings of the 6th Workshop on RF Superconductivity*, edited by R. M. Sundelin, CEBAF, Newport News, Va., p. 406 (1994).

[313] B. Yunn and R. M. Sundelin, in *Proceedings of the 1993 Particle Accelerator Conference*, Washington, DC, p. 1092 (1993). IEEE Cat. No. 3279-7.

[314] D. Metzger et al., in *Proceedings of the 1993 Particle Accelerator Conference*, Washington, DC, p. 1399 (1993). IEEE Cat. No. 3279-7.

[315] M. Pisharody et al., in *Proceedings of the 1995 Particle Accelerator Conference and International Conference on High Energy Accelerators*, Dallas, TX, p. 1720 (1996). Cat. No. 95CH35843.

[316] D. H. Priest and R. C. Talcott, *IRE Trans. Electron Devices*, **ED-8**:243 (1961).

[317] S. Michizono et al., *J. Vac. Sci. Technol. A, Vac. Surf. Films*, **10(4)**:1180 (1992).

[318] E. Haebel et al., in *Proceedings of the 7th Workshop on RF Superconductivity*, edited by B. Bonin, Gif-sur-Yvette, France, p. 707 (1995). CEA/Saclay 96 080/1.

[319] S. Isagawa, in *Proceedings of the 1987 IEEE Particle Accelerator Conference*, p. 1934 (1987). IEEE Cat. No. 87CH-2387-9.

[320] S. Mitsunobu, in *Proceedings of the 7th Workshop on RF Superconductivity*, edited by B. Bonin, Gif-sur-Yvette, France, p. 735 (1995). CEA/Saclay 96 080/1.

[321] H.-D. Gräf, in *Proceedings of the 5th Workshop on RF Superconductivity*, edited by D. Proch, DESY, Hamburg, Germany, p. 317 (1991). DESY-M-92-01.

[322] S. Simrock, in *Proceedings of the 6th Workshop on RF Superconductivity*, edited by R. M. Sundelin, CEBAF, Newport News, Va., p. 294 (1994).

[323] R. Legg, in *Proceedings of the 5th European Particle Accelerator Conference*, edited by S. Myers et al., Barcelona, Spain, IOPP Publishing, Bristol, pp. 130, 2013 (1996).

[324] A. Marziali and H. Schwettman, in *Proceedings of the 6th Workshop on RF Superconductivity*, edited by R. M. Sundelin, CEBAF, Newport News, Va., p. 782 (1994).

[325] J. Delayen, in *Proceedings of the 4th Workshop on RF Superconductivity*, edited by Y. Kojima, KEK, Tsukuba, Japan, p. 249 (1990). Rep. 89-21.

[326] J. Tückmantel, *CERN AT-RF (Int)*, **91**:99 (1991).

[327] A. Gamp, in *Proceedings of the 6th Workshop on RF Superconductivity*, edited by R. M. Sundelin, CEBAF, Newport News, Va., p. 492 (1994).

[328] D. Boussard et al., in *Proceedings of the 7th Workshop on RF Superconductivity*, edited by B. Bonin, Gif-sur-Yvette, France, p. 641 (1995). CEA/Saclay 96 080/1.

[329] G. Cavallari et al., in *Proceedings of the 3rd Workshop on RF Superconductivity*, edited by K. W. Shepard, Argonne Natl. Lab., Argonne, Il, p. 625 (1988). ANL-PHY-88-1.

[330] E. Nordberg et al., in *Proceedings of the 1993 Particle Accelerator Conference*, Washington, DC, p. 995 (1993). IEEE Cat. No. 3279-7.

[331] D. J. Liska et al., in *Proceedings of the 1992 Linac Conference*, edited by C. R. Hoffman, p. 163 (1992). AECL-10728.

[332] R. Peccei et al., eds., *Particle Physics, Perspectives and Opportunities*, World Scientific (1995).

[333] PEP-II, An Asymmetric B Factory, Conceptual Design Report. Technical Report SLAC-418, SLAC, June 1993.

[334] M.-P. Level et al., in *Proceedings of the 5th European Particle Accelerator Conference*, edited by S. Myers et al., Barcelona, Spain, IOPP Publishing, Bristol, p. 670 (1996).

[335] J. Kirchgessner, *Part. Accel.*, **46(1)**:151 (1995).

[336] W. Weingarten, in *Proceedings of the 7th Workshop on RF Superconductivity*, edited by B. Bonin, Gif-sur-Yvette, France, p. 23 (1995). CEA/Saclay 96 080/1.

[337] R. B. Palmer. Technical Report SLAC-PUB 4707, SLAC, 1988.

[338] K. Oide and K. Yokoya, *Phys. Rev. A, Gen. Phys.*, **40**:315 (1989).

[339] K. Akai, in *Proceedings of the 1995 Particle Accelerator Conference and International Conference on High Energy Accelerators*, Dallas, TX, p. 769 (1996). Cat. No. 95CH35843.

[340] F. Carminati et al. An Energy Amplifier For Cleaner and Inexhaustible Nuclear Energy Production Driven By a Particle Beam Accelerator. Technical Report, CERN, 1993. CERN/AT/93-47 (ET).

[341] G. P. Lawrence, in *Proceedings of the 1995 Particle Accelerator Conference and International Conference on High Energy Accelerators*, Dallas, TX, p. 35 (1996). Cat. No. 95CH35843.

REFERENCES

[342] R. Jameson, in *Proceedings of the 5th European Particle Accelerator Conference*, edited by S. Myers et al., Barcelona, Spain, IOPP Publishing, Bristol, p. 210 (1996).

[343] H. Lengeler, in *Proceedings of the 6th Workshop on RF Superconductivity*, edited by R. M. Sundelin, CEBAF, Newport News, Va., p. 338 (1994).

[344] B. Rusnak et al., in *Proceedings of the 1997 Particle Accelerator Conference* (1997). To be published.

[345] M. Tigner, *Nuovo Cimento*, **37**:1228 (1965).

[346] P. Emma, in *Proceedings of the 1995 Particle Accelerator Conference and International Conference on High Energy Accelerators*, Dallas, TX, p. 606 (1996). Cat. No. 95CH35843.

[347] P. B. Wilson, in *Applications of High-Power Microwaves*, edited by A. V. Gapanov-Grekhov and V. L. Granatstein, pp. 229–317. Artech House, Boston, 1994.

[348] G. Loew. International Linear Collider Technical Review Committee Report. Technical Report SLAC-R-95-471, SLAC, 1995.

[349] H. Padamsee, ed. *Proceedings of the 1st TESLA Workshop*, Cornell University, 1990. CLNS 90-1029.

[350] D. Proch, ed. *Proceedings of the 5th Workshop on RF Superconductivity*, DESY, Hamburg, Germany, 1991. DESY-M-92-01.

[351] R. M. Sundelin, ed. *Proceedings of the 6th Workshop on RF Superconductivity*, CEBAF, Newport News, Va., 1994.

[352] H. Padamsee, in *Proceedings of the 5th Workshop on RF Superconductivity*, edited by D. Proch, DESY, Hamburg, Germany, p. 963 (1991). DESY-M-92-01.

[353] R. Brinkmann, in *Proceedings of the 1995 Particle Accelerator Conference and International Conference on High Energy Accelerators*, Dallas, TX, p. 674 (1996). Cat. No. 95CH35843.

[354] R. Sundelin, in *Proceedings of the 1987 IEEE Particle Accelerator Conference*, p. 68 (1987). IEEE Cat. No. 87CH-2387-9.

[355] J. Rossbach, in *Proceedings of the 1995 Particle Accelerator Conference and International Conference on High Energy Accelerators*, Dallas, TX, p. 611 (1996). Cat. No. 95CH35843.

[356] S. Tazzari et al., in *Proceedings of the 6th Workshop on RF Superconductivity*, edited by R. M. Sundelin, CEBAF, Newport News, Va., p. 440 (1994).

[357] V. Balakin et al., in *Proceedings of the 12th International Conference on High Energy Accelerators*, Fermilab, p. 119 (1983).

[358] The NLC Design Group. Zeroth-order Design Report for the Next Linear Collider. Technical Report SLAC Report 474, SLAC, May 1996. This document is also available via http://www.slac.stanford.edu.

[359] H. Padamsee. Presented at Snowmass 96.

[360] J. Graber et al., in *Proceedings of the 1993 Particle Accelerator Conference*, Washington, DC, p. 892 (1993). IEEE Cat. No. 3279-7.

[361] D. A. Edwards, ed. The TESLA Test Facility Linac Design Report. Technical Report TESLA 95-01, DESY, 1995.

[362] H. Braun et al., in *Proceedings of the 1995 Particle Accelerator Conference and International Conference on High Energy Accelerators*, Dallas, TX, p. 716 (1996). Cat. No. 95CH35843.

[363] R. Bingham, in *Proceedings of the 5th European Particle Accelerator Conference*, edited by S. Myers et al., Barcelona, Spain, IOPP Publishing, Bristol, p. 120 (1996).

[364] R. B. Palmer et al., in *Proceedings of the 1995 Particle Accelerator Conference and International Conference on High Energy Accelerators*, Dallas, TX, p. 53 (1996). Cat. No. 95CH35843.

[365] The $\mu^+\mu^-$ Collider Collaboration. $\mu^+\mu^-$ Collider, A Feasibility Study. Technical Report BNL-52503, Brookhaven, 1996.

[366] S. Kurokawa et al., eds. *Proceedings of the Joint US-CERN-Japan International School, Frontiers of Accelerator Technology*. World Scientific, 1996.

[367] *Proceedings of the 5th Symposium on Accelerator Science and Technology*, KEK, 1984.

[368] *Proceedings of the 1987 IEEE Particle Accelerator Conference*, 1987. IEEE Cat. No. 87CH-2387-9.

[369] *Proceedings of the 1995 Particle Accelerator Conference and International Conference on High Energy Accelerators*, Dallas, TX, May 1996. Cat. No. 95CH35843.

[370] *Conference Record of the 1991 IEEE Particle Accelerator Conference*, May 1991.

[371] *Proceedings of the 1989 IEEE Particle Accelerator Conference*, 1989. IEEE Cat. 89CH2669-0.

REFERENCES

[372] V. Suller et al., eds. *Proceedings of the 4th European Particle Accelerator Conference*. World Scientific, 1994.

[373] E. H. Henke et al., eds. *Proceedings of the 1992 European Particle Accelerator Conference*. Editions Frontieres, 1992.

[374] *Proceedings of the 1972 Applied Superconductivity Conference*, Annapolis, 1972. IEEE. IEEE Pub. No. 72CH0682-5-TABSC.

[375] *Proceedings of the XIII International Symposium on Discharges and Electrical Insulation in Vacuum*, Paris, France, 1988.

[376] C. R. Hoffman, ed. *Proceedings of the 1992 Linac Conference*, 1992. AECL-10728.

[377] *Proceedings of the 1993 Particle Accelerator Conference*, Washington, DC, 1993. IEEE Cat. No. 3279-7.

[378] S. Myers et al., eds. *Proceedings of the 5th European Particle Accelerator Conference*, Barcelona, Spain, 1996. IOPP Publishing, Bristol.

Index

accelerators, heavy-ion, 6, 22, 25
   ATLAS, 18, 20, 23
   Australian National University, 23
   Bombay, 23
   Florida State University, 23
   JAERI, 5, 20, 23, 24
   Kansas State University, 23
   Legnaro, 20, 23
   New Delhi, 23
   Saclay, 20, 23
   Sao Paolo, 23
   SUNY, 18, 23
   University of Washington, 20, 23
acoustic mismatch theory, 212
aperture, 8, 11, 17, 342, 344, 443, 446

BCS theory, 67, 69, 70, 74, 90
bead pulling, 137
beam halo, 31, 344
   activation from, 456
beam instabilities, 12, 339, 341
   cumulative beam break-up, 346
   energy spread, 31, 343, 344
   growth rate, 346–348, 442
   longitudinal, 343, 381, 392–393, 396–397
   multibunch, 8, 12, 27, 343, 345, 442, 444–445
   multipass, 346, 348
   regenerative beam break-up, 345
   single bunch, 342, 343
   single pass, 342, 345
   stored current, maximum, 397
   transverse, 343, 347
beam loading, 381–401
   coupling, optimizing, 388, 389, 396
   frequency spectrum of beam, 384, 385
   fundamental theorem of, 333
   generator power required, 390–391, 396

   parameter, 387
beam tubes
   effect on fields, 41, 48, 129, 131, 350, 352
   making, 114

clean room assembly, 120, 123–125, 126, 127, 248, 249
coherence length, 8, 71, 74, 94
colliders, linear, 8, 13, 460, 464
   ac power, 460, 461
   background, of particles, 461
   beam power, 470
   beamstrahlung, 461
   BTC, 441, 442
   bunch charge, 461
   CLIC, 462, 473, 474
   comparisons, 466–469, 472, 473
   design flexibility, 470
   JLC, 462
   luminosity, 470
   NLC, 462, 465
   rf power, peak, 466
   RK-TBA, 462
   SBLC, 462
   SLC, 460
   spot size, 461
   TESLA, 344, 345, 462, 463, 465, 466, 470, 472
   TTF, 463, 473
   upgrade potential, 470–472
   VLEPP, 462
computer codes for cavity fields
   boundary conditions, 49, 51, 53, 54
   cavity geometry, specifying, 49
   CLANS, 350, 351–352
   higher order modes, 349–352
   MAFIA, 49
   material properties, 49, 53
   meshing, 49, 51, 52, 55
   SEAFISH, 350, 351–352

SUPERFISH, 41, 48, 49, 55, 51–55, 56, 351
SuperLANS, 41, 49, 53, 55, 56, 351
URMEL, 41, 48, 49, 51, 52, 55, 56, 349, 349–351
conductivity, electrical, 58, 59, 62, 72
conductivity, thermal, 10, 11, 58, 59, 62, 65, 75, 76, 208–210, 217
    thermal breakdown, influence on, 217
conductor, perfect, 80–82, 84
control systems, rf, 13, 426, 427
Cooper pairs, 8, 69–73, 75, 209
coupler, higher order mode, 3, 11, 28, 355
    absorbers, 359–361, 374, 376
    asymmetric cavities, 355
    beam-tube, 357, 374–376, 379, 446
    coaxial, 357, 361–373
    development, 369–371
    fundamental mode filter, 357, 362–367, 369
    polarized cavities, 356, 357
    waveguide, 17, 357–360
coupler, input, 3, 11, 28, 146, 148, 164, 403
    coaxial, 145, 404–407, 412, 421
    cooling, 405, 409
    development, 404
    HPP, 421, 422
    waveguide, 17, 404, 408–410, 413
coupler, transmitted power, 145, 146
coupling factor, 146, 148
    defined, 148
    external, 150
    measuring, 154, 155, 161
    optimizing, 154, 388, 388–389, 396
coupling, cell-to-cell, 130, 140
CP violation, 440, 441
crab cavity, 450–453
craters, 258, 260–264, 297, 309, 313
critical current density, 73, 93
critical magnetic field, 9
    lower, 96
    rf, 73, 99–101
    superheating, 9, 99, 100, 320
    thermodynamic, 91–93, 320
    upper, 98, 174, 175, 316

Debye temperature, 66

decay time of stored energy, 147, 161–162
deep drawing, 108, 109, 108–109, 110–112
defects, 9, 12
deflecting force of cavity fields, 341
density of electron states, 64, 68
detuning, of cavity, 389–392, 394–396
dipole mode, 41, 340
    calculating, 349, 352, 350–352
    excitation, 340–342

efficiency
    power transfer to the beam, 387
    refrigerator, 7, 48
electromagnetic field
    accelerating voltage, 42, 43, 340, 386–388, 392
    boundary conditions, 37, 38
    cavities, 39
    eigenvalue equation, 37–39
    peak electric, 9, 23, 43, 48, 102, 163
    peak magnetic, 9, 23, 43, 48
    transverse electric mode, 38–39, 41
    transverse magnetic mode, 38–39, 40, 41
    waveguides, 37
electron synchrotron cavity, historical foundation, 16, 17
electron–phonon interaction, 67, 68
emittance, 31
emittance growth, 8, 344
energy gap, 8, 69–72, 74, 93
    anisotropy effect, 89, 90
energy, stored in cavity, 156
equivalent circuit, 130, 149, 382, 383, 385, 382–385

Fermi energy, 63, 64, 71
Fermi surface, 68
Fermi velocity, 8, 63–65, 74
Fermi–Dirac distribution, 63–65
ferrite, microwave propteries, 378
field emission, 10–13, 228, 312, 227–313
    activation, 254, 271
    adsorbates, influence on, 252, 256–258, 271, 287
        resonant tunnelling theory, 257
    anode probe-hole studies, 252, 255

INDEX

current density, 239, 271
dc studies, 228
discharge, 228, 288, 296, 299
   dc, 265, 266
   simulations, 271
emission-free surface, 251, 254
emissive area, 233, 241
energy spectrum, electrons, 252, 253, 255
field-enhancement factor, 227, 233, 241
heat treatment, 290
helium processing, 269, 270, 287, 288
image-charge effect, 231–235
interface, influence of, 251
ionization of gas, 239, 271–272
melting, 271
MIM model, 252, 254, 255
mushroom cavity studies, 261, 262
onset, 281, 282
performance limits, 282, 283
rf processing, 228–229, 258, 259, 264, 265, 267, 268–271, 281, 299
simulations, 229, 230, 237, 272, 275, 272–279
statistical model, 283–285, 303, 305, 306, 313
thermometry, 235, 272
tip-on-tip model, 250
vacuum accidents, effect of, 309–312
whiskers, 233, 235, 238, 239, 245
work function, 231, 232
x rays, 229, 230
x-ray mapping, 235, 236
field emission sites, 227, 228, 235, 239
   activation, 240, 251, 252, 254
   active, 269
   dc scanning, 243, 244
   density of, 240, 242–244, 249, 290–293
   emissive area, 240, 241, 245
   field-enhancement factor, 240, 241, 245
   helium processed, 270
   intentionally introduced, 243, 246, 250–252
   locating, 272–279
   melting, 237, 267, 268
   particulate, 237, 238, 240, 246–248, 250, 271
   processed, 239, 240, 242, 246, 248, 254, 260
   whiskers, 245, 246, 271
flux vortex, 96, 97
fluxoid nucleation, 100
Fowler–Nordheim theory, 227, 230–233, 274, 277, 278
free-electron lasers, 14, 33
   DALINAC, 34
   JAERI, 34
   Stanford University, 33–34
   TJNAF, 34
free-electron model, 57–62

geometry factor, 45
Ginzburg–Landau parameter, 95
Ginzburg–Landau theory, 98

half-wave resonator, 26
harmonic number, 335, 385
heat treatment, 289–292
   annealing, 106, 110–112, 219
   hydrogen removal, 121, 177
helical resonator, 18, 19
helium, superfluid, 165–166
higher order mode
   calculating, 349, 349–352
   cryogenic losses, 338–339
   damping, 373, 376, 377
   excitation of, 331–341
      antiresonant, 338
      average, 337, 338
      resonant, 336, 338
   frequency spread, 347
   impedances, 361, 373, 377, 444, 445
   trapped modes, 12, 353, 356
hybrid mode, 42

impurities of niobium bulk, 107, 106–108
   carbon, 107
   hydrogen, 107, 175, 177
   interstitial, 58, 209
   nitrogen, 107
   oxygen, 107
   tantalum, 106, 107, 204
impurity scattering, of electrons, 58, 59, 62, 65

interdigital structures, 25
interface, niobium–helium
 film boiling, 212
 heat transfer coefficient, 210, 211, 214
 Kapitza conductance, 211, 212, 219
 nucleate boiling, 212
isotope effect, 68

lead film, 10, 18
linacs, recirculating, 31
 CEBAF, 32
 DALINAC, 32
 HEPL, 14–16, 31
 University of Illinois, 16
London equations, 80, 84, 85, 89
Lorentz-force detuning, 428, 430
loss factor
 dipole mode, 340, 341
 monopole mode, 333

Maxwell–Boltzmann distribution, 63, 64
mean free path
 electron, 8, 59, 63, 74, 79, 87, 88
 phonon, 75
mechanical properties of niobium
 elongation, 109, 113
 grain boundaries, 75
 grain size, 75, 106, 108–109, 111, 112, 219
 influence of heat treatment on, 218, 219
 recrystallization, 106, 111, 219
 tensile strength, 109
 work hardening, 109
 workability, 219
 yield strength, 109, 113, 220
Meissner effect, 66, 82–84
metallurgy, 105
microphonics, 13, 18, 427–429
microscope, scanning electron, 201, 202
momentum compaction, 392
monopole mode, 41
 calculating, 349, 350, 349–350
 excitation, 331–340
muffin-tin cavity, 16, 17, 179, 185, 187
multicell cavity
 choosing number of cells, 12

mode designations, 132, 133
mode solutions, 132, 133
multipacting, 4, 10, 11, 16, 17, 179–197, 312
 avoiding, 187, 196
 barrier, 179, 186
  hard, 181
  soft, 179, 184
 coaxial lines, 195, 196
 low-velocity structures, 10
 one-point, 185
 order, 185, 189
 parallel-plate geometry, 189–191, 193
 processing, 179, 184, 193
 resonance condition, 181, 186, 191, 193
 simulations, 192–193
 trajectory, 185, 187, 189
 two-point, 189
muon collider, 13, 474–476

Nb_3Sn, 315, 319–325
NbN, 315
niobium, high purity, 59, 218, 286
niobium–copper cavities, 318–320
niobium–copper composite, 24
niobium–copper films, 10, 315, 316
 characterization of, 316, 317
 defects, 204, 319
 fabrication, 316, 317

Panofsky–Wenzel theorem, 342
passband modes, 132–134
penetration depth, 8, 74, 81, 84, 86, 89, 90
perfect conductor, 80–82, 84
perturbation method
 cell-to-cell tuning, 134, 136, 137
 reviewed, 134–136
phase advance, in multicell modes, 133
phase lock, 160
phase stability, longitudinal, 381, 392–393, 396–397
phonon peak, 75, 76, 209
phonon radiation limit, 212
phonon scattering, 65
phonons, 58, 59, 65, 67, 75, 209
pill-box cavity, 3, 4, 48, 187

INDEX

modes, 40–41, 46
pondermotive oscillations, 428, 431
power
    dissipated by beam, 336, 338
    dissipated in HOMs, 336
    emitted through coupler, 155, 382
    measuring, 160
    reflected at coupler, 150, 153–154, 156, 161, 387
    transfer to beam, 387, 393
    processing, high power (HPP), 13, 15, 296, 297, 299–305, 307, 308, 312, 313
        apparatus, 298
        thermal breakdown, during, 307–309
        vacuum accidents, recovery from, 309–312
proton accelerators, 8, 453–458
purifying niobium, 10
    electron beam melting, 218
    from ore, 105, 107
    outgassing at high temperature, 217, 218
    postpurification, 15, 118, 119, 118–119, 212, 218, 286

Q switch, 200, 204, 205
quadrupole mode, 41
quality factor
    external, 147
    loaded, 147
    unloaded, 45
quarter-wave resonator, 5, 20, 24

re-entrant cavity, 48
reflection coefficient, 150, 153
refrigerator
    efficiency, 7, 48
    rf heat load, 48
residual resistance, 9, 75, 171, 172, 321, 322, 324
    defects, 171
    dielectic losses, 172
    hydrides, 175–177, 219, 281
    joint losses, 172
    oxides, 177
    trapped flux, 173, 174, 178, 316
rinsing, high pressure, 13, 15, 121–123, 125, 249, 286, 293, 295, 312
Robinson stability criterion, 392, 395, 397
RRR, 59, 62, 88, 220
    annealing, effect on, 106
    impurities, effect of, 107
    thermal breakdown, influence on, 215, 218, 220–222, 285, 286, 294, 295, 301, 303, 312
    thermal conductivity, influence on, 208
rust removal, 115

scanning electron microscope, 201, 202
scattering of electrons off impurities, 58, 59, 62, 65
secondary electron emission, 181–183, 183, 184, 197
    energy, 187
    surface adsorbates, 181, 182, 184, 197
separator cavity, historical foundations, 22
shielding dc magnetic field, 174
shunt impedance, 382
    dipole mode, 341, 342
    geometric, 6, 8
    monopole mode, 47, 48
simulating
    field emission heating, 272–279
    multipacting, 192–193
skin depth, 78
skin effect, anomalous, 79, 80
specific heat, 61–63, 65, 66, 69, 70, 92
spherical cavity, 55, 56, 188
spinning, 108, 111, 114
split-loop resonator, 19
split-ring resonator, 18, 20
spoke resonator, 26
stamping, 108, 109, 108–109, 110, 112
Standard Model, 439–441
starbursts, 258
    processed sites, 260, 261, 266–268
storage rings, 27
    beam tests, 18
    CESR, 18, 27
    HERA, 27
    LEP, 3, 4, 7, 27
    PETRA, 18

SPS, 27
TAR, 18
TEVATRON, 27
TRISTAN, 27
storage rings, high current, 442
  beam tests, 449
  CESR, 384, 399, 441, 442, 445, 447–449
  KEK-B, 441–443, 445, 447–451
  LHC, 441, 442, 445, 449, 450
  PEP-II, 441, 444, 445
  SOLEIL, 446
stored energy, in cavity, 156
superconductors
  high temperature, 66, 86, 92, 315, 323, 325–327
  Type I, 94, 95, 97, 98
  Type II, 94–99
superfluid helium, 165–166
surface preparation, 120
  chemical etching, 115, 120–121, 122, 123
    hydrogen contamination, 175
  degreasing, 115
  derusting, 115
  electropolishing, 121, 124
surface reactance, 86, 89, 177
surface resistance, 8, 72–75, 86, 87
  defined, 44
  frequency dependence, 74, 86, 89, 90
  lead, 87
  mean free path effect, 88, 89
  Nb_3Sn, 87, 321
  niobium, 87
  normal state, 79
synchronous bunch, 392
synchronous phase, 392
synchrotron radiation, 17, 460

tapered-helix structure, 20, 21
thermal breakdown, 9–13, 199–201, 312
  cures
    guided repair, 207
    thermal conductivity, 208, 209
    thin films on copper, 209
  defect-free case, 213, 216, 223, 224, 225, 223–225, 286
  defects, 199–201, 203, 204

  properties, 207, 215, 216, 286
  field emission induced, 229
  film boiling, *see also* interface, niobium–helium
  model, 205, 206
  performance limits, 220, 222, 282
  rf signature, 200
  simulations, 203, 206, 210, 213–218, 224, 286
thermometry, 145, 162, 167, 164–168, 169, 187, 277
  defects, 201–204
  field emission, 165, 166, 229, 230, 235–237, 256–259, 270, 272, 287, 290
  residual resistance, 171–173
  rotating arm, 164, 165, 166
  superfluid helium, 165–166
  thermal breakdown, 216, 224, 225
transit time factor, 42
trapped modes, 351
tuner, 425, 426, 432–434
  fast tuner, 426
  fine tuner, 425, 433
  magnetostrictive, 432
  piezoelectric, 432
  thermal, 434, 435
  tuning range, 431
  tuning sensitivity, 432
tuning, 13
  cell-to-cell, 120, 119–120, 129
  frequency of mode, 425
  measuring field profile, 129, 137, 138, 140, 143
tuning offset, 397
two-fluid model, 8, 72, 85, 90

voltage, beam-induced, 331–333, 335, 336, 386–398
voltage, deflecting, 342

wakefields, 342
  long-range, 12
  short-range, 8, 12, 343, 344
wall losses, 44, 45, 47, 48, 163, 382
welding
  electron-beam, 115, 116, 117, 115–117
  TIG, 117, 117
Wiedemann-Franz Law, 62, 65, 75

INDEX

window, 12, 404, 410, 412
  arcing, 411
  brazing and metallization, 415
  ceramic properties, 415
  coatings, 417
  coaxial
    conical, 407, 412
    cylindrical, 405, 412
    dc bias voltage, 417
    planar, 405
  cold, 410
  diagnostics, 418, 419
  location, 411
  multipacting, 411, 416–418
  performance, 421
  processing, 411, 413, 415, 416, 418–420
  waveguide, planar, 407, 408, 411, 413, 414

x-ray analysis, 202
x-ray mapping, 160, 162